SPACE – TIME – MATTER

HERMANN WEYL

TRANSLATED FROM THE GERMAN BY
HENRY L. BROSE

WITH FIFTEEN DIAGRAMS

HAWK PRESS

Published by

Hawk Press
4836/24, Ansari Road, Daryaganj
New Delhi – 110 002
Phones : 9643330713, 011-23278618, 011-35676207
E-mail : thehawkpress@gmail.com
www.thehawkpress.com

ISBN : 978-93-93971-02-9

FROM THE AUTHOR'S PREFACE TO THE FIRST EDITION

EINSTEIN'S Theory of Relativity has advanced our ideas of the structure of the cosmos a step further. It is as if a wall which separated us from Truth has collapsed. Wider expanses and greater depths are now exposed to the searching eye of knowledge, regions of which we had not even a presentiment. It has brought us much nearer to grasping the plan that underlies all physical happening.

Although very recently a whole series of more or less popular introductions into the general theory of relativity has appeared, nevertheless a systematic presentation was lacking. I therefore considered it appropriate to publish the following lectures which I gave in the Summer Term of 1917 at the *Eidgen. Technische Hochschule* in Zürich. At the same time it was my wish to present this great subject as an illustration of the intermingling of philosophical, mathematical, and physical thought, a study which is dear to my heart. This could be done only by building up the theory systematically from the foundations, and by restricting attention throughout to the principles. But I have not been able to satisfy these self-imposed requirements: the mathematician predominates at the expense of the philosopher.

The theoretical equipment demanded of the reader at the outset is a minimum. Not only is the special theory of relativity dealt with exhaustively, but even Maxwell's theory and analytical geometry are developed in their main essentials. This was a part of the whole scheme. The setting up of the Tensor Calculus—by means of which, alone, it is possible to express adequately the physical knowledge under discussion—occupies a relatively large amount of space. It is therefore hoped that the book will be found suit able for making physicists better acquainted with this mathematical instrument, and also that it will serve as a text-book for students and win their sympathy for the new ideas.

HERMANN WEYL

RIBBITZ IN MECKLENBURG
Easter, 1918

PREFACE TO THE THIRD EDITION

ALTHOUGH this book offers fruits of knowledge in a refractory shell, yet communications that have reached me have shown that to some it has been a source of comfort in troublous times. To gaze up from the ruins of the oppressive present towards the stars is to recognise the indestructible world of laws, to strengthen faith in reason, to realise the "harmonia mundi" that transfuses all phenomena, and that never has been, nor will be, disturbed.

My endeavour in this third edition has been to attune this harmony more perfectly. Whereas the second edition was a reprint of the first, I have now undertaken a thorough revision which affects Chapters II and IV above all. The discovery by Levi-Civita, in 1917, of the conception of infinitesimal parallel displacements suggested a renewed examination of the mathematical foundation of Riemann's geometry. The development of pure infinitesimal geometry in Chapter II, in which every step follows quite naturally, clearly, and necessarily, from the preceding one, is, I believe, the final result of this investigation as far as the essentials are concerned. Several shortcomings that were present in my first account in the *Mathematische Zeitschrift* (Bd. 2, 1918) have now been eliminated. Chapter IV, which is in the main devoted to Einstein's Theory of Gravitation has, in consideration of the various important works that have appeared in the meanwhile, in particular those that refer to the Principle of Energy-Momentum, been subjected to a very considerable revision. Furthermore, a new theory by the author has been added, which draws the physical inferences consequent on the extension of the foundations of geometry beyond Riemann, as shown in Chapter II, and represents an attempt to derive from world-geometry not only gravitational but also electromagnetic phenomena. Even if this theory is still only in its infant stage, I feel convinced that it contains no less truth than Einstein's Theory of Gravitation—whether this amount of truth is unlimited or, what is more probable, is bounded by the Quantum Theory.

I wish to thank Mr. Weinstein for his help in correcting the proof-sheets.

HERMANN WEYL

ACLA POZZOLI, NEAR SAMADEN
August, 1919

PREFACE TO THE FOURTH EDITION

IN this edition the book has on the whole preserved its general form, but there are a number of small changes and additions, the most important of which are: (1) A paragraph added to Chapter II in which the problem of space is formulated in conformity with the view of the Theory of Groups; we endeavour to arrive at an understanding of the inner necessity and uniqueness of Pythagorean space metrics based on a quadratic differential form. (2) We show that the reason that Einstein arrives necessarily at uniquely determined gravitational equations is that the scalar of curvature is the only invariant having a certain character in Riemann's space. (3) In Chapter IV the more recent experimental researches dealing with the general theory of relativity are taken into consideration, particularly the deflection of rays of light by the gravitational field of the sun, as was shown during the solar eclipse of 29th May, 1919, the results of which aroused great interest in the theory on all sides. (4) With Mie's view of matter there is contrasted another (*vide* particularly § 32 and § 36), according to which matter is a limiting singularity of the field, but charges and masses are force-fluxes in the field. This entails a new and more cautious attitude towards the whole problem of matter.

Thanks are due to various known and unknown readers for pointing out desirable modifications, and to Professor Nielsen (at Breslau) for kindly reading the proof-sheets.

HERMANN WEYL

ZÜRICH, *November*, 1920

TRANSLATOR'S NOTE

IN this rendering of Professor Weyl's book into English, pains have been taken to adhere as closely as possible to the original, not only as regards the general text, but also in the choice of English equivalents for technical expressions. For example, the word *affine* has been retained. It is used by Möbius in his *Der Barycentrische Calcul*, in which he quotes a Latin definition of the term as given by Euler. Veblen and Young have used the word in their *Projective Geometry*, so that it is not quite unfamiliar to English mathematicians. *Abbildung*, which signifies representation, is generally rendered equally well by transformation, inasmuch as it denotes a copy of certain elements of one space mapped out on, or expressed in terms of, another space. In some cases the German word is added in parenthesis for the sake of those who wish to pursue the subject further in original papers. It is hoped that the appearance of this English edition will lead to further efforts towards extending Einstein's ideas so as to embrace *all* physical knowledge. Much has been achieved, yet much remains to be done. The brilliant speculations of the latter chapters of this book show how vast is the field that has been opened up by Einstein's genius. The work of translation has been a great pleasure, and I wish to acknowledge here the courtesy with which suggestions concerning the type and the symbols have been received and followed by Messrs. Methuen & Co. Ltd. Acting on the advice of interested mathematicians and physicists I have used Clarendon type for the vector notation. My warm thanks are due to Professor G. H. Hardy of New College and Mr. T. W. Chaundy, M.A., of Christ Church, for valuable suggestions and help in looking through the proofs. Great care has been taken to render the mathematical text as perfect as possible.

HENRY L. BROSE

CHRIST CHURCH, OXFORD
December, 1921

CONTENTS

CHAPTER I

EUCLIDEAN SPACE. ITS MATHEMATICAL FORMULATION AND ITS RÔLE IN PHYSICS

CHAPTER II

The Metrical Continuum

CHAPTER III

Relativity of Space and Time

CHAPTER IV

The General Theory of Relativity

The formulæ are numbered anew for each chapter. Unless otherwise stated, references to formulæ are to those in the current chapter.

SPACE—TIME—MATTER

INTRODUCTION

SPACE and **time** are commonly regarded as the **forms** of existence of the real world, **matter** as its **substance**. A definite portion of matter occupies a definite part of space at a definite moment of time. It is in the composite idea of **motion** that these three fundamental conceptions enter into intimate relationship. Descartes defined the objective of the exact sciences as consisting in the description of all happening in terms of these three fundamental conceptions, thus referring them to motion. Since the human mind first wakened from slumber, and was allowed to give itself free rein, it has never ceased to feel the profoundly mysterious nature of time-consciousness, of the progression of the world in time,—of Becoming. It is one of those ultimate metaphysical problems which philosophy has striven to elucidate and unravel at every stage of its history. The Greeks made Space the subject-matter of a science of supreme simplicity and certainty. Out of it grew, in the mind of classical antiquity, the idea of pure science. Geometry became one of the most powerful expressions of that sovereignty of the intellect that inspired the thought of those times. At a later epoch, when the intellectual despotism of the Church, which had been maintained through the Middle Ages, had crumbled, and a wave of scepticism threatened to sweep away all that had seemed most fixed, those who believed in Truth clung to Geometry as to a rock, and it was the highest ideal of every scientist to carry on his science "***more geometrico***". Matter was imagined to be a substance involved in every change, and it was thought that every piece of matter could be measured as a quantity, and that its characteristic expression as a "substance" was the Law of Conservation of Matter which asserts that matter remains constant in amount throughout every change. This, which has hitherto represented our knowledge of space and matter, and which was in many quarters claimed by philosophers as *a priori* knowledge, absolutely general and necessary, stands today a tottering structure. First, the physicists in the persons of Faraday and Maxwell, proposed the "electromagnetic **field**" in contra-distinction to **matter**, as a reality of a different category. Then, during the last century, the mathematician, following a different line of thought, secretly undermined belief in the evidence of Euclidean Geometry. And now, in our time, there has been unloosed a cataclysm which has swept away space, time, and matter hitherto regarded as the firmest pillars of natural science, but only to make place for a view of things of wider scope, and entailing a deeper vision.

This revolution was promoted essentially by the thought of one man, Albert Einstein. The working-out of the fundamental ideas seems, at the present time, to have reached a certain conclusion; yet, whether or not we are already faced with a new state of affairs, we feel ourselves compelled to subject these new ideas to a close analysis. Nor is any retreat possible. The development of scientific thought may once again take us beyond the present achievement, but a return to the old narrow and restricted scheme is out of the question.

Philosophy, mathematics, and physics have each a share in the problems presented here. We shall, however, be concerned above all with the mathematical and physical aspect of these questions. I shall only touch lightly on the philosophical implications for the simple reason that in this direction nothing final has yet been reached, and that for my own part I am not in a position to give such answers to the epistemological questions involved as my conscience would allow me to uphold. The ideas to be worked out in this book are not the result of some speculative inquiry into the foundations of physical knowledge, but have been developed in the ordinary course of the handling of concrete physical problems—problems arising in the rapid development of science which has, as it were, burst its old shell, now become too narrow. This revision of fundamental principles was only undertaken later, and then only to the extent necessitated by the newly formulated ideas. As things are to-day, there is left no alternative but that the separate sciences should each proceed along these lines dogmatically, that is to say, should follow in good faith the paths along which they are led by reasonable motives proper to their own peculiar methods and special limitations. The task of shedding philosophic light on to these questions is none the less an important one, because it is radically different from that which falls to the lot of individual sciences. This is the point at which the philosopher must exercise his discretion. If he keep in view the boundary lines determined by the difficulties inherent in these problems, he may direct, but must not impede, the advance of sciences whose field of inquiry is confined to the domain of concrete objects.

Nevertheless I shall begin with a few reflections of a philosophical character. As human beings engaged in the ordinary activities of our daily lives, we find ourselves confronted in our acts of perception by material things. We ascribe a "real" existence to them, and we accept them in general as constituted, shaped, and coloured in such and such a way, and so forth, as they appear to us in our perception in "general," that is ruling out possible illusions, mirages, dreams, and hallucinations.

These material things are immersed in, and transfused by, a manifold, indefinite in outline, of analogous realities which unite to form a single ever-present world of space to which I, with my own body, belong. Let us here consider only these bodily objects, and not all the other things of a different category, with which we as ordinary beings are confronted; living creatures, persons, objects of daily use, values, such entities as state, right, language, etc. Philosophical reflection probably begins in every one of us who is endowed with an abstract turn of mind when he first becomes sceptical about the world-view of naïve realism to which I have briefly alluded.

It is easily seen that such a quality as "**green**" has an existence only as the correlate of the sensation "green" associated with an object given by perception, but that it is meaningless to attach it as a thing in itself to material things existing **in themselves**. This recognition of the **subjectivity of the qualities of sense** is found in Galilei (and also in Descartes and Hobbes) in a form closely related to the principle underlying the **constructive mathematical method of our modern physics which repudiates "qualities"**. According to this principle, colours are "really" vibrations of the æther, i.e. motions. In the field of philosophy Kant was the first to take the next decisive step towards the point of view that not only the qualities revealed by the senses, but also space and spatial characteristics have no objective significance in the absolute sense; in other words, that **space, too, is only a form of our perception**. In the realm of physics it is perhaps only the theory of relativity which has made it quite clear that the two essences, space and time, entering into our intuition have no place in the world constructed by mathematical physics. Colours are thus "really" not even æther-vibrations, but merely a series of values of mathematical functions in which occur four independent parameters corresponding to the three dimensions of space, and the one of time.

Expressed as a general principle, this means that the real world, and every one of its constituents with their accompanying characteristics, are, and can only be given as, intentional objects of acts of consciousness. The immediate data which I receive are the experiences of consciousness in just the form in which I receive them. They are not composed of the mere stuff of perception, as many Positivists assert, but we may say that in a sensation an object, for example, is actually physically present for me—to whom that sensation relates—in a manner known to every one, yet, since it is characteristic, it cannot be described more fully. Following Brentano, I shall call it the "**intentional object**". In experiencing perceptions I see this chair, for example. My attention is fully directed towards it. I "have" the perception, but it is only when I make this perception in turn the intentional object of a new inner perception (a free act of reflection enables me to do this) that I "know" something regarding it (and not the chair alone), and ascertain precisely what I remarked just above. In this second act the intentional object is immanent, i.e. like the act itself, it is a real component of my stream of experiences, whereas in the primary act of perception the object is transcendental, i.e. it is given in an experience of consciousness, but is not a real component of it. What is immanent is **absolute**, i.e. it is exactly what it is in the form in which I have it, and I can reduce this, its essence, to the axiomatic by acts of reflection. On the other hand, transcendental objects have only a **phenomenal** existence; they are appearances presenting themselves in manifold ways and in manifold "gradations". One and the same leaf seems to have such and such a size, or to be coloured in such and such a way, according to my position and the conditions of illumination. Neither of these modes of appearance can claim to present the leaf just as it is "in itself". Furthermore, in every perception there is, without doubt, involved the **thesis of reality** of the object appearing in it; the latter is, indeed, a fixed and lasting element of the general thesis of reality of the world. When, however, we pass from the natural view to the philosophical attitude, meditating upon perception, we no longer subscribe to this thesis. We simply affirm that something real is "supposed" in it. The meaning of such a supposition now becomes the problem which must be solved from the data of consciousness. In addition a justifiable ground for making it must be found. I do not by this in any way wish to imply that the view that the events of the world are a mere play of the consciousness produced by the ego, contains a higher degree of truth than naïve realism; on the contrary, we are only concerned in seeing clearly that the datum of consciousness is the starting-point at which we must place ourselves if we are to understand the absolute meaning as well as the right to the supposition of reality. In the field of logic we have an analogous case. A judgment, which I pronounce, affirms a certain set of circumstances; it takes them as true. Here, again, the philosophical question of the meaning of, and the justification for, this thesis of truth arises; here, again, the idea of objective truth is not denied, but becomes a problem which has to be grasped from what is given absolutely. "Pure consciousness" is the seat of that which is philosophically *a priori*. On the other hand, a philosophic examination of the thesis of truth must and will lead to the conclusion that none of these acts of perception, memory, etc., which present experiences from which I seize reality, gives us a conclusive right to ascribe to the perceived object an existence and a constitution as perceived. This right can always in its turn be over-ridden by rights founded on other perceptions, etc.

It is the nature of a real thing to be inexhaustible in content; we can get an ever deeper insight into this content by the continual addition of new experiences, partly in apparent contradiction, by bringing them into harmony with one another. In this interpretation, things of the real world are approximate ideas. From this arises the empirical character of all our knowledge of reality.*

Time is the primitive form of the stream of consciousness. It is a fact, however obscure and perplexing to our minds, that the contents of consciousness do not present themselves simply as being (such as conceptions, numbers, etc.), but as **being now** filling the form of the enduring present with a varying content. So that one does not say this **is** but this is **now**, yet now no more. If we project ourselves outside the stream of consciousness and represent its content as an object, it becomes an event happening in time, the separate stages of which stand to one another in the relations of **earlier** and **later**.

Just as time is the form of the stream of consciousness, so one may justifiably assert that space is the form of external material reality. All characteristics of material things as they are presented to us in the acts of external perception (e.g. colour) are endowed with the separateness of spatial extension, but it is only when we build up a single connected real world out of all our experiences that the spatial extension, which is a constituent of every perception, becomes a part of one and the same all-inclusive space. Thus space is the **form** of the external world. That is to say, every ma-

* *Vide* note 1.

terial thing can, without changing content, equally well occupy a position in Space different from its present one. This immediately gives us the property of the homogeneity of space which is the root of the conception, Congruence.

Now, if the worlds of consciousness and of transcendental reality were totally different from one another, or, rather, if only the passive act of perception bridged the gulf between them, the state of affairs would remain as I have just represented it, namely, on the one hand a consciousness rolling on in the form of a lasting present, yet spaceless; on the other, a reality spatially extended, yet timeless, of which the former contains but a varying appearance. Antecedent to all perception there is in us the experience of effort and of opposition, of being active and being passive. For a person leading a natural life of activity, perception serves above all to place clearly before his consciousness the definite point of attack of the action he wills, and the source of the opposition to it. As the doer and endurer of actions I become a single individual with a psychical reality attached to a body which has its place in space among the material things of the external world, and by which I am in communication with other similar individuals. Consciousness, without surrendering its immanence, becomes a piece of reality, becomes this particular person, namely myself, who was born and will die. Moreover, as a result of this, consciousness spreads out its web, in the form of time, over reality. Change, motion, elapse of time, becoming and ceasing to be, exist in time itself; just as my will acts on the external world through and beyond my body as a motive power, so the external world is in its turn **active** (as the German word "Wirklichkeit," reality, derived from "wirken" = to act, indicates). Its phenomena are related throughout by a **causal connection**. In fact physics shows that cosmic time and physical form cannot be dissociated from one another. The new solution of the problem of amalgamating space and time offered by the theory of relativity brings with it a deeper insight into the harmony of action in the world.

The course of our future line of argument is thus clearly outlined. What remains to be said of time, treated separately, and of grasping it mathematically and conceptually may be included in this introduction. We shall have to deal with space at much greater length. Chapter I will be devoted to a discussion of **Euclidean space** and its mathematical structure. In Chapter II will be developed those ideas which compel us to pass beyond the Euclidean scheme; this reaches its climax in the general space-conception of the metrical continuum (Riemann's conception of space). Following upon this Chapter III will discuss the problem mentioned just above of the **amalgamation** of Space and Time in the world. From this point on the results of mechanics and physics will play an important part, inasmuch as this problem by its very nature, as has already been remarked, comes into our view of the world as an active entity. The edifice constructed out of the ideas contained in Chapters II and III will then in the final Chapter IV lead us to Einstein's *General Theory of Relativity*, which, physically, entails a new Theory of **Gravitation**, and also to an extension of the latter which embraces electromagnetic phenomena in addition to gravitation. The revolutions which are brought about in our notions of Space and Time will of necessity affect the conception of matter too. Accordingly, all that has to be said about matter will be dealt with appropriately in Chapters III and IV.

To be able to apply mathematical conceptions to questions of Time we must postulate that it is theoretically possible to fix in Time, to any order of accuracy, an absolutely rigorous **now** (present) as a **point of Time**—i.e. to be able to indicate points of time, one of which will always be the earlier and the other the later. The following principle will hold for this "order-relation". If A is earlier than B and B is earlier than C, then A is earlier than C. Each two points of Time, A and B, of which A is the earlier, mark off a **length of time**; this includes every point which is later than A and earlier than B. The fact that Time is a form of our stream of experience is expressed in the idea of **equality**: the empirical content which fills the length of Time AB can in itself be put into any other time without being in any way different from what it is. The length of time which it would then occupy is equal to the distance AB. This, with the help of the principle of causality, gives us the following objective criterion in physics for equal lengths of time. If an absolutely isolated physical system (i.e. one not subject to external influences) reverts once again to exactly the same state as that in which it was at some earlier instant, then the same succession of states will be repeated in time and the whole series of events will constitute a cycle. In general such a system is called a **clock**. Each period of the cycle lasts **equally** long.

The mathematical fixing of time by **measuring** it is based upon these two relations, "earlier (or later) times" and "equal times". The nature of measurement may be indicated briefly as follows: Time is homogeneous, i.e. a single point of time can only be given by being specified individually. There is no inherent property arising from the general nature of time which may be ascribed to any one point but not to any other; or, every property logically derivable from these two fundamental relations belongs either to all points or to none. The same holds for time-lengths and point-pairs. A property which is based on these two relations and which holds for **one** point-pair must hold for every point-pair AB (in which A is earlier than B). A difference arises, however, in the case of three point-pairs. If any two time-points O and E are given such that O is earlier than E, it is possible to fix conceptually further time-points P by referring them to the unit-distance OE. This is done by constructing logically a relation t between three points such that for every two points O and E, of which O is the earlier, there is one and only one point P which satisfies the relation t between O, E and P, i.e. symbolically,

$$OP = t \cdot OE$$

(e.g. $OP = 2 \cdot OE$ denotes the relation $OE = EP$). **Numbers** are merely concise symbols for such relations as t, defined logically from the primary relations. P is the "time-point with the **abscissa** t **in the co-ordinate system** (taking OE as unit length)". Two different numbers t and t^* in the same co-ordinate system necessarily lead to two different points; for, otherwise, in consequence of the homogeneity of the continuum of time-lengths, the property expressed by

$$t \cdot AB = t^* \cdot AB,$$

since it belongs to the time-length $AB = OE$, must belong to **every** time-length, and hence the equations $AC = t \cdot AB$, $AC = t^* \cdot AB$ would both express the same relation, i.e. t would be equal to t^*. Numbers enable us to single out separate time-points relatively to a unit-distance OE out of the time-continuum by a conceptual, and hence objective and precise, process. But the objectivity of things conferred by the exclusion of the ego and its data derived directly from intuition, is not entirely satisfactory; the co-ordinate system which can only be specified by an individual act (and then only approximately) remains as an inevitable residuum of this elimination of the percipient.

It seems to me that by formulating the principle of measurement in the above terms we see clearly how mathematics has come to play its rôle in exact natural science. *An essential feature of measurement is the difference between the "determination" of an object by individual specification and the determination of the same object by some conceptual means.* The latter is only possible relatively to objects which must be defined directly. That is why a **theory of relativity** is perforce always involved in measurement. The general problem which it proposes for an arbitrary domain of objects takes the form: (1) What must be given such that relatively to it (and to any desired order of precision) one can single out conceptually a single arbitrary object P from the continuously extended domain of objects under consideration? That which has to be given is called the **co-ordinate system**, the conceptual definition is called the **co-ordinate** (or abscissa) of P in the co-ordinate system. Two different co-ordinate systems are completely equivalent for an objective standpoint. There is no property, that can be fixed conceptually, which applies to one co-ordinate system but not to the other; for in that case too much would have been given directly. (2) What relationship exists between the co-ordinates of one and the same arbitrary object P in two different co-ordinate systems?

In the realm of time-points, with which we are at present concerned, the answer to the first question is that the co-ordinate system consists of a time-length OE (giving the origin and the unit of measure). The answer to the second question is that the required relationship is expressed by the formula of transformation

$$t = at' + b \qquad (a > 0)$$

in which a and b are constants, whilst t and t' are the co-ordinates of the same arbitrary point P in an "unaccented" and "accented" system respectively. For all possible pairs of co-ordinate systems the characteristic numbers, a and b, of the transformation may be any real numbers with the limitation that a must always be positive. The aggregate of transformations constitutes a **group**, as their nature would imply, i.e.,

1. "identity" $t = t'$ is contained in it.

2. Every transformation is accompanied by its reciprocal in the group, i.e. by the transformation which exactly cancels its effect. Thus, the inverse of the transformation (a, b), viz. $t = at' + b$, is $\left(\frac{1}{a}, -\frac{b}{a}\right)$, viz. $t' = \frac{1}{a}t - \frac{b}{a}$.

3. If two transformations of a group are given, then the one which is produced by applying these two successively also belongs to the group. It is at once evident that, by applying the two transformations

$$t = at' + b \qquad t' = a't'' + b'$$

in succession, we get

$$t = a_1 t'' + b_1$$

where $a_1 = a \cdot a'$ and $b_1 = (ab') + b$; and if a and a' are positive, so is their product.

The theory of relativity discussed in Chapters III and IV proposes the problem of relativity, not only for time-points, but for the physical world in its entirety. We find, however, that this problem is solved once a solution has been found for it in the case of the two forms of this world, space and time. By choosing a co-ordinate system for space and time, we may also fix the physically real content of the world conceptually in all its parts by means of numbers.

All beginnings are obscure. Inasmuch as the mathematician operates with his conceptions along strict and formal lines, he, above all, must be reminded from time to time that the origins of things lie in greater depths than those to which his methods enable him to descend. Beyond the knowledge gained from the individual sciences, there remains the task of **comprehending**. In spite of the fact that the views of philosophy sway from one system to another, we cannot dispense with it unless we are to convert knowledge into a meaningless chaos.

CHAPTER I

EUCLIDEAN SPACE. ITS MATHEMATICAL FORMULATION AND ITS RÔLE IN PHYSICS

§ 1. Deduction of the Elementary Conceptions of Space from that of Equality

JUST as we fixed the present moment ("now") as a geometrical point in time, so we fix an exact "here," a point in space, as the first element of continuous spatial extension, which, like time, is infinitely divisible. Space is not a one-dimensional continuum like time. The principle by which it is continuously extended cannot be reduced to the simple relation of "earlier" or "later". We shall refrain from inquiring what relations enable us to grasp this continuity conceptually. On the other hand, space, like time, is a **form** of phenomena. Precisely the same content, identically the same thing, still remaining what it is, can equally well be at some place in space other than that at which it is actually. The new portion of Space $\mathbf{S}'$ then occupied by it is equal to that portion $\mathbf{S}$ which it actually occupied. $\mathbf{S}$ and $\mathbf{S}'$ are said to be **congruent**. To every point P of $\mathbf{S}$ there corresponds one definite **homologous** point P' of $\mathbf{S}'$ which, after the above displacement to a new position, would be surrounded by exactly the same part of the given content as that which surrounded P originally. We shall call this "transformation" (in virtue of which the point P' corresponds to the point P) a **congruent transformation**. Provided that the appropriate subjective conditions are satisfied the given material thing would seem to us after the displacement exactly the same as before. There is reasonable justification for believing that a rigid body, when placed in two positions successively, realises this idea of the equality of two portions of space; by a **rigid** body we mean one which, however it be moved or treated, can always be made to appear the same to us as before, if we take up the appropriate position with respect to it. I shall evolve the scheme of geometry from the conception of equality combined with that of continuous connection—of which the latter offers great difficulties to analysis—and shall show in a superficial sketch how all fundamental conceptions of geometry may be traced back to them. My real object in doing so will be to single out **translations** among possible congruent transformations. Starting from the conception of translation I shall then develop Euclidean geometry along strictly axiomatic lines.

First of all the **straight line**. Its distinguishing feature is that it is determined by two of its points. Any *other* line can, even when two of its points are kept fixed, be brought into another position by a congruent transformation (the test of straightness).

Thus, if A and B are two different points, the straight line $g = AB$ includes every point which becomes transformed into itself by all those congruent transformations which transform AB into themselves. (In familiar language, the straight line lies evenly between its points.) Expressed kinematically, this is tantamount to saying that we regard the straight line as an axis of rotation. It is homogeneous and a linear continuum just like time. Any arbitrary point on it divides it into two parts, two "rays". If B lies on one of these parts and C on the other, then A is said to be between B and C and the points of one part lie to the right of A, the points of the other part to the left. (The choice as to which is right or left is determined arbitrarily.) The simplest fundamental facts which are implied by the conception "between" can be formulated as exactly and completely as a geometry which is to be built up by deductive processes demands. For this reason we endeavour to trace back all conceptions of continuity to the conception "between," i.e. to the relation "A is a point of the straight line BC and lies between B and C" (this is the reverse of the real intuitional relation). Suppose A' to be a point on g to the right of A, then A' also divides the line g into two parts. We call that to which A belongs the left-hand side. If, however, A' lies to the left of A the position is reversed. With this convention, analogous relations hold not only for A and A' but also for **any** two points of a straight line. The points of a straight line are ordered by the terms left and right in precisely the same way as points of time by the terms earlier and later.

Left and right are equivalent. There is one congruent transformation which leaves A fixed, but which interchanges the two halves into which A divides the straight line. Every finite portion of straight line AB may be superposed upon itself in such a way that it is reversed (i.e. so that B falls on A, and A falls on B). On the other hand, a congruent transformation which transforms A into itself, and all points to the right of A into points to the right of A, and all points to the left of A into points to the left of A, leaves every point of the straight line undisturbed. The homogeneity of the straight line is expressed in the fact that the straight line can be placed upon itself in such a way that any point A of it can be transformed into any other point A' of it, and that the half to the right of A can be transformed into the half to the right of A', and likewise for the portions to the left of A and A' respectively (this implies a mere translation of the straight line). If we now introduce the equation $AB = A'B'$ for the points of the straight line by interpreting it as meaning that AB is transformed into the straight line $A'B'$ by a translation, then the same things hold for this conception as for time. These same circumstances enable us to introduce numbers, and to establish a reversible and single correspondence between the points of a straight line and real numbers by using a unit of length OE.

Let us now consider the group of congruent transformations which leaves the straight line g fixed, i.e. transforms every point of g into a point of g again.

We have called particular attention to rotations among these as having the property of leaving not only g as a whole, but also every single point of g unmoved in position. How can translations in this group be distinguished from twists?

I shall here outline a preliminary argument in which not only the straight line, but also the plane is based on a property of rotation.

Two rays which start from a point O form an **angle**. Every angle can, when inverted, be superposed exactly upon itself, so that one arm falls on the other, and *vice versa*. Every **right** angle is congruent with its complementary angle. Thus, if h is a straight line perpendicular to g at the point A, then there is one rotation about g ("inversion") which interchanges the two halves into which h is divided by A. All the straight lines which are perpendicular to g at A together form the **plane** E through A perpendicular to g. Each pair of these perpendicular straight lines may be produced from any other by a rotation about g.

If g is inverted, and placed upon itself in some way, so that A is transformed into itself, but so that the two halves into which A divides g are interchanged, then the plane E of necessity coincides with itself. The plane may also be defined by taking this property in conjunction with that of symmetry of rotation. Two congruent tables of revolution (i.e. symmetrical with respect to rotations) are

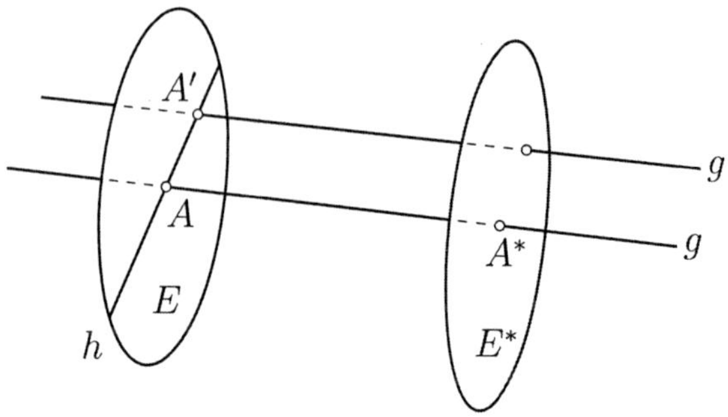

FIG. 1.

plane if, by means of inverting one, so that its axis is vertical in the opposite direction, and placing it on the other, the two table-surfaces can be made to coincide. The plane is homogeneous. The point A on E which appears as the centre in this example is in no way unique among the points of E. A straight line g' passes through each one A' of them in such a way that E is made up of all straight lines through A' perpendicular to g'. The straight lines g' which are perpendicular to E at its points A' respectively form a group of **parallel** straight lines. The straight line g with which we started is in no wise unique among them. The straight lines of this group occupy the whole of space in such a way that only one straight line of the group passes through each point of space. This in no way depends on the point A of the straight line g, at which the above construction was performed.

If A^* is any point on g, then the plane which is erected normally to g at A^* cuts not only g perpendicularly, but also **all** straight lines of the group of parallels. All such normal planes E^* which are erected at all points A^* on g form a group of parallel planes.

These also fill space continuously and uniquely. We need only take another small step to pass from the above framework of space to the rectangular system of co-ordinates. We shall use it here, however, to fix the conception of spatial translation.

Translation is a congruent transformation which transforms not only g but every straight line of the group of parallels into itself. There is one and only one translation which transfers the arbitrary point A on g to the arbitrary point A^* on the same straight line.

I shall now give an alternate method of arriving at the conception of translation. The chief characteristic of translation is that all points are of equal importance in it, and that the behaviour of a point during translation does not allow any objective assertion to be made about it, which could not equally well be made of any other point (this means that the points of space for a given translation can only be distinguished by specifying each one singly ["that one there"], whereas in the case of rotation, for example, the points on the axis are distinguished by the property that they preserve their positions). By using this as a basis we get the following definition of translation, which is quite independent of the conception of rotation. Let the arbitrary point P be transformed into P' by a congruent transformation: we shall call P and P' connected points. A second congruent transformation which has the property of again transforming every pair of connected points into connected points, is to be called **interchangeable** with the first transformation. A congruent transformation is then called a translation, if it gives rise to interchangeable congruent transformations, which transform the arbitrary point A into the arbitrary point B. The statement that two congruent transformations I and II are interchangeable signifies (as is easily proved from the above definition) that the congruent transformation resulting from the successive application of I and II is identical with that which results when these two transformations are performed in the reverse order. It is a fact that one translation (and, as we shall see, **only** one) exists, which transforms the arbitrary point A into the arbitrary point B. Moreover, not only is it a fact that, if $\mathbf{T}$ denote a translation and A and B any two points, there is, according to our definition, a congruent transformation, interchangeable with $\mathbf{T}$, which transforms A into B, but also that the particular **translation** which transforms A into B has the required property. A translation is therefore interchangeable with all other translations, and a congruent transformation which is interchangeable with all translations is also necessarily a translation. From this it follows that the congruent transformation which results from successively performing two translations, and also the "inverse" of a translation (i.e. that transformation which exactly reverses or neutralises the original translation) is itself a translation. Translations possess the "group" property.* There is no translation which transforms A into A except **identity**, in which every point remains undisturbed. For if such a translation were to transform P into P', then, according to definition, there must be a congruent transformation, which transforms A into P and simultaneously A into P'; P and P' must therefore be identical points. Hence there cannot be two different translations both of which transform A into another point B.

As the conception of translation has thus been defined independently of that of rotation, the translational view of the straight line and plane may thus be formed in contrast with the above

* *Vide* note 2.

view based on rotations. Let **a** be a translation which transfers the point A_0 to A. This same translation will transfer A_1 to a point A_2, A_2 to A_3, etc. Moreover, through it A_0 will be derived from a certain point A_{-1}, A_{-1} from A_{-2}, etc. This does not yet give us the whole straight line, but only a series of equidistant points on it. Now, if n is a natural number (integer), a translation $\frac{\mathbf{a}}{n}$ exists which, when repeated n times, gives **a**. If, then, starting from the point A_0 we use $\frac{\mathbf{a}}{n}$ in the same way as we just now used **a** we shall obtain an array of points on the straight line under construction, which will be n times as dense.

If we take all possible whole numbers as values of n this array will become denser in proportion as n increases, and all the points which we obtain finally fuse together into a linear continuum, in which they become embedded, giving up their individual existences (this description is founded on our intuition of continuity). We may say that the straight line is derived from a point by an infinite repetition of the same infinitesimal translation and its inverse. A plane, however, is derived by translating one straight line, g, along another, h. If g and h are two different straight lines passing through the point A_0, then if we apply to g all the translations which transform h into itself, all straight lines which thus result from g together form the **common** plane of g and h.

We succeed in introducing logical order into the structure of geometry only if we first narrow down the general conception of congruent transformation to that of translation, and use this as an axiomatic foundation (§§ 2 and 3). By doing this, however, we arrive at a geometry of translation alone, viz. affine geometry within the limits of which the general conception of congruence has later to be re-introduced (§ 4). Since intuition has now furnished us with the necessary basis we shall in the next paragraph enter into the region of deductive mathematics.

§ 2. The Foundations of Affine Geometry

For the present we shall use the term vector to denote a translation or a displacement **a** in the space. Later we shall have occasion to attach a wider meaning to it. The statement that the displacement **a** transfers the point P to the point Q ("transforms" P into Q) may also be expressed by saying that Q is the end-point of the vector **a** whose starting-point is at P. If P and Q are any two points then there is one and only one displacement **a** which transfers P to Q. We shall call it the vector defined by P and Q, and indicate it by $\overrightarrow{PQ}$.

The translation **c** which arises through two successive translations **a** and **b** is called the sum of **a** and **b**, i.e. $\mathbf{c} = \mathbf{a} + \mathbf{b}$. The definition of summation gives us: (1) the meaning of multiplication (repetition) and of the division of a vector by an integer; (2) the purport of the operation which transforms the vector **a** into its inverse $-\mathbf{a}$; (3) the meaning of the nil-vector **0**, viz. "identity," which leaves all points fixed, i.e. $\mathbf{a}+\mathbf{0} = \mathbf{a}$ and $\mathbf{a}+(-\mathbf{a}) = \mathbf{0}$. It also tells us what is conveyed by the symbols $\pm\frac{m\mathbf{a}}{n} = \lambda\mathbf{a}$, in which m and n are any two natural numbers (integers) and λ denotes the fraction $\pm\frac{m}{n}$. By taking account of the postulate of continuity this also gives us the significance of $\lambda\mathbf{a}$, when λ is **any** real number. The following system of axioms may be set up for affine geometry:—

I. Vectors

Two vectors $\mathbf{a}$ and $\mathbf{b}$ uniquely determine a vector $\mathbf{a}+\mathbf{b}$ as their sum. A number λ and a vector $\mathbf{a}$ uniquely define a vector $\lambda\mathbf{a}$, which is "λ times $\mathbf{a}$" (multiplication). These operations are subject to the following laws:—

(α) Addition—

(1) $\mathbf{a}+\mathbf{b}=\mathbf{b}+\mathbf{a}$ (Commutative Law).

(2) $(\mathbf{a}+\mathbf{b})+\mathbf{c}=\mathbf{a}+(\mathbf{b}+\mathbf{c})$ (Associative Law).

(3) If $\mathbf{a}$ and $\mathbf{c}$ are any two vectors, then there is one and only one value of $\mathbf{x}$ for which the equation $\mathbf{a}+\mathbf{x}=\mathbf{c}$ holds. It is called the difference between $\mathbf{c}$ and $\mathbf{a}$ and signifies $\mathbf{c}-\mathbf{a}$ (Possibility of Subtraction).

(β) Multiplication—

(1) $(\lambda+\mu)\mathbf{a}=(\lambda\mathbf{a})+(\mu\mathbf{a})$ (First Distributive Law).

(2) $\lambda(\mu\mathbf{a})=(\lambda\mu)\mathbf{a}$ (Associative Law).

(3) $1\cdot\mathbf{a}=\mathbf{a}$.

(4) $\lambda(\mathbf{a}+\mathbf{b})=(\lambda\mathbf{a})+(\lambda\mathbf{b})$ (Second Distributive Law).

For rational multipliers λ, μ, the laws (β) follow from the axioms of addition if multiplication by such factors be **defined** from addition. In accordance with the principle of continuity we shall also make use of them for any arbitrary real numbers, but we purposely formulate them as separate axioms because they cannot be derived in the general form from the axioms of addition by logical reasoning alone. By refraining from reducing multiplication to addition we are enabled through these axioms to banish continuity, which is so difficult to fix precisely, from the logical structure of geometry. The law (β) 4 comprises the theorems of similarity.

(γ) The "Axiom of Dimensionality," which occupies the next place in the system, will be formulated later.

II. Points and Vectors

1. Every pair of points A and B determines a vector $\mathbf{a}$; expressed symbolically $\overrightarrow{AB} = \mathbf{a}$. If A is any point and $\mathbf{a}$ any vector, there is one and only one point B for which $\overrightarrow{AB} = \mathbf{a}$.

2. If $\overrightarrow{AB} = \mathbf{a}$, $\overrightarrow{BC} = \mathbf{b}$, then $\overrightarrow{AC} = \mathbf{a} + \mathbf{b}$.

In these axioms two fundamental categories of objects occur, viz. points and vectors; and there are three fundamental relations, those expressed symbolically by—

$$\mathbf{a} + \mathbf{b} = \mathbf{c} \qquad \mathbf{b} = \lambda\mathbf{a} \qquad \overrightarrow{AB} = \mathbf{a}. \tag{1}$$

All conceptions which may be defined from (1) by logical reasoning alone belong to affine geometry. The doctrine of affine geometry is composed of all theorems which can be deduced logically from the axioms (1), and it can thus be erected deductively on the axiomatic basis (1) and (2). The axioms are not all logically independent of one another for the axioms of addition for vectors (Iα, 2 and 3) follow from those (II) which govern the relations between points and vectors. It was our aim, however, to make the vector-axioms I suffice in themselves, so that we should be able to deduce from them all those facts which involve vectors exclusively (and not the relations between vectors and points).

From the axioms of addition Iα we may conclude that a definite vector $\mathbf{0}$ exists which, for every vector $\mathbf{a}$, satisfies the equation $\mathbf{a} + \mathbf{0} = \mathbf{a}$. From the axioms II it further follows that $\overrightarrow{AB}$ is equal to this vector $\mathbf{0}$ when, and only when, the points A and B coincide.

If O is a point and $\mathbf{e}$ is a vector differing from $\mathbf{0}$, the end-points of all vectors OP which have the form $\xi\mathbf{e}$ (ξ being an arbitrary real number) form a **straight line**. This explanation gives the translational or affine view of straight lines the form of an exact definition which rests solely upon the fundamental conceptions involved in the system of affine axioms. Those points P for which the abscissa ξ is positive form one-half of the straight line through O, those for which ξ is negative form the other half. If we write $\mathbf{e}_1$ in place of $\mathbf{e}$, and if $\mathbf{e}_2$ is another vector, which is not of the form $\xi\mathbf{e}_1$, then the end-points P of all vectors $\overrightarrow{OP}$ which have the form $\xi_1\mathbf{e}_1+\xi_2\mathbf{e}_2$ form a **plane E** (in this way the plane is derived affinely by sliding one straight line along another). If we now displace the plane **E** along a straight line passing through O but not lying on **E**, the plane passes through all space. Accordingly, if $\mathbf{e}_3$ is a vector not expressible in the form $\xi_1\mathbf{e}_1 + \xi_2\mathbf{e}_2$, then every vector can be represented in one and only one way as a linear combination of $\mathbf{e}_1$, $\mathbf{e}_2$, and $\mathbf{e}_3$, viz.

$$\xi_1\mathbf{e}_1 + \xi_2\mathbf{e}_2 + \xi_3\mathbf{e}_3.$$

We thus arrive at the following set of definitions:—

A finite number of vectors $\mathbf{e}_1, \mathbf{e}_2, \ldots, \mathbf{e}_h$ is said to be **linearly independent** if

$$\xi_1\mathbf{e}_1 + \xi_2\mathbf{e}_2 + \cdots + \xi_h\mathbf{e}_h \tag{2}$$

only vanishes when all the co-efficients ξ vanish simultaneously. With this assumption all vectors of the form (2) together constitute a so-called ***h*-dimensional linear vector-manifold** (or simply vector field); in this case it is the one mapped out by the vectors $\mathbf{e}_1, \mathbf{e}_2, \ldots, \mathbf{e}_h$. An h-dimensional linear vector-manifold **M** can be characterised without referring to its particular base $\mathbf{e}$, as follows:—

(1) The two fundamental operations, viz. addition of two vectors and multiplication of a vector by a number do not transcend

the manifold, i.e. the sum of two vectors belonging to **M** as also the product of such a vector and any real number also lie in **M**.

(2) There are h linearly independent vectors in **M**, but every $h+1$ are linearly dependent on one another.

From the property (2) (which may be deduced from our original definition with the help of elementary results of linear equations) it follows that h, the dimensional number, is as such characteristic of the manifold, and is not dependent on the special vector base by which we map it out. The dimensional axiom which was omitted in the above table of axioms may now be formulated.

There are n linearly independent vectors, but every $n+1$ are linearly dependent on one another,
or: The vectors constitute an n-dimensional linear manifold. If $n = 3$ we have affine geometry of space, if $n = 2$ plane geometry, if $n = 1$ geometry of the straight line. In the deductive treatment of geometry it will, however, be expedient to leave the value of n undetermined, and to develop an "n-dimensional geometry" in which that of the straight line, of the plane, and of space are included as special cases. For we see (at present for affine geometry, later on for **all** geometry) that there is nothing in the mathematical structure of space to prevent us from exceeding the dimensional number 3. In the light of the mathematical uniformity of space as expressed in our axioms, its special dimensional number 3 appears to be accidental, so that a systematic deductive theory cannot be restricted by it. We shall revert to the idea of an n-dimensional geometry, obtained in this way, in the next paragraph.* We must first complete the definitions outlined.

If O is an arbitrary point, then the sum-total of all the end-

* *Vide* note 3.

points P of vectors, the origin of which is at O and which belong to an h-dimensional vector field $\mathbf{M}$ as represented by (2), occupy fully **an h-dimensional point-configuration**. We may, as before, say that it is **mapped out** by the vectors $\mathbf{e}_1$, $\mathbf{e}_2$, $\ldots$, $\mathbf{e}_h$, which start from O. The one-dimensional configuration of this type is called a straight line, the two-dimensional a plane. The point O does not play a unique part in this linear configuration. If O' is any other point of it, then $\overrightarrow{O'P}$ traverses the same vector manifold $\mathbf{M}$ if all possible points of the linear aggregate are substituted for P in turn.

If we measure off all vectors of the manifold $\mathbf{M}$ firstly from the point O and then from any other arbitrary point O' the two resulting linear point aggregates are said to be **parallel** to one another. The definition of parallel planes and parallel straight lines is contained in this. That part of the h-dimensional linear assemblage which results when we measure off all the vectors (2) from O, subject to the limitation

$$0 \leq \xi_1 \leq 1, \qquad 0 \leq \xi_2 \leq 1, \quad \ldots, \qquad 0 \leq \xi_h \leq 1,$$

will be called the h-dimensional **parallelepiped** which has its origin at O and is mapped out by the vectors $\mathbf{e}_1$, $\mathbf{e}_2$, $\ldots$, $\mathbf{e}_h$. (The one-dimensional parallelepiped is called *distance*, the two-dimensional one is called *parallelogram*. None of these conceptions is limited to the case $n = 3$, which is presented in ordinary experience.)

A point O in conjunction with n linear independent vectors $\mathbf{e}_1$, $\mathbf{e}_2$, $\ldots$, $\mathbf{e}_h$ will be called a co-ordinate system (**C**). Every vector $\mathbf{x}$ can be presented in one and only one way in the form

$$\mathbf{x} = \xi_1\mathbf{e}_1 + \xi_2\mathbf{e}_2 + \cdots + \xi_n\mathbf{e}_n. \tag{3}$$

The numbers ξ_i will be called its **components** in the co-ordinate system (**C**). If P is any arbitrary point and if $\overrightarrow{OP}$ is equal to the vector (3), then the ξ_i are called the **co-ordinates** of P. All co-ordinate systems are equivalent in affine geometry. There is no property of this geometry which distinguishes one from another. If

$$O' \mid \mathbf{e}_1', \mathbf{e}_2', \ldots, \mathbf{e}_n'$$

denote a second co-ordinate system, equations

$$\mathbf{e}_i' = \sum_{k=1}^{n} \alpha_k^i \mathbf{e}_k \tag{4}$$

will hold in which the α_k^i form a number system which must have a non-vanishing determinant (since the $\mathbf{e}_i'$ are linearly independent). If ξ_i are the components of a vector $\mathbf{x}$ in the first co-ordinate system and ξ_i' the components of the same vector in the second co-ordinate system, then the relation

$$\xi_i = \sum_{k=1}^{n} \alpha_i^k \xi_k' \tag{5}$$

holds; this is easily shown by substituting the expressions (4) in the equation

$$\sum_i \xi_i \mathbf{e}_i = \sum_i \xi_i' \mathbf{e}_i'.$$

Let $\alpha_1, \alpha_2, \ldots, \alpha_n$ be the co-ordinates of O' in the first co-ordinate system. If x_i are the co-ordinates of any arbitrary point in the first system and x_i' its co-ordinates in the second, the equations

$$x_i = \sum_{k=1}^{n} \alpha_i^k x_k' + \alpha_i \tag{6}$$

hold. For $x_i - \alpha_i$ are the components of

$$\overrightarrow{O'P} = \overrightarrow{OP} - \overrightarrow{OO'}$$

in the first system; x'_i are the components of $\overrightarrow{O'P}$ in the second. Formulæ (6) which give the transformation for the co-ordinates are thus linear. Those (viz. 5) which transform the vector components are easily derived from them by cancelling the terms α_i which do not involve the variables. An analytical treatment of affine geometry is possible, in which every vector is represented by its components and every point by its co-ordinates. The geometrical relations between points and vectors then express themselves as relations between their components and co-ordinates respectively of such a kind that they are not destroyed by linear arbitrary transformations.

Formulæ (5) and (6) may also be interpreted in another way. They may be regarded as a mode of representing an affine **transformation** in a definite co-ordinate system. A transformation, i.e. a rule which assigns a vector $\mathbf{x}'$ to every vector $\mathbf{x}$ and a point P' to every point P, is called linear or affine if the fundamental affine relations (1) are not disturbed by the transformation: so that if the relations (1) hold for the original points and vectors they also hold for the transformed points and vectors:

$$\mathbf{a}' + \mathbf{b}' = \mathbf{c}' \qquad \mathbf{b}' = \lambda\mathbf{a}' \qquad \overrightarrow{A'B'} = \mathbf{a}' - \mathbf{b}'$$

and if in addition no vector differing from $\mathbf{0}$ transforms into the vector $\mathbf{0}$. Expressed in other words this means that two points are transformed into one and the same point only if they are themselves identical. Two figures which are formed from one another by an affine transformation are said to be affine. From the point of view of affine geometry they are identical. There can be no affine property possessed by the one which is not possessed by the other. The conception of linear transformation thus plays the same part in affine geometry as congruence plays in general geometry; hence its fundamental importance. In affine transformations linearly independent vectors become transformed into linearly independent vectors again; likewise an h-dimensional linear configuration into a like configuration; parallels into parallels; a co-ordinate system $O \mid \mathbf{e}_1, \ \mathbf{e}_2, \ \ldots, \mathbf{e}_n$ into a new co-ordinate system $O' \mid \mathbf{e}'_1, \ \mathbf{e}'_2, \ \ldots, \mathbf{e}'_n$.

Let the numbers α^i_k, α_i, have the same meaning as above. The vector (3) is changed by the affine transformation into

$$\mathbf{x}' = \xi_1 \mathbf{e}'_1 + \xi_2 \mathbf{e}'_2 + \cdots + \xi_n \mathbf{e}'_n.$$

If we substitute in this the expressions for $\mathbf{e}'_i$ and use the original co-ordinate system $O \mid \mathbf{e}_1, \ \mathbf{e}_2, \ \ldots, \mathbf{e}_n$ to picture the affine transformation, then, interpreting ξ_i as the components of any vector and ξ'_i as the components of its transformed vector,

$$\xi'_i = \sum_{k=1}^{n} \alpha^k_i \xi_k. \tag{5'}$$

If P becomes P', the vector $\overrightarrow{OP}$ becomes $\overrightarrow{O'P'}$, and it follows from this that if x_i are the co-ordinates of P and x'_i those of P', then

$$x'_i = \sum_{k=1}^{n} \alpha^k_i x_k + \alpha_i.$$

In analytical geometry it is usual to characterise linear configurations by linear equations connecting the co-ordinates of the "current" point (variable). This will be discussed in detail in the next paragraph. Here we shall just add the fundamental conception of "linear forms" upon which this discussion is founded. A function $L(\mathbf{x})$, the argument $\mathbf{x}$ of which assumes the value of every vector in turn, these values being real numbers only, is called a **linear form**, if it has the functional properties

$$L(\mathbf{a}+\mathbf{b}) = L(\mathbf{a}) + L(\mathbf{b}); \qquad L(\lambda\mathbf{a}) = \lambda \cdot L(\mathbf{a}).$$

In a co-ordinate system $\mathbf{e}_1, \mathbf{e}_2, \ldots, \mathbf{e}_n$ each of the n vector-components ξ_i of $\mathbf{x}$ is such a linear form. If $\mathbf{x}$ is defined by (3), then any arbitrary linear form L satisfies

$$L(\mathbf{x}) = \xi_1 L(\mathbf{e}_1) + \xi_2 L(\mathbf{e}_2) + \cdots + \xi_n L(\mathbf{e}_n).$$

Thus if we put $L(\mathbf{e}_i) = a_i$, the linear form, expressed in terms of components, appears in the form

$$a_1\xi_1 + a_2\xi_2 + \cdots + a_n\xi_n \quad \text{(the } a_i\text{'s are its constant co-efficients).}$$

Conversely, every expression of this type gives a linear form. A number of linear forms $L_1, L_2, L_3, \ldots, L_h$ are linearly independent, if no constants λ_i exist, for which the identity-equation holds:

$$\lambda_1 L_1(\mathbf{x}) + \lambda_2 L_2(\mathbf{x}) + \ldots + \lambda_h L_h(\mathbf{x}) = 0$$

except $\lambda_i = 0$. $n+1$ linear forms are *always* linearly interdependent.

§ 3. The Conception of n-dimensional Geometry. Linear Algebra. Quadratic Forms

To recognise the perfect mathematical harmony underlying the laws of space, we must discard the particular dimensional number

$n = 3$. Not only in geometry, but to a still more astonishing degree in physics, has it become more and more evident that as soon as we have succeeded in unravelling fully the natural laws which govern reality, we find them to be expressible by mathematical relations of surpassing simplicity and architectonic perfection. It seems to me to be one of the chief objects of mathematical instruction to develop the faculty of perceiving this simplicity and harmony, which we cannot fail to observe in the theoretical physics of the present day. It gives us deep satisfaction in our quest for knowledge. Analytical geometry, presented in a compressed form such as that I have used above in exposing its principles, conveys an idea, even if inadequate, of this perfection of form. But not only for this purpose must we go beyond the dimensional number $n = 3$, but also because we shall later require four-dimensional geometry for concrete physical problems such as are introduced by the theory of relativity, in which Time becomes added to Space in a four-dimensional geometry.

We are by no means obliged to seek illumination from the mystic doctrines of spiritists to obtain a clearer vision of multi-dimensional geometry. Let us consider, for instance, a homogeneous mixture of the four gases, hydrogen, oxygen, nitrogen, and carbon dioxide. An arbitrary quantum of such a mixture is specified if we know how many grams of each gas are contained in it. If we call each such quantum a vector (we may bestow names at will) and if we interpret addition as implying the union of two quanta of the gases in the ordinary sense, then all the axioms I of our system referring to vectors are fulfilled for the dimensional number $n = 4$, provided we agree also to talk of negative quanta of gas. One gram of pure hydrogen, one gram of oxygen, one gram of nitrogen, and one gram of carbon dioxide are four "vectors," independent of one another from which all other gas quanta may be built up linearly; they thus form a co-ordinate system. Let us take another example. We have five parallel horizontal bars upon each of which a small bead slides. A definite condition of this primitive "adding-machine" is defined if the position of each of the five beads upon its respective rod is known. Let us call such a condition a "point" and every simultaneous displacement of the five beads a "vector," then all of our axioms are satisfied for the dimensional number $n = 5$. From this it is evident that constructions of various types may be evolved which, by an appropriate disposal of names, satisfy our axioms. Infinitely more important than these somewhat frivolous examples is the following one which shows that **our axioms characterise the basis of our operations in the theory of linear equations**. If α_i and α are given numbers,

$$\alpha_1 x_1 + \alpha_2 x_2 + \ldots + \alpha_n x_n = 0 \tag{7}$$

is usually called a **homogeneous** linear equation in the unknowns x_i, whereas

$$\alpha_1 x_1 + \alpha_2 x_2 + \ldots + \alpha_n x_n = \alpha \tag{8}$$

is called a **non-homogeneous** linear equation. In treating the theory of linear homogeneous equations, it is found useful to have a short name for the system of values of the variables x_i; we shall call it "vector". In carrying out calculations with these vectors, we shall define the sum of the two vectors

$$(a_1, a_2, \ldots, a_n) \quad \text{and} \quad (b_1, b_2, \ldots, b_n)$$

to be the vector

$$(a_1 + b_1, a_2 + b_2, \ldots, a_n + b_n)$$

and λ times the first vector to be

$$(\lambda a_1, \lambda a_2, \ldots, \lambda a_n).$$

The axioms I for vectors are then fulfilled for the dimensional number n.

$$\begin{aligned}\mathbf{e}_1 &= (1, 0, 0, \ldots, 0),\\ \mathbf{e}_2 &= (0, 1, 0, \ldots, 0),\\ &\ldots\ldots\ldots\ldots\ldots\ldots\\ \mathbf{e}_n &= (0, 0, 0, \ldots, 1)\end{aligned}$$

form a system of independent vectors. The components of any arbitrary vector $(x_1, x_2, \ldots, x_n)$ in this co-ordinate system are the numbers x_i themselves. The fundamental theorem in the solution of linear homogeneous equations may now be stated thus:—

$$\text{if} \quad L_1(\mathbf{x}), \quad L_2(\mathbf{x}), \quad \ldots, \quad L_h(\mathbf{x})$$

are h linearly independent linear forms, the solutions $\mathbf{x}$ of the equations

$$L_1(\mathbf{x}) = 0, \quad L_2(\mathbf{x}) = 0, \quad \ldots, \quad L_h(\mathbf{x}) = 0$$

form an $(n - h)$-dimensional linear vector manifold.

In the theory of non-homogeneous linear equations we shall find it advantageous to denote a system of values of the variables x_i a "point". If x_i and x_i' are two systems which are solutions of equation (8), their difference

$$x_1' - x_1, \quad x_2' - x_2, \quad \ldots, \quad x_n' - x_n$$

is a solution of the corresponding homogeneous equation (7). We shall, therefore, call this difference of two systems of values of the variables x_i a "vector," viz. the "vector" defined by the two "points" (x_i) and (x'_i); we make the above conventions for the addition and multiplication of these vectors. **All the axioms then hold.** In the particular co-ordinate system composed of the vectors $\mathbf{e}_i$ given above, and having the "origin" $O = (0, 0, \ldots, 0)$, the co-ordinates of a point (x_i) are the numbers x_i themselves. The fundamental theorem concerning linear equations is: those points which satisfy h independent linear equations, form a point-configuration of $n - h$ dimensions.

In this way we should not only have arrived quite naturally at our axioms without the help of geometry by using the theory of linear equations, but we should also have reached the wider conceptions which we have linked up with them. In some ways, indeed, it would appear expedient (as is shown by the above formulation of the theorem concerning homogeneous equations) to build up the theory of linear equations upon an axiomatic basis by starting from the axioms which have here been derived from geometry. A theory developed along these lines would then hold for any domain of operations, for which these axioms are fulfilled, and not only for a "system of values in n variables". It is easy to pass from such a theory which is more conceptual, to the usual one of a more formal character which operates from the outset with numbers x_i by taking a definite co-ordinate system as a basis, and then using in place of vectors and points their components and co-ordinates respectively.

It is evident from these arguments that the whole of affine geometry merely teaches us that space is a **region of three dimensions in linear quantities** (the meaning of this statement will be sufficiently clear without further explanation). All the separate facts of intuition which were mentioned in § 1 are simply disguised forms of this one truth. Now, if on the one hand it is very satisfactory to be able to give a common ground in the theory of knowledge for the many varieties of statements concerning space, spatial configurations, and spatial relations which, taken together, constitute geometry, it must on the other hand be emphasised that this demonstrates very clearly with what little right mathematics may claim to expose the intuitional nature of space. Geometry contains no trace of that which makes the space of intuition what it **is** in virtue of its own entirely distinctive qualities which are not shared by "states of addition-machines" and "gas-mixtures" and "systems of solutions of linear equations". It is left to metaphysics to make this "comprehensible" or indeed to show why and in what sense it is incomprehensible. We as mathematicians have reason to be proud of the wonderful insight into the knowledge of space which we gain, but, at the same time, we must recognise with humility that our conceptual theories enable us to grasp only one aspect of the nature of space, that which, moreover, is most formal and superficial.

To complete the transition from affine geometry to complete metrical geometry we yet require several conceptions and facts which occur in linear algebra and which refer to **bilinear and quadratic forms**. A function $Q(\mathbf{x}, \mathbf{y})$ of two arbitrary vectors $\mathbf{x}$ and $\mathbf{y}$ is called a bilinear form if it is a linear form in $\mathbf{x}$ as well as in $\mathbf{y}$. If in a certain co-ordinate system ξ_i are the components of $\mathbf{x}$, η_i those of $\mathbf{y}$, then an equation

$$Q(\mathbf{x}, \mathbf{y}) = \sum_{i,k=1}^{n} a_{ik}\xi_i\eta_k$$

with constant co-efficients a_{ik} holds. We shall call the form "non-degenerate" if it vanishes identically in $\mathbf{y}$ only when the vector $\mathbf{x} = \mathbf{0}$. This happens when, and only when, the homogeneous equations

$$\sum_{i=1}^{n} a_{ik}\xi_i = 0$$

have a single solution $\xi_i = 0$ or when the determinant $|a_{ik}| \neq 0$. From the above explanation it follows that this condition, viz. the non-vanishing of the determinant, persists for arbitrary linear transformations. The bilinear form is called **symmetrical** if $Q(\mathbf{y}, \mathbf{x}) = Q(\mathbf{x}, \mathbf{y})$. This manifests itself in the co-efficients by the symmetrical property $a_{ki} = a_{ik}$. Every bilinear form $Q(\mathbf{x}, \mathbf{y})$ gives rise to a **quadratic form** which depends on only one variable vector $\mathbf{x}$

$$Q(\mathbf{x}) = Q(\mathbf{x}, \mathbf{x}) = \sum_{i,k=1}^{n} a_{ik}\xi_i\xi_k.$$

In this way every quadratic form is derived in general from one, and only one, **symmetrical** bilinear form. The quadratic form $Q(\mathbf{x})$ which we have just formed may also be produced from the symmetrical form

$$\tfrac{1}{2}\{Q(\mathbf{x}, \mathbf{y}) + Q(\mathbf{y}, \mathbf{x})\}$$

by identifying $\mathbf{x}$ with $\mathbf{y}$.

To prove that one and the same quadratic form cannot arise from two different symmetrical bilinear forms, one need merely show that a symmetrical bilinear form $Q(\mathbf{x}, \mathbf{y})$ which satisfies the equation $Q(\mathbf{x}, \mathbf{x})$ identically for $\mathbf{x}$, vanishes identically. This, however, immediately results from the relation which holds for every symmetrical bilinear form

$$Q(\mathbf{x}+\mathbf{y},\mathbf{x}+\mathbf{y}) = Q(\mathbf{x},\mathbf{x}) + 2Q(\mathbf{x},\mathbf{y}) + Q(\mathbf{y},\mathbf{y}). \qquad (9)$$

If $Q(\mathbf{x})$ denotes any arbitrary quadratic form then $Q(\mathbf{x},\mathbf{y})$ is always to signify the symmetrical bilinear form from which $Q(\mathbf{x})$ is derived (to avoid mentioning this in each particular case). When we say that a quadratic form is non-degenerate we wish to convey that the above symmetrical bilinear form is non-degenerate. A quadratic form is **positive definite** if it satisfies the inequality $Q(\mathbf{x}) > 0$ for every value of the vector $\mathbf{x} \neq \mathbf{0}$. Such a form is certainly non-degenerate, for no value of the vector $\mathbf{x} \neq \mathbf{0}$ can make $Q(\mathbf{x},\mathbf{y})$ vanish identically in $\mathbf{y}$, since it gives a positive result for $\mathbf{y} = \mathbf{x}$.

§ 4. The Foundations of Metrical Geometry

To bring about the transition from affine to metrical geometry we must once more draw from the fountain of intuition. From it we obtain for three-dimensional space the definition of the **scalar product** of two vectors **a** and **b**. After selecting a definite vector as a unit we measure out the length of **a** and the length (negative or positive as the case may be) of the perpendicular projection of **b** upon **a** and multiply these two numbers with one another. This means that the lengths of not only parallel straight lines may be compared with one another (as in affine geometry) but also such as are arbitrarily inclined to one another. The following rules hold for scalar products:—

$$\lambda\mathbf{a}\cdot\mathbf{b} = \lambda(\mathbf{a}\cdot\mathbf{b}) \qquad (\mathbf{a}+\mathbf{a}')\cdot\mathbf{b} = (\mathbf{a}\cdot\mathbf{b}) + (\mathbf{a}'\cdot\mathbf{b})$$

and analogous expressions with reference to the second factor; in addition, the commutative law $\mathbf{a}\cdot\mathbf{b} = \mathbf{b}\cdot\mathbf{a}$. The scalar product of $\mathbf{a}$ with $\mathbf{a}$ itself, viz. $\mathbf{a}\cdot\mathbf{a}=\mathbf{a}^2$, is always positive except when $\mathbf{a}=\mathbf{0}$, and is equal to the square of the length of $\mathbf{a}$. These laws signify that the scalar product of two arbitrary vectors, i.e. $\mathbf{x}\cdot\mathbf{y}$ is a symmetrical bilinear form, and that the quadratic form which arises from it is positive definite. We thus see that not the length, but the square of the length of a vector depends in a simple rational way on the vector itself; it is a quadratic form. This is the real content of Pythagoras' Theorem. The scalar product is nothing more than the symmetrical bilinear form from which this quadratic form has been derived. We accordingly formulate the following:—

METRICAL AXIOM: *If a unit vector* $\mathbf{e}$*, differing from zero, be chosen, every two vectors* $\mathbf{x}$ *and* $\mathbf{y}$ *uniquely determine a number* $(\mathbf{x}\cdot\mathbf{y})=Q(\mathbf{x},\mathbf{y})$*; the latter, being dependent on the two vectors, is a symmetrical bilinear form.* The quadratic form $(\mathbf{x}\cdot\mathbf{x})=Q(\mathbf{x})$ which arises from it is positive definite. $Q(\mathbf{e})=1$.

We shall call Q the **metrical groundform**. We then have that *an affine transformation which, in general, transforms the vector* $\mathbf{x}$ *into* $\mathbf{x}'$ *is a congruent one if it leaves the metrical groundform unchanged:*—

$$Q(\mathbf{x}')=Q(\mathbf{x}). \tag{10}$$

*Two geometrical figures which can be transformed into one another by a congruent transformation are congruent.** The conception of congruence is **defined** in our axiomatic scheme by these statements. If we have a domain of operation in which the axioms of § 2 are fulfilled, we can choose any arbitrary positive definite

*We take no notice here of the difference between direct congruence and mirror congruence (lateral inversion). It is present even in affine transformations, in n-dimensional space as well as 3-dimensional space.

quadratic form in it, "promote" it to the position of a fundamental metrical form, and, using it as a basis, define the conception of congruence as was just now done. This form then endows the affine space with metrical properties and Euclidean geometry in its entirety now holds for it. The formulation at which we have arrived is not limited to any special dimensional number.

It follows from (10), in virtue of relation (9) of § 3, that for a congruent transformation the more general relation

$$Q(\mathbf{x}', \mathbf{y}') = Q(\mathbf{x}, \mathbf{y})$$

holds.

Since the conception of congruence is defined by the metrical groundform it is not surprising that the latter enters into all formulæ which concern the measure of geometrical quantities. Two vectors $\mathbf{a}$ and $\mathbf{a}'$ are congruent if, and only if,

$$Q(\mathbf{a}) = Q(\mathbf{a}').$$

We could accordingly introduce $Q(\mathbf{a})$ as a measure of the vector $\mathbf{a}$. Instead of doing this, however, we shall use the positive square root of $Q(\mathbf{a})$ for this purpose and call it the length of the vector $\mathbf{a}$ (this we shall adopt as our definition) so that the further condition is fulfilled that the length of the sum of two parallel vectors pointing in the same direction is equal to the sum of the lengths of the two single vectors. If $\mathbf{a}$, $\mathbf{b}$ as well as $\mathbf{a}'$, $\mathbf{b}'$ are two pairs of vectors, all of length unity, then the figure formed by the first two is congruent with that formed by the second pair, if, and only if, $Q(\mathbf{a}, \mathbf{b}) = Q(\mathbf{a}', \mathbf{b}')$.

In this case again we do not introduce the number $Q(\mathbf{a}, \mathbf{b})$ itself as a measure of the **angle**, but a number θ which is related to it by the transcendental function cosine thus:—

$$\cos\theta = Q(\mathbf{a}, \mathbf{b})$$

so as to be in agreement with the theorem that the numerical measure of an angle composed of two angles in the same plane is the sum of the numerical values of these angles. The angle which is formed from any two arbitrary vectors $\mathbf{a}$ and $\mathbf{b}$ ($\neq \mathbf{0}$) is then calculated from

$$\cos\theta = \frac{Q(\mathbf{a}, \mathbf{b})}{\sqrt{Q(\mathbf{a}, \mathbf{a}) \cdot Q(\mathbf{b}, \mathbf{b})}}. \tag{11}$$

In particular, two vectors $\mathbf{a}$, $\mathbf{b}$ are said to be **perpendicular** to one another if $Q(\mathbf{a}, \mathbf{b}) = 0$. This reminder of the simplest metrical formulæ of analytical geometry will suffice.

The angle defined by (11) which has been formed by two vectors is shown always to be real by the inequality

$$Q^2(\mathbf{a}, \mathbf{b}) \leq Q(\mathbf{a}) \cdot Q(\mathbf{b}) \tag{12}$$

which holds for every quadratic form Q which is ≥ 0 for all values of the argument. It is most simply deduced by forming

$$Q(\lambda\mathbf{a} + \mu\mathbf{b}) = \lambda^2 Q(\mathbf{a}) + 2\lambda\mu Q(\mathbf{a}, \mathbf{b}) + \mu^2 Q(\mathbf{b}) \geq 0.$$

Since this quadratic form in λ and μ cannot assume both positive and negative values its "discriminant" $Q^2(\mathbf{a}, \mathbf{b}) - Q(\mathbf{a}) \cdot Q(\mathbf{b})$ cannot be positive.

A number, n, of independent vectors form a **Cartesian coordinate system** if for every vector

$$\begin{aligned} \mathbf{x} &= x_1\mathbf{e}_1 + x_2\mathbf{e}_2 + \ldots + x_n\mathbf{e}_n \\ Q(\mathbf{x}) &= x_1^2 + x_2^2 + \ldots + x_n^2 \end{aligned} \tag{13}$$

holds, i.e. if

$$Q(\mathbf{e}_i, \mathbf{e}_j) = \begin{cases} 1 & (i = k), \\ 0 & (i \neq k). \end{cases}$$

From the standpoint of metrical geometry all co-ordinate systems are of equal value. A proof (appealing directly to our geometrical sense) of the theorem that such systems exist will now be given not only for a "definite" but also for any arbitrary non-degenerate quadratic form, inasmuch as we shall find later in the theory of relativity that it is just the "indefinite" case that plays the decisive rôle. We enunciate as follows:—

Corresponding to every non-degenerate quadratic form Q a co-ordinate system $\mathbf{e}_i$ can be introduced such that

$$Q(\mathbf{x}) = \epsilon_1 x_1^2 + \epsilon_2 x_2^2 + \cdots + \epsilon_n x_n^2 \quad (\epsilon_i = \pm 1). \tag{14}$$

Proof.—Let us choose any arbitrary vector $\mathbf{e}_1$ for which $Q(\mathbf{e}_1) \neq 0$. By multiplying it by an appropriate positive constant we can arrange so that $Q(\mathbf{e}_1) = \pm 1$. We shall call a vector $\mathbf{x}$ for which $Q(\mathbf{e}_1, \mathbf{x}) = 0$ **orthogonal** to $\mathbf{e}_1$. If $\mathbf{x}^*$ is a vector which is orthogonal to $\mathbf{e}_1$, and if x_1 is any arbitrary number, then

$$\mathbf{x} = x_1 \mathbf{e}_1 + x^* \tag{15}$$

satisfies Pythagoras' Theorem:—

$$Q(\mathbf{x}) = x_1^2 Q(\mathbf{e}_1) + 2x_1 Q(\mathbf{e}_1, \mathbf{x}^*) + Q(\mathbf{x}^*) = \pm x_1^2 + Q(\mathbf{x}^*).$$

The vectors orthogonal to $\mathbf{e}_1$ constitute an $(n-1)$-dimensional linear manifold, in which $Q(\mathbf{x})$ is a non-degenerate quadratic form. Since our theorem is self-evident for the dimensional number $n = 1$, we may assume that it holds for $(n-1)$ dimensions (proof by successive induction from the case $(n-1)$ to that of n). According to this, $n-1$ vectors $\mathbf{e}_2, \ldots, \mathbf{e}_n$, orthogonal to $\mathbf{e}_1$ exist, such that for

$$\mathbf{x}^* = x_2\mathbf{e}_2 + \cdots + x_n\mathbf{e}_n$$

the relation

$$Q(\mathbf{x}^*) = \pm x_2^2 \pm \ldots \pm x_n^2$$

holds. This enables $Q(\mathbf{x})$ to be expressed in the required form. Then

$$Q(\mathbf{e}_i) = \epsilon_i \qquad Q(\mathbf{e}_i, \mathbf{e}_k) = 0 \quad (i \neq k).$$

These relations result in all the $\mathbf{e}_i$'s being independent of one another and in each vector $\mathbf{x}$ being representable in the form (13). They give

$$x_i = \epsilon_i \cdot Q(\mathbf{e}_i, \mathbf{x}). \tag{16}$$

An important corollary is to be made in the "indefinite" case. The numbers r and s attached to the ϵ_i's, and having positive and negative signs respectively, are uniquely determined by the quadratic form: it may be said to have r positive and s negative dimensions. (s may be called the inertial index of the quadratic form, and the theorem just enunciated is known by the name "Law of Inertia". The classification of surfaces of the second order depends on it.) The numbers r and s may be characterised invariantly thus:—

There are r mutually orthogonal vectors $\mathbf{e}$, for which $Q(\mathbf{e}) > 0$; but for a vector $\mathbf{x}$ which is orthogonal to these and not equal to $\mathbf{0}$, it necessarily follows that $Q(\mathbf{x}) < 0$. Consequently there cannot be more than r such vectors. A corresponding theorem holds for s.

r vectors of the required type are given by **those** r fundamental vectors $\mathbf{e}_i$ of the co-ordinate system upon which the expression (14) is founded, to **which** the positive signs ϵ_i correspond. The corresponding components x_i $(i = 1, 2, 3, \ldots, r)$ are definite linear forms of $\mathbf{x}$ [cf. (16)]: $x_i = L_i(\mathbf{x})$. If, now, $\mathbf{e}_i$ $(i = 1, 2, \ldots, r)$ is any system of vectors which are mutually orthogonal to one another, and satisfy the condition $Q(\mathbf{e}_i) > 0$, and if $\mathbf{x}$ is a vector orthogonal to these $\mathbf{e}_i$, we can set up a linear combination

$$\mathbf{y} = \lambda_1\mathbf{e}_1 + \ldots + \lambda_r\mathbf{e}_r + \mu\mathbf{x}$$

in which not all the co-efficients vanish and which satisfies the r homogeneous equations

$$L_1(\mathbf{y}) = 0, \quad \ldots, \quad L_r(\mathbf{y}) = 0.$$

It is then evident from the form of the expression that $Q(\mathbf{y})$ must be negative unless $\mathbf{y} = \mathbf{0}$. In virtue of the formula

$$Q(\mathbf{y}) - \{\lambda_1^2 Q(\mathbf{e}_1) + \cdots + \lambda_r^2 Q(\mathbf{e}_r)\} = \mu^2 Q(\mathbf{x})$$

it then follows that $Q(\mathbf{x}) < 0$ except in the case in which if $\mathbf{y} = \mathbf{0}$, $\lambda_1 = \cdots = \lambda_r$ also $= 0$. But then, by hypothesis, μ must $\neq 0$, i.e. $\mathbf{x} = \mathbf{0}$.

In the theory of relativity the case of a quadratic form with one negative and $n - 1$ positive dimensions becomes important. In three-dimensional space, if we use affine co-ordinates,

$$-x_1^2 + x_2^2 + x_3^2 = 0$$

is the equation of a cone having its vertex at the origin and consisting of two sheets, as expressed by the negative sign of x_1^2, which are only connected with one another at the origin of co-ordinates. This division into two sheets allows us to draw a distinction between past and future in the theory of relativity. We shall endeavour to describe this by an elementary analytical method here instead of using characteristics of continuity.

Let Q be a non-degenerate quadratic form having only one negative dimension. We choose a vector, for which $Q(\mathbf{e}) = -1$. We shall call these vectors $\mathbf{x}$, which are not zero and for which $Q(\mathbf{x}) \leq 0$ "negative vectors". According to the proof just given for the Theorem of Inertia, no negative vector can satisfy the equation $Q(\mathbf{e}, \mathbf{x}) = 0$. Negative vectors thus belong to one of two classes or "cones" according as $Q(\mathbf{e}, \mathbf{x}) < 0$ or > 0; $\mathbf{e}$ itself belongs to the former class, $-\mathbf{e}$ to the latter. A negative vector $\mathbf{x}$ lies "inside" or "on the sheet" of its cone according as $Q(\mathbf{x}) < 0$ or $= 0$. To show that the two cones are independent of the choice of the vector $\mathbf{e}$, one must prove that, from $Q(\mathbf{e}) = Q(\mathbf{e}') = -1$, and $Q(\mathbf{x}) \leq 0$, it follows that the sign of $\dfrac{Q(\mathbf{e}', \mathbf{x})}{Q(\mathbf{e}, \mathbf{x})}$ is the same as that of $-Q(\mathbf{e}, \mathbf{e}')$.

Every vector $\mathbf{x}$ can be resolved into two summands

$$\mathbf{x} = x\mathbf{e} + \mathbf{x}^*$$

such that the first is proportional and the second ($\mathbf{x}^*$) is orthogonal to $\mathbf{e}$. One need only take $\mathbf{x} = -Q(\mathbf{e}, \mathbf{x})$ and we then get

$$Q(\mathbf{x}) = -x^2 + Q(\mathbf{x}^*);$$

$Q(\mathbf{x}^*)$ is, as we know, necessarily ≥ 0. Let us denote it by Q^*.

The equation

$$Q^* = x^2 + Q(\mathbf{x}) = Q^2(\mathbf{e}, \mathbf{x}) + Q(\mathbf{x})$$

then shows that Q^* is a quadratic form (degenerate), which satisfies the identity or inequality, $Q^*(\mathbf{x}) \geq 0$. We now have

$$Q(\mathbf{x}) = -x^2 + Q^*(\mathbf{x}) \leq 0, \qquad Q(\mathbf{e}') = -e'^2 + Q^*(\mathbf{e}') < 0,$$
$$\{x = -Q(\mathbf{e}, \mathbf{x})\}; \qquad \{e' = -Q(\mathbf{e}, \mathbf{e})\}.$$

From the inequality (12) which holds for Q^*, it follows that

$$\{Q^*(\mathbf{e}', \mathbf{x})\}^2 \leq Q^*(\mathbf{e}') \cdot Q^*(\mathbf{x}) < e'^2 x^2;$$

consequently

$$-Q(\mathbf{e}', \mathbf{x}) = e'x - Q^*(\mathbf{e}', \mathbf{x})$$

has the same sign as the first summand $e'x$.

Let us now revert to the case of a definitely positive metrical groundform with which we are at present concerned. If we use a Cartesian co-ordinate system to represent a congruent transformation, the co-efficients of transformation α_i^k in formula (5′), § 2, will have to be such that the equation

$$\xi_1'^2 + \xi_2'^2 + \cdots + \xi_n'^2 = \xi_1^2 + \xi_2^2 + \cdots + \xi_n^2$$

is identically satisfied by the ξ's. This gives the "conditions for orthogonality"

$$\sum_{r=1}^{n} \alpha_r^i \alpha_r^j = \begin{cases} 1 & (i = j), \\ 0 & (i \neq j). \end{cases} \tag{17}$$

They signify that the transition to the inverse transformation converts the co-efficients α_i^k into α_k^i:—

$$\xi_i = \sum_{k=1}^{n} \alpha_k^i \xi_k'.$$

It furthermore follows that the determinant $\Delta = |\alpha_i^k|$ of a congruent transformation is identical with that of its inverse, and since their product must equal 1, $\Delta = \pm 1$. The positive or the negative sign would occur according as the congruence is real or inverted as in a mirror ("lateral inversion").

Two possibilities present themselves for the analytical treatment of metrical geometry. **Either** one imposes no limitation upon the affine co-ordinate system to be used: the problem is then to develop a theory of invariance with respect to arbitrary linear transformations, in which, however, in contra-distinction to the case of affine geometry, we have a definite invariant quadratic form, viz. the metrical groundform

$$Q(\mathbf{x}) = \sum_{i,k=1}^{n} g_{ik}\xi_i\xi_k$$

once and for all as an absolute datum. **Or**, we may use Cartesian co-ordinate systems from the outset: in this case, we are concerned with a theory of invariance for orthogonal transformations, i.e. linear transformations, in which the co-efficients satisfy the secondary conditions (17). We must here follow the first course so as to be able to pass on later to generalisations which extend beyond the limits of Euclidean geometry. This plan seems advisable from the algebraic point of view, too, since it is easier to gain a survey of those expressions which remain unchanged for **all** linear transformations than of those which are only invariant for orthogonal transformations (a class of transformations which are subjected to secondary limitations not easy to define).

We shall here develop the Theory of Invariance as a "Tensor Calculus" along lines which will enable us to express in a convenient mathematical form, not only geometrical laws, but also all physical laws.

§ 5. Tensors

Two linear transformations,

$$\xi^i = \sum_k \alpha^i_k \bar{\xi}^k, \qquad \left(|\alpha^i_k| \neq 0\right) \tag{18}$$

$$\eta_i = \sum_k \breve{\alpha}^k_i \bar{\eta}_k, \qquad \left(|\breve{\alpha}^k_i| \neq 0\right) \tag{18'}$$

in the variables ξ and η respectively, leading to the variables $\bar{\xi}$, $\bar{\eta}$ are said to be **contra-gredient** to one another, if they make the bilinear form $\sum_i \eta_i \xi^i$ transform into itself, i.e.

$$\sum_i \eta_i \xi^i = \sum_i \bar{\eta}_i \bar{\xi}^i. \tag{19}$$

Contra-gredience is thus a reversible relationship. If the variables ξ, η are transformed into $\bar{\xi}$, $\bar{\eta}$ by one pair of contra-gredient transformations A, $\breve{A}$, and then $\bar{\xi}$, $\bar{\eta}$ into $\bar{\bar{\xi}}$, $\bar{\bar{\eta}}$ by a second pair B, $\breve{B}$ it follows from

$$\sum_i \eta_i \xi^i = \sum_i \bar{\eta}_i \bar{\xi}^i = \sum_i \bar{\bar{\eta}}_i \bar{\bar{\xi}}^i$$

that the two transformations combined, which transform ξ directly into $\bar{\bar{\xi}}$, and η into $\bar{\bar{\eta}}$ are likewise contra-gredient. The co-efficients of two contra-gredient substitutions satisfy the conditions

$$\sum_r \alpha^r_i \breve{\alpha}^k_r = \delta^k_i = \begin{cases} 1 & (i = k), \\ 0 & (i \neq k). \end{cases} \tag{20}$$

If we substitute for the ξ's in the left-hand member of (19) their values in terms of $\bar{\xi}$ obtained from (18), it becomes evident that the equations (18′) are derived by reduction from

$$\bar{\eta}_i = \sum_k \alpha_i^k \eta_k. \tag{21}$$

There is thus one and only one contra-gredient transformation corresponding to every linear transformation. For the same reason as (21)

$$\bar{\xi}^i = \sum_k \breve{\alpha}_k^i \xi^k$$

holds. By substituting these expressions and (21) in (19), we find that the co-efficients, in addition to satisfying the conditions (20), satisfy

$$\sum_r \alpha_r^i \breve{\alpha}_k^r = \delta_k^i.$$

An orthogonal transformation is one which is contra-gredient to itself. If we subject a linear form in the variables ξ_i to any arbitrary linear transformation the co-efficients become transformed contragrediently to the variables, or they assume a "contra-variant" relationship to these, as it is sometimes expressed.

In an affine co-ordinate system $O \mid \mathbf{e}_1, \mathbf{e}_2, \ldots, \mathbf{e}_n$ we have up to the present characterised a displacement $\mathbf{x}$ by the uniquely defined components ξ^i given by the equation

$$\mathbf{x} = \xi^1\mathbf{e}_1 + \xi^2\mathbf{e}_2 + \cdots + \xi^n\mathbf{e}_n.$$

If we pass over into another affine co-ordinate system $\bar{O} \mid \bar{\mathbf{e}}_1, \bar{\mathbf{e}}_2, \ldots, \bar{\mathbf{e}}_n$, whereby

$$\bar{\mathbf{e}}_i = \sum_k \alpha_i^k \mathbf{e}_k,$$

the components of $\mathbf{x}$ undergo the transformation

$$\xi^i = \sum_k \alpha_k^i \bar{\xi}^k$$

as is seen from the equation

$$\mathbf{x} = \sum_i \xi^i \mathbf{e}_i = \sum_i \bar{\xi}^i \bar{\mathbf{e}}_i.$$

These components thus transform themselves contra-grediently to the fundamental vectors of the co-ordinate system, and are related contra-variantly to them; they may thus be more precisely termed the **contra-variant components** of the vector $\mathbf{x}$. In **metrical** space, however, we may also characterise a displacement in relation to the co-ordinate system by the values of its scalar product with the fundamental vectors $\mathbf{e}_i$ of the co-ordinate system

$$\xi_i = (\mathbf{x} \cdot \mathbf{e}_i).$$

In passing over into another co-ordinate system these quantities transform themselves—as is immediately evident from their definition—"co-grediently" to the fundamental vectors (just like the latter themselves), i.e. in accordance with the equations

$$\bar{\xi}_i = \sum_k \alpha_i^k \xi_k;$$

they behave "co-variantly". We shall call them the **co-variant components** of the displacement. The connection between co-variant and contra-variant components is given by the formulæ

$$\xi_i = \sum_k (\mathbf{e}_i \cdot \mathbf{e}_k) \xi^k = \sum_k g_{ik} \xi^k \tag{22}$$

or by their inverses (which are derived from them by simple resolution) respectively

$$\xi^i = \sum_k g^{ik}\xi_k. \tag{22'}$$

In a Cartesian co-ordinate system the co-variant components coincide with the contra-variant components. It must again be emphasised that the contra-variant components alone are at our disposal in affine space, and that, consequently, wherever in the following pages we speak of the components of a displacement without specifying them more closely, the contra-variant ones are implied.

Linear forms of one or two arbitrary displacements have already been discussed above. We can proceed from two arguments to three or more. Let us take, for example, a trilinear form $A(\mathbf{x}, \mathbf{y}, \mathbf{z})$. If in an arbitrary co-ordinate system we represent the two displacements $\mathbf{x}$, $\mathbf{y}$ by their contra-variant components, $\mathbf{z}$ by its co-variant components, i.e. ξ^i, η^i, and ζ_i respectively, then A is algebraically expressed as a trilinear form of these three series of variables with definite number-co-efficients

$$\sum_{i,j,k} a^l_{ik}\xi^i\eta^k\zeta_l. \tag{23}$$

Let the analogous expression in a different co-ordinate system, indicated by bars, be

$$\sum_{i,j,k} \bar{a}^l_{ik}\bar{\xi}^i\bar{\eta}^k\bar{\zeta}_l. \tag{23'}$$

A connection between the two algebraic trilinear forms (23) and (23′) then exists, by which the one resolves into the other if the two series of variables ξ, η are transformed contra-grediently to the fundamental vectors, but the series ζ co-grediently to the latter. This relationship enables us to calculate the co-efficient $\bar{a}^l_{ik}$ of A in the second co-ordinate system if the co-efficients a^l_{ik} and also the transformation co-efficient α^k_i leading from one co-ordinate system to the other are known. We have thus arrived at the conception of the "r-fold co-variant, s-fold contra-variant tensor of the $(r + s)$th degree": it is not confined to metrical geometry but only assumes the space to be affine. We shall now give an explanation of this tensor *in abstracto*. To simplify our expressions we shall take special values for the numbers r and s as in the example quoted above: $r = 2$, $s = l$, $r + s = 3$. We then enunciate:—

A trilinear form of three series of variables which is dependent on the co-ordinate system is called a doubly co-variant, singly contra-variant tensor of the third degree if the above relationship is as follows. The expressions for the linear form in any two co-ordinate systems, viz.:—

$$\sum a^l_{ik}\xi^i\eta^k\zeta_l, \qquad \sum \bar{a}^l_{ik}\bar{\xi}^i\bar{\eta}^k\bar{\zeta}_l$$

resolve into one another, if two of the series of variables (viz. the first two ξ and η) are transformed contra-grediently to the fundamental vectors of the co-ordinate system and the third co-grediently to the same. The co-efficients of the linear form are called the components of the tensor in the co-ordinate system in question. Furthermore, they are called co-variant in the indices, i, k, which are associated with the variables to be transformed contra-grediently, and contra-variant in the others (here only the one index l).

The terminology is based upon the fact that the co-efficients of a uni-linear form behave co-variantly if the variables are transformed contra-grediently, but contra-variantly if they are transformed co-grediently. Co-variant indices are always attached as suffixes to the co-efficients, contra-variant ones written at the top of the co-efficients. Variables with lowered indices are always to be transformed co-grediently to the fundamental vectors of the co-ordinate system, those with raised indices are to be transformed contra-grediently to the same. A tensor is fully known if its components in a co-ordinate system are given (assuming, of course, that the co-ordinate system itself is given); these components may, however, be prescribed arbitrarily. The tensor calculus is concerned with setting out the properties and relations of tensors, which are independent of the co-ordinate system. *In an extended sense a quantity in geometry and physics will be called a tensor if it defines uniquely a Linear algebraic form depending on the co-ordinate system in the manner described above; and conversely the tensor is fully characterised if this form is given.* For example, a little earlier we called a function of three displacements which depended linearly and homogeneously on each of their arguments a tensor of the third degree—one which is twofold co-variant and singly contra-variant. This was possible in **metrical** space. In this space, indeed, we are at liberty to represent this quantity by a "none" fold, single, twofold or threefold co-variant tensor. In affine space, however, we should only have been able to express it in the last form as a co-variant tensor of the third degree.

We shall illustrate this general explanation by some examples in which we shall still adhere to the standpoint of affine geometry alone.

1. If we represent a displacement **a** in an arbitrary co-ordinate system by its (contra-variant) components a^i and assign to it the linear form

$$a^1\xi_1 + a^2\xi_2 + \cdots + a^n\xi_n$$

having the variables ξ_i in this co-ordinate system, we get a contravariant tensor of the first order.

From now on we shall no longer use the term "vector" as being synonymous with "displacement" but to signify a "tensor of the first order," so that we shall say, **displacements are contravariant vectors**. The same applies to the **velocity** of a moving point, for it is obtained by dividing the infinitely small displacement which the moving point suffers during the time-element dt by dt (in the limiting case when $dt \to 0$). The present use of the word vector agrees with its usual significance which includes not only displacements but also every quantity which, after the choice of an appropriate unit, can be represented uniquely by a displacement.

2. It is usually claimed that **force** has a geometrical character on the ground that it may be represented in this way. In opposition, however, to this representation there is another which, we nowadays consider, does more justice to the physical nature of force, inasmuch as it is based on the conception of **work**. In modern physics the conception work is gradually usurping the conception of force, and is claiming a more decisive and fundamental rôle. We shall define the **components of a force** in a co-ordinate system $O \mid \mathbf{e}_i$ to be those numbers p_i which denote how much work it performs during each of the virtual displacements $\mathbf{e}_i$ of its point of application. These numbers completely characterise the force. The work performed during the arbitrary displacement

$$\mathbf{x} = \xi^1\mathbf{e}_1 + \xi^2\mathbf{e}_2 + \cdots + \xi^n\mathbf{e}_n$$

of its point of application is then $= \sum_i p_i\xi^i$. Hence it follows that for two definite co-ordinate systems the relation

$$\sum_i p_i\xi^i = \sum_i \bar{p}_i\bar{\xi}^i$$

holds, if the variables ξ^i, as signified by the upper indices, are transformed contra-grediently with respect to the co-ordinate system. According to this view, then, **forces are co-variant vectors**. The connection between this representation of forces and the usual one in which they are displacements will be discussed when we pass from affine geometry, with which we are at present dealing, to metrical geometry. The components of a co-variant vector become transformed co-grediently to the fundamental vectors in passing to a new co-ordinate system.

Additional Remarks.—Since the transformations of the components a^i of a co-variant vector and of the components b^i of a contra-variant vector are contra-gredient to one another, $\sum_i a_ib^i$ is a definite number which is defined by these two vectors and is independent of the co-ordinate system. This is our first example of an invariant tensor operation. Numbers or **scalars** are to be classified as tensors of zero order in the system of tensors.

It has already been explained under what conditions a bilinear form of two series of variables is called **symmetrical** and what makes a symmetrical bilinear form non-degenerate. A bilinear form $F(\xi, \eta)$ is called **skew-symmetrical** if the interchange of the two sets of variables converts it into its negative, i.e. merely changes its sign

$$F(\eta, \xi) = -F(\xi, \eta).$$

This property is expressed in the co-efficients a_{ik} by the equations $a_{ki} = -a_{ik}$. These properties persist if the two sets of variables are subjected to the same linear transformations. The property of being skew-symmetrical, symmetrical or (symmetrical and) non-degenerate, possessed by co-variant or contra-variant tensors of the second order is thus independent of the co-ordinate system.

Since the bilinear unit form resolves into itself after a contragredient transformation of the two series of variables there is among the **mixed** tensors of the second order (i.e. those which are simply co-variant and simply contra-variant) one, called the unit tensor, which has the components $\delta_i^k = \begin{matrix} 1 \ (i = k) \\ 0 \ (i \neq k) \end{matrix}$ in every co-ordinate system.

3. The metrical structure underlying Euclidean space assigns to every two displacements

$$\mathbf{x} = \sum_i \xi^i \mathbf{e}_i \qquad \mathbf{y} = \sum_i \eta^i \mathbf{e}_i$$

a number which is independent of the co-ordinate system and is their scalar product

$$\mathbf{x} \cdot \mathbf{y} = \sum_{i,k} g_{ik} \xi^i \eta^k \qquad g_{ik} = (\mathbf{e}_i \cdot \mathbf{e}_k).$$

Hence the bilinear form on the right depends on the co-ordinate system in such a way that a co-variant tensor of the second order is given by it, viz. the **fundamental metrical tensor**. The metrical structure is fully characterised by it. It is symmetrical and non-degenerate.

4. A **linear vector transformation** makes any displacement $\mathbf{x}$ correspond linearly to another displacement, $\mathbf{x}'$, i.e. so that the sum $\mathbf{x}' + \mathbf{y}'$ corresponds to the sum $\mathbf{x} + \mathbf{y}$ and the product $\lambda \mathbf{x}'$

to the product $\lambda\mathbf{x}$. In order to be able to refer conveniently to such linear vector transformations, we shall call them **matrices**. If the fundamental vectors $\mathbf{e}_i$ of a co-ordinate system become

$$\mathbf{e}'_i = \sum_k \alpha_i^k \mathbf{e}_k$$

as a result of the transformation it will in general convert the arbitrary displacement

$$\mathbf{x} = \sum_i \xi^i \mathbf{e}_i \quad \text{into} \quad \mathbf{x}' = \sum_i \xi^i \mathbf{e}'_i = \sum_{i,k} \alpha_i^k \xi^i \mathbf{e}_k. \tag{24}$$

We may, therefore, characterise the matrix in the particular co-ordinate system chosen by the bilinear form

$$\sum_{i,k} a_i^k \xi^i \eta_k.$$

It follows from (24) that the relation

$$\sum_{i,k} \bar{a}_i^k \bar{\xi}^i \mathbf{e}_k = \sum_{i,k} a_i^k \xi^i \mathbf{e}_k \ (= \mathbf{x}')$$

holds between two co-ordinate systems (we have used the same terminology as above) if

$$\sum_i \bar{\xi}^i \bar{\mathbf{e}}_i = \sum_i \xi^i \mathbf{e}_i \ (= \mathbf{x});$$

thus

$$\sum_{i,k} \bar{a}_i^k \bar{\xi}^i \bar{\eta}_k = \sum_{i,k} a_i^k \xi^i \eta_k$$

if the η^i are transformed co-grediently to the fundamental vectors and the ξ^i are transformed contra-grediently to them (the latter remark about the transformations of the variables is self-evident so that in future we shall simply omit it in similar cases). In this way matrices are represented as tensors of the second order. In particular, the unit tensor corresponds to "identity" which assigns to every displacement $\mathbf{x}$ itself.

As was shown in the examples of force and metrical space it often happens that the representation of geometrical or physical quantities by a tensor becomes possible only after a unit measure has been chosen: this choice can only be made by specifying it in each particular case. If the unit measure is altered the representative tensors must be multiplied by a universal constant, viz. the ratio of the two units of measure.

The following criterion is manifestly equivalent to this exposition of the conception tensor. *A linear form in several series of variables, which is dependent on the co-ordinate system, is a tensor if in every case it assumes a value independent of the co-ordinate system (a) whenever the components of an arbitrary contra-variant vector are substituted for every contra-gredient series of variables, and (b) whenever the components of an arbitrary co-variant vector are substituted for a co-gredient series.*

If we now return from **affine** to **metrical** geometry, we see from the arguments at the beginning of the paragraph that the difference between co-variants and contra-variants which affects the tensors themselves in affine geometry shrinks to a mere difference in the mode of representation.

Instead of talking of co-variant, mixed, and contra-variant *tensors* we shall hence find it more convenient here to talk only of the co-variant, mixed, and contra-variant *components* of a tensor. After the above remarks it is evident that the transition from one tensor to another which has a different character of co-variance may be formulated simply as follows. If we interpret the contra-gredient variables in a tensor as the contra-variant components of an arbitrary displacement, and the co-gredient variables as co-variant components of an arbitrary displacement, the tensor becomes transformed into a linear form of several arbitrary displacements which is independent of the co-ordinate system. By representing the arguments in terms of their co-variant or contra-variant components in any way which suggests itself as being appropriate we pass on to other representations of the same tensor. From the purely algebraic point of view the conversion of a co-variant index into a contra-variant one is performed by substituting new ξ_i's for the corresponding variables ξ^i in the linear form in accordance with (22). The invariant nature of this process depends on the circumstance that this substitution transforms contra-gredient variables into co-gredient ones. The converse process is carried out according to the inverse equations(22′). The components themselves are changed (on account of the symmetry of the g_{ik}'s) from contra-variants to co-variants, i.e. the indices are "lowered" according to the rule:

$$\text{Substitute } a_i = \sum_j g_{ij} a^j \text{ for } a^i$$

irrespective of whether the numbers a^i carry any other indices or not: the raising of the index is effected by the inverse equations.

If, in particular, we apply these remarks to the fundamental metrical tensor, we get

$$\sum_{i,k} g_{ik} \xi^i \eta^k = \sum_i \xi^i \eta_i = \sum_k \xi_k \eta^k = \sum_{i,k} g^{ik} \xi_i \eta_k .$$

Thus its mixed components are the numbers δ^i_k, its contra-variant components are the co-efficients g^{ik} of the equations (22′), which are the inverse of (22). It follows from the symmetry of the tensor that these as well as the g_{ik}'s satisfy the condition of symmetry $g^{ki} = g^{ik}$.

With regard to notation we shall adopt the convention of denoting the co-variant, mixed, and contra-variant components of the same tensor by similar letters, and of indicating by the position of the index at the top or bottom respectively whether the components are contra-variant or co-variant with respect to the index, as is shown in the following example of a tensor of the second order:

$$\sum_{i,k} a_{ik}\xi^i\eta^k = \sum_{i,k} a^i_k\xi_i\eta^k = \sum_{i,k} a^k_i\xi^i\eta_k = \sum_{i,k} a^{ik}\xi_i\eta_k$$

(in which the variables with lower and upper indices are connected in pairs by (22)).

In metrical space it is clear, from what has been said, that the difference between a co-variant and a contra-variant vector disappears: in this case we can represent a force, which, according to our view, is by nature a co-variant vector, as a contra-variant vector, too, i.e. by a displacement. For, as we represented it above by the linear form $\sum_i p_i\xi^i$ in the contra-gredient variables ξ^i, we can now transform the latter by means of (22′) into one having co-gredient variables ξ_i, viz. $\sum_i p^i\xi_i$. We then have

$$\sum_i p^i\xi_i = \sum_{i,k} g_{ik}p^i\xi^k = \sum_{i,k} g_{ik}p^k\xi^i = \sum_i p_i\xi^i;$$

the representative displacement $\mathbf{p}$ is thus defined by the fact that the work which the force performs during an arbitrary displacement is equal to the scalar product of the displacements $\mathbf{p}$ and $\mathbf{x}$.

In a Cartesian co-ordinate system in which the fundamental tensor has the components

$$g_{ik} = \begin{cases} 1 & (i = k), \\ 0 & (i \neq k), \end{cases}$$

the connecting equations (22) are simply: $\xi_i = \xi^i$. If we confine ourselves to the use of Cartesian co-ordinate systems, the difference between co-variants and contra-variants ceases to exist, not only for tensors but also for the tensor components. It must, however, be mentioned that the conceptions which have so far been outlined concerning the fundamental tensor g_{ik} assume only that it is symmetrical and non-degenerate, whereas the introduction of a Cartesian co-ordinate system implies, in addition, that the corresponding quadratic form is definitely positive. This entails a difference. In the Theory of Relativity the time co-ordinate is added as a fully equivalent term to the three-space co-ordinates, and the measure-relation which holds in this four-dimensional manifold is not based on a definite form but on an indefinite one (Chapter III). In this manifold, therefore, we shall not be able to introduce a Cartesian co-ordinate system if we restrict ourselves to real co-ordinates; but the conceptions here developed which are to be worked out in detail for the dimensional number $n = 4$ may be applied without alteration. Moreover, the algebraic simplicity of this calculus advises us against making exclusive use of Cartesian co-ordinate systems, as we have already mentioned at the end of § 4. Above all, finally, it is of great importance for later extensions which take us beyond Euclidean geometry that the affine view should even at this stage receive full recognition independently of the metrical one.

Geometrical and physical quantities are scalars, vectors, and tensors: this expresses the mathematical constitution of the space in which these quantities exist. The mathematical symmetry which this conditions is by no means restricted to geometry but, on the contrary, attains its full validity in physics. As natural phenomena take place in a metrical space this tensor calculus is the natural mathematical instrument for expressing the uniformity underlying them.

§ 6. Tensor Algebra. Examples

Addition of Tensors.—The multiplication of a linear form, bilinear form, trilinear form ... by a number, likewise the addition of two linear forms, or of two bilinear forms ... always gives rise to a form of the same kind. Vectors and tensors may thus be multiplied by a number (a scalar), and two or more tensors of the same order may be added together. These operations are carried out by multiplying the components by the number in question or by addition, respectively. Every set of tensors of the same order contains a unique tensor **0**, of which all the components vanish, and which, when added to any tensor of the same order, leaves the latter unaltered. The state of a physical system is described by specifying the values of certain scalars and tensors.

The fact that a tensor which has been derived from them by mathematical operations and is an invariant (i.e. dependent upon them alone and not upon the choice of the co-ordinate system) is equal to zero is what, in general, the expression of a physical law amounts to.

Examples.—The motion of a point is represented analytically by giving the position of the moving-point or of its co-ordinates, respectively, as functions of the time t. The derivatives $\frac{dx_i}{dt}$ are the contra-variant components u^i of the vector "velocity". By multiplying it by the mass m of the moving-point, m being a scalar which serves to express the inertia of matter, we get the "impulse" (or "momentum"). By adding the impulses of several points of mass or of those, respectively, of which one imagines a rigid body to be composed in the mechanics of point-masses, we get the total impulse of the point-system or of the rigid body. In the case of continuously extended matter we must supplant these sums by integrals. The fundamental law of motion is

$$\frac{dG^i}{dt} = p^i; \quad G^i = mu^i \tag{25}$$

where G^i denote the contra-variant components of the impulse of a mass-point and p^i denote those of the force.

Since, according to our view, force is primarily a co-variant vector, this fundamental law is possible only in a metrical space, but not in a purely affine one. The same law holds for the total impulse of a rigid body and for the total force acting on it.

Multiplication of Tensors.—By multiplying together two linear forms $\sum_i a_i\xi^i$, $\sum_i b_i\eta^i$ in the variables ξ and η, we get a bilinear form

$$\sum_{i,k} a_i b_k \xi^i \eta^k$$

and hence from the two vectors a and b we get a tensor c of the second order, i.e.

$$a_i b_k = c_{ik}. \tag{26}$$

Equation (26) represents an invariant relation between the vectors a and b and the tensor c, i.e. if we pass over to a new co-ordinate system precisely the same equations hold for the components (distinguished by a bar) of these quantities in this new co-ordinate system, i.e.

$$\bar{a}_i \bar{b}_k = \bar{c}_{ik}.$$

In the same way we may multiply a tensor of the first order by one of the second order (or generally, a tensor of any order by a tensor of any order). By multiplying

$$\sum_i a_i \xi^i \text{ by } \sum_{i,k} b_i^k \eta^i \zeta_k$$

in which the Greek letters denote variables which are to be transformed contra-grediently or co-grediently according as the indices are raised or lowered, we derive the trilinear form

$$\sum_{i,k,l} a_i b_k^l \xi^i \eta^k \zeta_l$$

and, accordingly, by multiplying the two tensors of the first and second order, a tensor c of the third order, i.e.

$$a_i \cdot b_k^l = c_{ik}^l.$$

This multiplication is performed on the components by merely multiplying each component of one tensor by each component of the other, as is evident above. It must be noted that the co-variant components (with respect to the index l, for example) of the resultant tensor of the third order, i.e. $c_{ik}^l = a_i b_k^l$, are given by: $c_{ikl} = a_i b_{kl}$. It is thus immediately permissible in such multiplication formulæ to transfer an index on both sides of the equation from below to above or *vice versa*.

Examples of Skew-symmetrical and Symmetrical Tensors.—If two vectors with the contra-variant components a^i, b^i, are multiplied first in one order and then in the reverse order, and if we then subtract the one result from the other, we get a skew-symmetrical tensor c of the second order with the contra-variant components

$$c^{ik} = a^i b^k - a^k b^i.$$

This tensor occurs in ordinary vector analysis as the "vectorial product" of the two vectors a and b. By specifying a certain direction of twist in three-dimensional space, it becomes possible to establish a reversible one-to-one correspondence between these tensors and the vectors. (This is impossible in four-dimensional space for the obvious reason that, in it, a skew-symmetrical tensor of the second order has six independent components, whereas a vector has only four; similarly in the case of spaces of still higher dimensions.) In three-dimensional space the above method of representation is founded on the following. If we use only Cartesian co-ordinate systems and introduce in addition to a and b an arbitrary displacement ξ, the determinant

$$\begin{vmatrix} a^1 & a^2 & a^3 \\ b^1 & b^2 & b^3 \\ \xi^1 & \xi^2 & \xi^3 \end{vmatrix} = c^{23}\xi^1 + c^{31}\xi^2 + c^{12}\xi^3$$

becomes multiplied by the determinant of the co-efficients of transformation, when we pass from one co-ordinate system to another. In the case of orthogonal transformations this determinant $= \pm 1$. If we confine our attention to "proper" orthogonal transformations, i.e. such for which this determinant $= +1$ the above linear form in the ξ's remains unchanged. Accordingly, the formulæ

$$c^{23} = c_1^* \qquad c^{31} = c_2^* \qquad c^{12} = c_3^*$$

express a relation between the skew-symmetrical tensor c and a vector c^*, this relation being invariant for proper orthogonal transformations. The vector c^* is perpendicular to the two vectors a and b, and its magnitude (according to elementary formulæ of analytical geometry) is equal to the area of the parallelogram of which the sides are a and b. It may be justifiable on the ground of economy of expression to replace skew-symmetrical tensors by vectors in ordinary vector analysis, but in some ways it hides the essential feature; it gives rise to the well-known "swimming-rules" in electrodynamics, which in no wise signify that there is a unique direction of twist in the space in which electrodynamics events occur; they become necessary only because the magnetic intensity of field is regarded as a vector, whereas it is, in reality, a skew-symmetrical tensor (like the so-called vectorial product of two vectors). If we had been given one more space-dimension, this error could never have occurred.

In mechanics the skew-symmetrical tensor product of two vectors occurs—

1. As moment of momentum (angular momentum) about a point O. If there is a point-mass at P and if ξ^1, ξ^2, ξ^3 are the components of $\overrightarrow{OP}$ and u^i are the (contra-variant) components of the velocity of the points at the moment under consideration, and m its mass, the momentum of momentum is defined by

$$L^{ik} = m(u^i\xi^k - u^k\xi^i).$$

The moment of momentum of a rigid body about a point O is the sum of the moments of momentum of each of the point-masses of the body.

2. As the **turning-moment (torque) of a force**. If the latter acts at the point P and if p^i are its contra-variant components,

the torque is defined by

$$q^{ik} = p^i \xi^k - p^k \xi^i.$$

The turning-moment of a system of forces is obtained by simple addition. In addition to (25) the law

$$\frac{dL^{ik}}{dt} = q^{ik} \tag{27}$$

holds for a point-mass as well as for a rigid body free from constraint. The turning-moment of a rigid body about a fixed point O is governed by the law (27) alone.

A further example of a skew-symmetrical tensor is the **rate of rotation** (angular velocity) of a rigid body about the fixed point O. If this rotation about O brings the point P in general to P', the vector $\overrightarrow{OP'}$ is produced, and hence also PP', by a linear transformation from $\overrightarrow{OP}$. If ξ^i are the components of $\overrightarrow{OP}$, $\delta\xi^i$ those of PP', v^i_k the components of this linear transformation (matrix), we have

$$\delta\xi^i = \sum_k v^i_k \xi^k. \tag{28}$$

We shall concern ourselves here only with infinitely small rotations. They are distinguished among infinitesimal matrices by the additional property that, regarded as an identity in ξ,

$$\delta\left(\sum_i \xi_i \xi^i\right) = \delta\left(\sum_{i,k} g_{ik} \xi^i \xi^k\right) = 0.$$

This gives

$$\sum_i \xi_i \, \delta\xi^i = 0.$$

By inserting the expressions (28) we get

$$\sum_{i,k} v^i_k \xi_i \xi^k = \sum_{i,k} v_{ik} \xi^i \xi^k = 0.$$

This must be identically true in the variables ξ_i, and hence

$$v_{ki} + v_{ik} = 0$$

i.e. the tensor which has v_{ik} for its co-variant components is skew-symmetrical.

A rigid body in motion experiences an infinitely small rotation during an infinitely small element of time δt. We need only to divide by δt the infinitesimal rotation-tensor v just formed to derive (in the limit when $\delta t \to 0$) the skew-symmetrical tensor "angular velocity," which we shall again denote by v. If u^i signify the contra-variant components of the velocity of the point P, and u_i its co-variant components in the formulæ (28), the latter resolves into the fundamental formula of the kinematics of a rigid body, viz.

$$u_i = \sum_k v_{ik} \xi^k. \tag{29}$$

The existence of the "instantaneous axis of rotation" follows from the circumstance that the linear equations

$$\sum_k v_{ik} \xi^k = 0$$

with the skew-symmetrical co-efficients v_{ik} always have solutions **in the case** $n = 3$, which differ from the trivial one $\xi^1 = \xi^2 = \xi^3 = 0$. One usually finds angular velocity, too, represented as a vector.

Finally the "moment of inertia" which presents itself in the rotation of a body offers a simple example of a symmetrical tensor of the second order.

If a point-mass of mass m is situated at the point P to which the vector $\overrightarrow{OP}$ starting from the centre of rotation O and with the components ξ^i leads, we call the symmetrical tensor of which the contra-variant components are given by $m\xi^i\xi^k$ (multiplication!), the "inertia of rotation" of the point-mass (with respect to the centre of rotation O). The inertia of rotation T^{ik} of a point-system or body is defined as the sum of these tensors formed separately for each of its points P. This definition is different from the usual one, but is the correct one if the intention of regarding the velocity of rotation as a skew-symmetrical tensor and not as a vector is to be carried out, as we shall presently see. The tensor T^{ik} plays the same part with regard to a rotation about O as that of the scalar m in translational motion.

Contraction.—If a_i^k are the mixed components of a tensor of the second order, $\sum_i a_i^i$ is an invariant. Thus, if $\bar{a}_i^k$ are the mixed components of the same tensor after transformation to a new co-ordinate system, then

$$\sum_i a_i^i = \sum_i \bar{a}_i^i.$$

Proof.—The variables ξ^i, η_i of the bilinear form

$$\sum_{i,k} a_i^k \xi^i \eta_k$$

must be subjected to the contra-gredient transformations

$$\xi^i = \sum_k \alpha_k^i \bar{\xi}^k, \qquad \eta_i = \sum_k \breve{\alpha}_i^k \bar{\eta}_k$$

if we wish to bring them into the form

$$\sum_{i,k} \bar{a}_i^k \bar{\xi}^i \bar{\eta}_k.$$

From this it follows that

$$\bar{a}_r^r = \sum_{i,k} a_i^k \alpha_r^i \breve{\alpha}_k^r,$$

and

$$\begin{aligned}\sum_r \bar{a}_r^r &= \sum_{i,k} \left(a_i^k \sum_r \alpha_r^i \breve{\alpha}_k^r \right) \\ &= \sum_i a_i^i \quad \text{by } (20').\end{aligned}$$

The invariant $\sum_i a_i^i$ which has been formed from the components a_i^k of a matrix is called the **trace (spur) of the matrix**.

This theorem enables us immediately to carry out a general operation on tensors, called "contraction," which takes a second place to multiplication. By making a definite upper index in the mixed components of a tensor coincide with a definite lower one, and summing over this index, we derive from the given tensor a new one the order of which is two less than that of the original one, e.g. we get from the components a_{hik}^{lm} of a tensor of the fifth order a tensor of the third order, thus:—

$$\sum_r a_{hir}^{lr} = c_{hi}^l. \tag{30}$$

The connection expressed by (30) is an invariant one, i.e. it preserves its form when we pass over to a new co-ordinate system, viz.

$$\sum_r \bar{a}^{lr}_{hir} = \bar{c}^{l}_{hi}. \tag{31}$$

To see this we only need the help of two arbitrary contra-variant vectors ξ^i, η^i and a co-variant one ζ_i. By means of them we form the components,

$$\sum_{h,i,l} a^{lm}_{hik} \xi^h \eta^i \zeta_l = f^m_k,$$

of a mixed tensor of the second order: to this we apply the theorem

$$\sum_r f^r_r = \sum_r \bar{f}^r_r$$

which was just proved. We then get the formula

$$\sum_{h,i,l} c^l_{hi} \xi^h \eta^i \zeta_l = \sum_{h,i,l} \bar{c}^l_{hi} \bar{\xi}^h \bar{\eta}^i \bar{\zeta}_l$$

in which the c's are defined by (30), the $\bar{c}$'s by (31). The $\bar{c}^l_{hi}$'s are thus, in point of fact, the components of the same tensor of the third order in the new system, of which the components in the old one $= c^l_{ih}$.

Examples of this process of contraction have been met with in abundance in the above. Wherever summation took place with respect to certain indices, the summation index appeared twice in the general member of summation, once above and once below the co-efficient: each such summation was an example of contraction. For example, in formula (29): by multiplication of v_{ik} with ξ^i one can form the tensor $v_{ik}\xi^l$ of the third order; by making k coincide with l and summing for k, we get the contracted tensor of the first order u_i. If a matrix A transforms the arbitrary displacement **x**

into $\mathbf{x}' = A(\mathbf{x})$, and if a second matrix B transforms this $\mathbf{x}'$ into $\mathbf{x}'' = -B(\mathbf{x}')$, a combination BA results from the two matrices, which transforms $\mathbf{x}$ directly into $\mathbf{x}'' = BA(\mathbf{x})$. If A has the components a_i^k and B components b_i^k, the components of the combined matrix BA are

$$c_i^k = \sum_r b_i^r a_r^k.$$

Here, again, we have the case of multiplication followed by contraction.

The process of contraction may be applied simultaneously for several pairs of indices. From the tensors of the 1st, 2nd, 3rd, ... order with the co-variant components a_i, a_{ik}, a_{ikl}, ..., we thus get, in particular, the invariants

$$\sum_i a_i a^i, \qquad \sum_{i,k} a_{ik} a^{ik}, \qquad \sum_{i,k,l} a_{ikl} a^{ikl}, \quad \ldots.$$

If, as is here assumed, the quadratic form corresponding to the fundamental metrical tensor is definitely positive, these invariants are all positive, for, in a Cartesian co-ordinate system they disclose themselves directly as the sums of the squares of the components. Just as in the simplest case of a vector, the square root of these invariants may be termed the measure or the magnitude of the tensor of the 1st, 2nd, 3rd, ... order.

We shall at this point make the convention, once and for all, that if an index occurs twice (once above and once below) in a term of a formula to which indices are attached, this is always to signify that summation is to be carried out with respect to the index in question, and we shall not consider it necessary to put a summation sign in front of it.

The operations of addition, multiplication, and contraction only require affine geometry: they are not based upon a "fundamental metrical tensor". The latter is only necessary for the process of passing from co-variant to contra-variant components and the reverse.

Euler's Equations for a Spinning Top

As an exercise in tensor calculus, we shall deduce Euler's equations for the motion of a rigid body under no forces about a fixed point O. We write the fundamental equations (27) in the co-variant form

$$\frac{dL_{ik}}{dt} = 0$$

and multiply them, for the sake of briefness, by the contra-variant components w^{ik} of an arbitrary skew-symmetrical tensor which is constant (independent of the time), and apply contraction with respect to i and k. If we put H_{ik} equal to the sum

$$\sum_m m u_i \xi^k$$

which is to be taken over all the points of mass, we get

$$\tfrac{1}{2} L_{ik} w^{ik} = H_{ik} w^{ik} = H,$$

an invariant, and we can compress our equation into

$$\frac{dH}{dt} = 0. \tag{32}$$

If we introduce the expressions (29) for u_i, and the tensor of inertia T, then

$$H_{ik} = v_{ir} T^r_k. \tag{33}$$

We have hitherto assumed that a co-ordinate system which is fixed in **space** has been used. The components T of inertia then change with the distribution of matter in the course of time. If, however, in place of this we use a co-ordinate system which is fixed in the **body**, and consider the symbols so far used as referring to the components of the corresponding tensors with respect to this co-ordinate system, whereas we distinguish the components of the same tensors with respect to the co-ordinate system fixed in space by a horizontal bar, the equation (32) remains valid on account of the invariance of H. The T_i^k's are now constants; on the other hand, however, the w^{ik}'s vary with the time. Our equation gives us

$$\frac{dH_{ik}}{dt} \cdot w^{ik} + H_{ik} \cdot \frac{dw^{ik}}{dt} = 0. \tag{34}$$

To determine $\frac{dw^{ik}}{dt}$, we choose two arbitrary vectors fixed in the body, of which the co-variant components in the co-ordinate system attached to the body are ξ_i and η_i respectively. These quantities are thus constants, but their components $\bar{\xi}_i$, $\bar{\eta}_i$ in the space co-ordinate system are functions of the time. Now,

$$w^{ik}\xi_i\eta_k = \bar{w}^{ik}\bar{\xi}_i\bar{\eta}_k,$$

and hence, differentiating with respect to the time

$$\frac{dw^{ik}}{dt} \cdot \xi_i\eta_k = \bar{w}^{ik}\left(\frac{d\bar{\xi}_i}{dt} \cdot \bar{\eta}_k + \bar{\xi}_i \cdot \frac{d\bar{\eta}_k}{dt}\right). \tag{35}$$

By formula (29)

$$\frac{d\bar{\xi}_i}{dt} = \bar{v}_{ir}\bar{\xi}^r = \bar{v}_i^r\bar{\xi}_r.$$

We thus get for the right-hand side of (35)

$$\bar{w}^{ik}(\bar{v}_i^r\bar{\xi}_r\bar{\eta}_k + \bar{v}_k^r\bar{\xi}_i\bar{\eta}_r),$$

and as this is an invariant, we may remove the bars, obtaining

$$\xi_i\eta_k\frac{dw^{ik}}{dt} = w^{ik}(\xi_r\eta_k v_i^r + \xi_i\eta_r v_k^r).$$

This holds identically in ξ and η; thus if the H^{ik} are arbitrary numbers,

$$H_{ik}\frac{dw^{ik}}{dt} = w^{ik}(v_i^r H_{rk} + v_k^r H_{ir}).$$

If we take the H_{ik}'s to be the quantities which we denoted above by this symbol, the second term of (34) is determined, and our equation becomes

$$\left\{\frac{dH_{ik}}{dt} + (v_i^r H_{rk} + v_k^r H_{ir})\right\} w^{ik} = 0,$$

which is an identity in the skew-symmetrical tensor w^{ik}; hence

$$\frac{d(H_{ik} - H_{ki})}{dt} + \begin{bmatrix} v_i^r H_{rk} + v_k^r H_{ir} \\ -v_k^r H_{ri} + v_i^r H_{kr} \end{bmatrix} = 0.$$

We shall now substitute the expression (33) for H_{ik}. Since, on account of the symmetry of T_{ik},

$$v_k^r H_{ir} (= v_k^r v_i^s T_{ro})$$

is also symmetrical in i and k, the two last terms of the sum in the square brackets destroy one another. If we now put the symmetrical tensor

$$v_i^r v_{kr} = g_{rs} v_i^r v_k^s = (v, v)_{ik}$$

we finally get our equations into the form

$$\frac{d}{dt}(v_{ir}T_k^r - v_{kr}T_i^r) = (v,v)_{ir}T_k^r - (v,v)_{kr}T_i^r.$$

It is well known that we may introduce a Cartesian co-ordinate system composed of the three principal axes of inertia, so that in these

$$g_{ik} = \begin{cases} 1 & (i = k), \\ 0 & (i \neq k), \end{cases} \qquad \text{and} \qquad T_{ik} = 0 \quad (\text{for } i \neq k).$$

If we then write T_1 in place of T_1^1, and do the same for the remaining indices, our equations in this co-ordinate system assume the simple form

$$(T_i + T_k)\frac{dv_{ik}}{dt} = (T_k - T_i)(v,v)_{ik}.$$

These are the differential equations for the components v_{ik} of the unknown angular velocity—equations which, as is known, may be solved in elliptic functions of t. The principal moments of inertia T_i which occur here are connected with those, T_i^*, given in accordance with the usual definitions by the equations

$$T_1^* = T_2 + T_3, \qquad T_2^* = T_3 + T_1, \qquad T_3^* = T_1 + T_2.$$

The above treatment of the problem of rotation may, in contradistinction to the usual method, be transposed, word for word, from three-dimensional space to multi-dimensional spaces. This is, indeed, irrelevant in practice. On the other hand, the fact that we have freed ourselves from the limitation to a definite dimensional number and that we have formulated physical laws in such a way that the dimensional number appears **accidental** in them, gives us an assurance that we have succeeded fully in grasping them mathematically.

The study of tensor-calculus* is, without doubt, attended by conceptual difficulties—over and above the apprehension inspired by indices, which must be overcome. From the formal aspect, however, the method of reckoning used is of extreme simplicity; it is much easier than, e.g., the apparatus of elementary vector-calculus. There are two operations, multiplication and contraction; then putting the components of two tensors with totally different indices alongside of one another; the identification of an upper index with a lower one, and, finally, summation (not expressed) over this index. Various attempts have been made to set up a standard terminology in this branch of mathematics involving only the vectors themselves and not their components, analogous to that of vectors in vector analysis. This is highly expedient in the latter, but very cumbersome for the much more complicated framework of the tensor calculus. In trying to avoid continual reference to the components we are obliged to adopt an endless profusion of names and symbols in addition to an intricate set of rules for carrying out calculations, so that the balance of advantage is considerably on the negative side. An emphatic protest must be entered against these orgies of formalism which are threatening the peace of even the technical scientist.

* *Vide* note 4.

§ 7. Symmetrical Properties of Tensors

It is obvious from the examples of the preceding paragraph that symmetrical and skew-symmetrical tensors of the second order, wherever they are applied, represent entirely different kinds of quantities. Accordingly the character of a quantity is not in general described fully, if it is stated to be a tensor of such and such an order, but **symmetrical characteristics** have to be added.

A linear form of several series of variables is called **symmetrical** if it remains unchanged after any two of these series of variables are interchanged, but is called **skew-symmetrical** if this converts it into its negative, i.e. reverses its sign. A symmetrical linear form does not change if the series of variables are subjected to any permutations among themselves; a skew-symmetrical one does not change if an even permutation is carried out with the series of variables, but changes its sign if the permutation is odd. The co-efficients a_{ikl} of a symmetrical trilinear form (we purposely choose three again as an example) satisfy the conditions

$$a_{ikl} = a_{kli} = a_{lik} = a_{kil} = a_{lki} = a_{ilk}.$$

Of the co-efficients of a skew-symmetrical tensor only those which have three different indices can be $\neq 0$ and they satisfy the equations

$$a_{ikl} = a_{kli} = a_{lik} = -a_{kil} = -a_{lki} = -a_{ilk}.$$

There can consequently be no (non-vanishing) skew-symmetrical forms of more than n series of variables in a domain of n variables. Just as a symmetrical bilinear form may be entirely replaced by the quadratic form which is derived from it by identifying the two series of variables, so a symmetrical trilinear form is uniquely determined by the cubical form of a single series of variables with the co-efficients a_{ikl}, which is derived from the trilinear form by the same process. If in a skew-symmetrical trilinear form

$$F = \sum_{i,k,l} a_{ikl} \xi^i \eta^k \zeta^l$$

we perform the 3! permutations on the series of variables ξ, η, ζ, and prefix a positive or negative sign to each according as the permutation is even or odd, we get the original form six times. If they are all added together, we get the following scheme for them:—

$$F = \frac{1}{3!} \sum a_{ikl} \begin{vmatrix} \xi^i & \xi^k & \xi^l \\ \eta^i & \eta^k & \eta^l \\ \zeta^i & \zeta^k & \zeta^l \end{vmatrix}. \tag{36}$$

In a linear form the property of being symmetrical or skew-symmetrical is not destroyed if each series of variables is subjected to the same linear transformation. Consequently, a meaning may be attached to the terms **symmetrical** and **skew-symmetrical**, **co-variant** or **contra-variant** tensors. But these expressions have no meaning in the domain of mixed tensors. We need spend no further time on symmetrical tensors, but must discuss skew-symmetrical co-variant tensors in somewhat greater detail as they have a very special significance.

The components ξ^i of a displacement determine the direction of a straight line (positive or negative) as well as its magnitude. If ξ^i and η^i are any two linearly independent displacements, and if they are marked out from any arbitrary point O, they trace out a plane. The ratios of the quantities

$$\xi^i \eta^k - \xi^k \eta^i = \xi^{ik}$$

define the "position" of this plane (a "direction" of the plane) in the same way as the ratios of the ξ^i fix the position of a straight line (its "direction"). The ξ^{ik} are each $= 0$ if, and only if, the two displacements ξ^i, η^i are linearly dependent; in this case they do not map out a two-dimensional manifold. When two linearly independent displacements ξ^i and η^i trace out a plane, a definite sense of rotation is implied, viz. the sense of the rotation about O in the plane which for a turn $< 180°$ brings ξ to coincide with η; also a definite measure (quantity), viz. the area of the parallelogram enclosed by ξ and η. If we mark off two displacements ξ, η from an arbitrary point O, and two ξ_*, η_* from an arbitrary point O_*, then the position, the sense of rotation, and the magnitude of the plane marked out are identical in each if, and only if, the ξ^{ik}'s of the one pair coincide with those of the other, i.e.

$$\xi^i\eta^k - \xi^k\eta^i = \xi^i_*\eta^k_* - \xi^k_*\eta^i_*.$$

So that just as the ξ^i's determine the direction and length of a straight line, so the ξ^{ik}'s determine the sense and surface area of a plane; the completeness of the analogy is evident.

To express this we may call the first configuration a **one-dimensional space-element**, the second a **two-dimensional space-element**. Just as the square of the magnitude of a one-dimensional space-element is given by the invariant

$$\xi_i\xi^i = g_{ik}\xi^i\xi^k = Q(\xi)$$

so the square of the magnitude of the two-dimensional space-element is given, in accordance with the formulæ of analytical geometry, by

$$\tfrac{1}{2}\xi^{ik}\xi_{ik};$$

for which we may also write

$$\begin{aligned}\xi_i\eta_k(\xi^i\eta^k - \xi^k\eta^i) &= (\xi_i\xi^i)(\eta^k\eta_k) - (\xi_i\eta^i)(\xi^k\eta_k) \\ &= Q(\xi)\cdot Q(\eta) - Q^2(\xi,\eta).\end{aligned}$$

In the same sense the determinants

$$\xi^{ikl} = \begin{vmatrix} \xi^i & \xi^k & \xi^l \\ \eta^i & \eta^k & \eta^l \\ \zeta^i & \zeta^k & \zeta^l \end{vmatrix}$$

which are derived from three independent displacements ξ, η, ζ, are the components of a **three-dimensional space-element**, the magnitude of which is given by the square root of the invariant

$$\tfrac{1}{3!}\xi^{ikl}\xi_{ikl}.$$

In three-dimensional space this invariant is

$$\xi_{123}\xi^{123} = g_{1i}g_{2k}g_{3l}\xi^{ikl}\xi^{123},$$

and since $\xi^{ikl} = \pm\xi^{123}$, according as ikl is an even or an odd permutation of 123, it assumes the value

$$g\cdot(\xi^{123})^2$$

where g is the determinant of the co-efficients g_{ik} of the fundamental metrical form. The volume of the parallelepiped thus becomes

$$= \sqrt{g}\cdot\begin{vmatrix} \xi^1 & \xi^2 & \xi^3 \\ \eta^1 & \eta^2 & \eta^3 \\ \zeta^1 & \zeta^2 & \zeta^3 \end{vmatrix}$$ (taking the absolute, i.e. positive value of the determinants).

This agrees with the elementary formulæ of analytical geometry. In a space of more than three dimensions we may similarly pass on to four-dimensional space-elements, etc.

Just as a co-variant tensor of the first order assigns a number linearly (and independently of the co-ordinate system) to every one-dimensional space-element (i.e. displacement), so a skew-symmetrical co-variant tensor of the second order assigns a number to every two-dimensional space-element, a skew-symmetrical tensor of the third order to each three-dimensional space-element, and so on: this is immediately evident from the form in which (36) is expressed. For this reason we consider it justifiable to call the co-variant skew-symmetrical tensors simply **linear tensors**. Among operations in the domain of linear tensors we shall mention the two following ones:—

$$a_i b_k - a_k b_i = c_{ik}, \tag{37}$$

$$a_i b_{kl} - a_k b_{li} + a_l b_{ik} = c_{ikl}. \tag{38}$$

The former produces a linear tensor of the second order from two linear tensors of the first order; the latter produces a linear tensor of the third order from one of the first and one of the second.

Sometimes conditions of symmetry more complicated than those considered heretofore occur. In the realm of quadrilinear forms $F(\xi, \eta, \xi', \eta')$ those play a particular part which satisfy the conditions

$$F(\eta, \xi, \xi', \eta') = F(\xi, \eta, \eta', \xi') = -F(\xi, \eta, \xi', \eta'), \tag{39_1}$$

$$F(\xi', \eta', \xi, \eta) = F(\xi, \eta, \xi', \eta'), \tag{39_2}$$

$$F(\xi, \eta, \xi', \eta') + F(\xi, \xi', \eta', \eta) + F(\xi, \eta', \eta, \xi') = 0. \tag{39_3}$$

For it may be shown that for every quadratic form of an arbitrary two-dimensional space-element

$$\xi^{ik} = \xi^i \eta^k - \xi^k \eta^i$$

there is one and only one quadrilinear form F which satisfies these conditions of symmetry, and from which the above quadratic form is derived by identifying the second pair of variables ξ', η' with the first pair ξ, η. We must consequently use co-variant tensors of the fourth order having the symmetrical properties (39) if we wish to represent functions which stand in quadratic relationship with an element of surface.

The **most general form of the condition of symmetry** for a tensor F of the fifth order of which the first, second, and fourth series of variables are contra-gredient, the third and fifth co-gredient (we are taking a particular case) are

$$\sum_S e_S F_S = 0$$

in which S signifies all permutations of the five series of variables in which the contra-gredient ones are interchanged among themselves and likewise the co-gredient ones; F_S denotes the form which results from F after the permutation S; e_S is a system of definite numbers, which are assigned to the permutations S. The summation is taken over all the permutations S. The kind of symmetry underlying a definite type of tensors expresses itself in one or more of such conditions of symmetry.

§ 8. Tensor Analysis. Stresses

Quantities which describe how the state of a spatially extended physical system varies from point to point have not a distinct value but only one "for each point": in mathematical language they are "functions of the place or point". According as we are dealing with a scalar, vector, or tensor, we speak of a scalar, vector, or tensor **field**.

Such a field is given if a scalar, vector, or tensor of the proper type is assigned to every point of space or to a definite region of it. If we use a definite co-ordinate system the value of the scalar quantities or of the components of the vector or tensor quantities respectively, appear in the co-ordinate system as functions of the co-ordinates of a variable point in the region under consideration.

Tensor analysis tells us how, by differentiating with respect to the space co-ordinates, a new tensor can be derived from the old one in a manner entirely independent of the co-ordinate system. This method, like tensor algebra, is of extreme simplicity. Only one operation occurs in it, viz. **differentiation**.

If

$$\phi = f(x_1, x_2, \ldots, x_n) = f(x)$$

denotes a given scalar field, the change of ϕ corresponding to an infinitesimal displacement of the variable point, in which its co-ordinates x_i suffer changes dx_i respectively, is given by the total differential

$$df = \frac{\partial f}{\partial x_1}\,dx_1 + \frac{\partial f}{\partial x_2}\,dx_2 + \cdots + \frac{\partial f}{\partial x_n}\,dx_n.$$

This formula signifies that if the Δx_i are first taken as the components of a finite displacement and the Δf are the corresponding changes in f, then the difference between

$$\Delta f \quad \text{and} \quad \sum_i \frac{\partial f}{\partial x_i}\,\Delta x_i$$

does not only decrease absolutely to zero with the components of the displacement, but also relatively to the amount of the displacement, the measure of which may be defined as $|\Delta x_1| + |\Delta x_2| + \cdots + |\Delta x_n|$. We link up the linear form

$$\sum_i \frac{\partial f}{\partial x_i} \xi^i$$

in the variables ξ^i to this differential. If we carry out the same construction in another co-ordinate system (with horizontal bars over the co-ordinates), it is evident from the meaning of the term differential that the first linear form passes into the second, if the ξ^i's are subjected to the transformation which is contra-gredient to the fundamental vectors. Accordingly

$$\frac{\partial f}{\partial x_1}, \quad \frac{\partial f}{\partial x_2}, \quad \ldots, \quad \frac{\partial f}{\partial x_n}$$

are the co-variant components of a vector which arises from the scalar field ϕ in a manner independent of the co-ordinate system. In ordinary vector analysis it occurs as the **gradient** and is denoted by the symbol grad ϕ.

This operation may immediately be transposed from a scalar to any arbitrary tensor field. If, e.g., $f^h_{ik}(x)$ are components of a tensor field of the third order, contra-variant with respect to h, but co-variant with respect to i and k, then

$$f^h_{ik} \xi_h \eta^i \zeta^k$$

is an invariant, if we take ξ_h as standing for the components of an arbitrary but constant co-variant vector (i.e. independent of its position), and η^i, ζ^i each as standing for the components of a similar contra-variant vector in turn. The change in this invariant due to an infinitesimal displacement with components dx_i is given by

$$\frac{\partial f^h_{ik}}{\partial x_l}\,\xi_h \eta^i \zeta^k\,dx_l$$

hence

$$f^h_{ikl} = \frac{\partial f^h_{ik}}{\partial x_l}$$

are the components of a tensor field of the fourth order, which arises from the given one in a manner independent of the co-ordinate system. **Just this is the process of differentiation**; as is seen, it raises the order of the tensor by 1. We have still to remark that, on account of the circumstance that the fundamental metrical tensor is independent of its position, one obtains the components of the tensor just formed, for example, which are contra-variant with respect to the index k, by transposing the index k under the sign of differentiation to the top, viz. $\dfrac{\partial f^{hki}}{\partial x_l}$. The change from co-variant to contra-variant is interchangeable with differentiation. Differentiation may be carried out purely formally by imagining the tensor in question multiplied by a vector having the co-variant components

$$\frac{\partial}{\partial x_1}, \quad \frac{\partial}{\partial x_2}, \quad \ldots, \quad \frac{\partial}{\partial x_n} \tag{40}$$

and treating the differential quotient $\dfrac{\partial f}{\partial x_i}$ as the symbolic product of f and $\dfrac{\partial}{\partial x_i}$. The symbolic vector (40) is often encountered in mathematical literature under the mysterious name "nabla-vector".

Examples.—The vector with the co-variant components u_i gives rise to the tensor of the second order $\frac{\partial u_i}{\partial x_k} = u_{ik}$. From this we form

$$\frac{\partial u_i}{\partial x_k} - \frac{\partial u_k}{\partial x_i}. \tag{41}$$

These quantities are the co-variant components of a linear tensor of the second order. In ordinary vector analysis it occurs (with the signs reversed) as "**rotation**" (rot, spin or **curl**). On the other hand the quantities

$$\tfrac{1}{2}\left(\frac{\partial u_i}{\partial x_k} + \frac{\partial u_k}{\partial x_i}\right)$$

are the co-variant components of a symmetrical tensor of the second order. If the vector u represents the velocity of continuously extended moving matter as a function of its position, the vanishing of this tensor at a point signifies that the immediate neighbourhood of the point moves as a rigid body; it thus merits the name **distortion tensor**. Finally by contracting u_k^i we get the scalar

$$\frac{\partial u^i}{\partial x_i}$$

which is known in vector analysis as "**divergence**" (div.).

By differentiating and contracting a tensor of the second order having mixed components S_i^k we derive the vector

$$\frac{\partial S_i^k}{\partial x_k}.$$

If v_{ik} are the components of a linear tensor field of the second order, then, analogously to formula (38) in which we substitute v

or b and the symbolic vector "differentiation" for a, we get the linear tensor of the third order with the components

$$\frac{\partial v_{kl}}{\partial x_i} + \frac{\partial v_{li}}{\partial x_k} + \frac{\partial v_{ik}}{\partial x_l}. \tag{42}$$

Tensor (41), i.e. the curl, vanishes if v_i is the gradient of a scalar field; tensor (42) vanishes if v_{ik} is the curl of a vector u_i.

Stresses.—An important example of a tensor field is offered by the stresses occurring in an elastic body; it is, indeed, from this example that the name "tensor" has been derived. When tensile or compressional forces act at the surface of an elastic body, whilst, in addition, "volume-forces" (e.g. gravitation) act on various portions of the matter within the body, a state of equilibrium establishes itself, in which the forces of cohesion called up in the matter by the distortion balance the impressed forces from without. If we imagine any portion J of the matter cut out of the body and suppose it to remain coherent after we have removed the remaining portion, the impressed volume forces will not of themselves keep this piece of matter in a state of equilibrium. They are, however, balanced by the compressional forces acting on the surface Ω of the portion J, which are exerted on it by the portion of matter removed. We have actually, if we do not take the atomic (granular) structure of matter into account, to imagine that the forces of cohesion are only active in direct contact, with the consequence that the action of the removed portion upon J must be representable by superficial forces such as pressure: and indeed, if $\mathbf{S}\,do$ is the pressure acting on an element of surface do ($\mathbf{S}$ here denotes the pressure per unit surface), $\mathbf{S}$ can depend only upon the place at which the element of surface do happens to be and on the inward normal n of this element of surface with respect to J, which characterises the "position" of do. We shall write $\mathbf{S}_n$ for $\mathbf{S}$ to emphasise this connection between $\mathbf{S}$ and n. If $-n$ denotes the normal in a direction reversed to that of n, it follows from the equilibrium of a small infinitely thin disc, that

$$\mathbf{S}_{-n} = -\mathbf{S}_n. \tag{43}$$

We shall use Cartesian co-ordinates x_1, x_2, x_3. The compressional forces per unit of area at a point, which act on an element of surface situated at the same point, the inward normals of which coincide with the direction of the positive x_1-, x_2-, x_3-axis respectively will be denoted by $\mathbf{S}_1$, $\mathbf{S}_2$, $\mathbf{S}_3$. We now choose any three positive numbers a_1, a_2, a_3, and a positive number ϵ, which is to converge to the value 0 (whereas the a_i remain fixed). From the point O under consideration we mark off in the direction of the positive co-ordinate axes the distances

$$OP_1 = \epsilon a_1, \qquad OP_2 = \epsilon a_2, \qquad OP_3 = \epsilon a_3$$

and consider the infinitesimal tetrahedron $OP_1P_2P_3$ having OP_2P_3, OP_3P_1, OP_1P_2 as walls and $P_1P_2P_3$ as its "roof". If f is the superficial area of the roof and a_1, a_2, a_3 are the direction cosines of its inward normals n, then the areas of the walls are

$$-f \cdot a_1 (= \tfrac{1}{2}\epsilon^2 a_2 a_3), \qquad -f \cdot a_2, \qquad -f \cdot a_3.$$

The sum of the pressures on the walls and the roof becomes for evanescent values of ϵ:

$$f\big\{\mathbf{S}_n - (a_1\mathbf{S}_1 + a_2\mathbf{S}_2 + a_3\mathbf{S}_3)\big\}.$$

The magnitude of f is of the order ϵ^2: but the volume force acting upon the volume of the tetrahedron is only of the order of magnitude ϵ^3. Hence, owing to the condition for equilibrium, we must have

$$\mathbf{S}_n = (a_1\mathbf{S}_1 + a_2\mathbf{S}_2 + a_3\mathbf{S}_3).$$

With the help of (43) this formula may be extended immediately to the case in which the tetrahedron is situated in any of the remaining 7 octants. If we call the components of $\mathbf{S}_i$ with respect to the co-ordinate axes S_{i1}, S_{i2}, S_{i3}, and if ξ^i, η^i are the components of any two arbitrary displacements of length 1, then

$$\sum_{i,k} S_{ik}\xi^i\eta^k \tag{44}$$

is the component, in the direction η, of the compressional force which is exerted on an element of surface of which the inner normal is ξ. The bilinear form (44) has thus a significance independent of the co-ordinate system, and the S_{ik}'s are the components of a "stress" tensor field. We shall continue to operate in rectangular co-ordinate systems so that we shall not have to distinguish between co-variant and contra-variant quantities.

We form the vector $\mathbf{S}_1'$ having components S_{1i}, S_{2i}, S_{3i}. The component of $\mathbf{S}_1'$ in the direction of the inward normal n of an element of surface is then equal to the x_1-component of $\mathbf{S}_n$. The x_1-component of the total pressure which acts on the surface Ω of the detached portion of matter J is therefore equal to the surface integral of the normal components of $\mathbf{S}_1'$ and this, by Gauss's Theorem, is equal to the volume integral

$$-\int_J \operatorname{div} \mathbf{S}_1' \cdot dV.$$

The same holds for the x_2 and the x_3 component. We have thus to form the vector $\mathbf{p}$ having the components

$$p_i = -\sum_k \frac{\delta S_i^k}{\delta x_k}$$

(this is performed, as we know, according to an invariant law). The compressional forces $\mathbf{S}$ are then equivalent to a volume force having the direction and intensity given by $\mathbf{p}$ per unit volume in the sense that, for every dissociated portion of matter J,

$$\int_\Omega \mathbf{S}_n \, do = \int_J \mathbf{p} \, dV. \tag{45}$$

If $\mathbf{k}$ is the impressed force per unit volume, the first condition of equilibrium for the piece of matter considered coherent after being detached is

$$\int_J (\mathbf{p} + \mathbf{k}) \, dV = \mathbf{0},$$

and as this must hold for every portion of matter

$$\mathbf{p} + \mathbf{k} = \mathbf{0}. \tag{46}$$

If we choose an arbitrary origin O and if $\mathbf{r}$ denote the radius vector to the variable point P, and the square bracket denote the "vectorial" product, the second condition for equilibrium, the equation of moments, is

$$\int_\Omega [\mathbf{r}, \mathbf{S}_n] \, do + \int_J [\mathbf{r}, \mathbf{k}] \, dV = \mathbf{0},$$

and since (46) holds generally we must have, besides (45),

$$\int_\Omega [\mathbf{r}, \mathbf{S}_n] \, do = \int_J [\mathbf{r}, \mathbf{p}] \, dV.$$

The x_1 component of $[\mathbf{r}, \mathbf{S}_n]$ is equal to the component of $x_2\mathbf{S}'_3 - x_3\mathbf{S}'_2$ in the direction of n. Hence, by Gauss's theorem, the x_1 component of the left-hand member is

$$-\int_J \operatorname{div}(x_2\mathbf{S}'_3 - x_3\mathbf{S}'_2)\, dV.$$

Hence we get the equation

$$\operatorname{div}(x_2\mathbf{S}'_3 - x_3\mathbf{S}'_2) = -(x_2p_3 - x_3p_2).$$

But the left-hand member

$$\begin{aligned} &= (x_2 \operatorname{div} \mathbf{S}'_3 - x_3 \operatorname{div} \mathbf{S}'_2) + (\mathbf{S}'_3 \cdot \operatorname{grad} x_2 - \mathbf{S}'_2 \cdot \operatorname{grad} x_3) \\ &= -(x_2p_3 - x_3p_2) + (S_{23} - S_{32}). \end{aligned}$$

Accordingly, if we form the x_2 and x_3 components in addition to the x_1 component, this condition of equilibrium gives us

$$S_{23} = S_{32}, \qquad S_{31} = S_{13}, \qquad S_{12} = S_{21},$$

i.e. the symmetry of the **stress-tensor S**. For an arbitrary displacement having the components ξ^i,

$$\frac{\sum S_{ik}\xi^i\xi^k}{\sum g_{ik}\xi^i\xi^k}$$

is the component of the pressure per unit surface for the component in the direction ξ, which acts on an element of surface placed at right angles to this direction. (We may here again use any arbitrary affine co-ordinate system.) **The stresses are fully**

equivalent to a volume force of which the density p is calculated according to the invariant formulæ

$$-p_i = \frac{\delta S_i^k}{\delta x_k}. \tag{47}$$

In the case of a pressure p which is equal in all directions

$$S_i^k = p \cdot \delta_i^k, \qquad p_i = -\frac{\delta p}{\delta x_i}.$$

As a result of the foregoing reasoning we have formulated in exact terms the conception of stress alone, and have discovered how to represent it mathematically. To set up the fundamental laws of the theory of elasticity it is, in addition, necessary to find out how the stresses depend on the distortion brought about in the matter by the impressed forces. There is no occasion for us to discuss this in greater detail.

§ 9. Stationary Electromagnetic Fields

Hitherto, whenever we have spoken of mechanical or physical things, we have done so for the purpose of showing in what manner their spatial nature expresses itself: namely, that its laws manifest themselves as invariant tensor relations. This also gave us an opportunity of demonstrating the importance of the tensor calculus by giving concrete examples of it. It enabled us to prepare the ground for later discussions which will grapple with physical theories in greater detail, both for the sake of the theories themselves and for their important bearing on the problem of time. In this connection the **theory of the electromagnetic field**, which is the most perfect branch of physics at present known, will be of the highest importance. It will here only be considered in so far as time does not enter into it, i.e. we shall confine our attention to conditions which are stationary and invariable in time.

Coulomb's Law for electrostatics may be enunciated thus. If any charges of electricity are distributed in space with the density ρ they exert a force

$$\mathbf{K} = e \cdot \mathbf{E} \tag{48}$$

upon a point-charge e, whereby

$$\mathbf{E} = -\int \frac{\rho \cdot \mathbf{r}}{4\pi r^3}\, dV. \tag{49}$$

$\mathbf{r}$ here denotes the vector $\overrightarrow{OP}$ which leads from the "point of emergence O" at which $\mathbf{E}$ is to be determined, to the "current point" or source, with respect to which the integral is taken: r is its length and dV is the element of volume. The force is thus composed of two factors, the charge e of the small testing body, which depends on its condition alone, and of the "intensity of field" $\mathbf{E}$, which on the contrary is determined solely by the given distribution of the charges in space. We picture in our minds that even if we do not observe the force acting on a testing body, an "electric field" is called up by the charges distributed in space, this field being described by the vector $\mathbf{E}$; the action on a point-charge e expresses itself in the force (48). We may derive $\mathbf{E}$ from a potential $-\phi$ in accordance with the formulæ

$$\mathbf{E} = \operatorname{grad} \phi, \qquad -4\pi\phi = \int \frac{\rho}{r}\, dV. \tag{50}$$

From (50) it follows (1) that $\mathbf{E}$ is an irrotational (and hence lamellar) vector, and (2) that the flux of $\mathbf{E}$ through any closed surface is equal to the charges enclosed by this surface, or that the electricity is the source of the electric field; i.e. in formulæ

$$\operatorname{curl}\mathbf{E} = \mathbf{0}, \qquad \operatorname{div}\mathbf{E} = \rho. \tag{51}$$

Inversely, Coulomb's Law arises out of these simple differential laws if we add the condition that the field $\mathbf{E}$ vanish at infinite distances. For if we put $\mathbf{E} = \operatorname{grad}\phi$ from the first of the equations (51), we get from the second, to determine ϕ, Poisson's equation $\Delta\phi = \rho$, the solution of which is given by (50).

Coulomb's Law deals with "**action at a distance**". The intensity of the field at a point is expressed by it depending on the charges at all other points, near or far, in space. In contradistinction from this the far simpler formulæ (51) express laws relating to "infinitely near" action. As a knowledge of the values of a function in an arbitrarily small region surrounding a point is sufficient to determine the differential quotient of the function at the point, the values of ρ and $\mathbf{E}$ at a point and in its immediate neighbourhood are brought into connection with one another by (51). We shall regard these laws of infinitely near action as the true expression of the uniformity of action in nature, whereas we look upon (49) merely as a mathematical result following logically from it. In the light of the laws expressed by (51) which have such a simple intuitional significance we believe that we **understand** the source of Coulomb's Law. In doing this we do indeed bow to dictates of the theory of knowledge. Even Leibniz formulated the postulate of continuity, of infinitely near action, as a general principle, and could not, for this reason, become reconciled to Newton's Law of Gravitation, which entails action at a distance and which corresponds fully to that of Coulomb. The mathematical clearness and the simple meaning of the laws(51) are additional factors to be taken into account. In building up the theories of physics we notice repeatedly that once we have succeeded in bringing to light the uniformity of a certain group of phenomena it may be expressed in formulæ of perfect mathematical harmony. After all, from the physical point of view, Maxwell's theory in its later form bears uninterrupted testimony to the stupendous fruitfulness which has resulted through passing from the old idea of action at a distance to the modern one of infinitely near action.

The field exerts on the charges which produce it a force of which the density per unit volume is given by the formula

$$\mathbf{p} = \rho\mathbf{E}. \tag{52}$$

This is the rigorous interpretation of the equation (48).

If we bring a test charge (on a small body) into the field, it also becomes one of the field-producing charges, and formula (48) will lead to a correct determination of the field **E** existing before the test charge was introduced, only if the test charge e is so weak that its effect on the field is imperceptible. This is a difficulty which permeates the whole of experimental physics, viz. that by introducing a measuring instrument the original conditions which are to be measured become disturbed. This is, to a large extent, the source of the errors to the elimination of which the experimenter has to apply so much ingenuity.

The fundamental law of mechanics: mass $\times$ acceleration $=$ force, tells us how masses move under the influence of given forces (the initial velocities being given). Mechanics does not, however, teach us what is force; this we learn from physics. *The fundamental law of mechanics is a blank form which acquires a concrete content only when the conception of force occurring in it is filled in by physics.* The unfortunate attempts which have been made to develop mechanics as a branch of science distinct in itself have, in consequence, always sought help by resorting to an explanation in *words* of the fundamental law: force **signifies** mass × acceleration. In the present case of electrostatics, i.e. for the particular category of physical phenomena, we recognise what is force, and how it is determined according to a definite law by (52) from the phase-quantities charge and field. If we regard the charges as being given, the field equations (51) give the relation in virtue of which the charges determine the field which they produce. With regard to the charges, it is known that they are bound to matter. The modern theory of electrons has shown that this can be taken in a perfectly rigorous sense. Matter, is composed of elementary quanta, electrons, which have a definite invariable mass, and, in addition, a definite invariable charge. Whenever new charges appear to spring into existence, we merely observe the separation of positive and negative elementary charges which were previously so close together that the "action at a distance" of the one was fully compensated by that of the other. In such processes, accordingly, just as much positive electricity "arises" as negative. The laws thus constitute a cycle. The distribution of the elementary quanta of matter provided with charges fixed once and for all (and, in the case of non-stationary conditions, also their velocities) determine the field. The field exerts upon charged matter a ponderomotive force which is given by (52). The force determines, in accordance with the fundamental law of mechanics, the acceleration, and hence the distribution and velocity of the matter at the following moment. **We require this whole network of theoretical considerations to arrive at an experimental means of verification,**—if we assume that what we directly observe is the motion of matter. (Even this can be admitted only conditionally.) We cannot merely test a single law detached from this theoretical fabric! The connection between direct experience and the objective element behind it, which reason seeks to grasp conceptually in a theory, is not so simple that every single statement of the theory has a meaning which may be verified by direct intuition. We shall see more and more clearly in the sequel that Geometry, Mechanics, and Physics form an inseparable theoretical whole in this way. We must never lose sight of this totality when we enquire whether these sciences interpret rationally the reality which proclaims itself in all subjective experiences of consciousness, and which itself transcends consciousness: that is, truth forms a **system**. For the rest, the physical world-picture here described in its first outlines is characterised by the dualism of **matter** and **field**, between which there is a reciprocal action. Not till the advent of the theory of relativity was this dualism overcome, and, indeed, in favour of a physics based solely on fields (cf. § 24).

The ponderomotive force in the electric field was traced back to stresses even by Faraday. If we use a rectangular system of co-ordinates x_1, x_2, x_3 in which E_1, E_2, E_3 are the components of the electrical intensity of field, the x_i component of the force-density is

$$p_i = \rho E_i = E\left(\frac{\partial E_1}{\partial x_1} + \frac{\partial E_2}{\partial x_2} + \frac{\partial E_3}{\partial x_3}\right).$$

By a simple calculation which takes account of the irrotational property of $\mathbf{E}$ we discover from this that the components p_i of the force-density are derived by the formulæ (47) from the stress tensor, the components S_{ik} of which are tabulated in the following quadratic scheme

$$\begin{vmatrix} \frac{1}{2}(E_2^2+E_3^2-E_1^2) & -E_1E_2 & -E_1E_3 \\ -E_2E_1 & \frac{1}{2}(E_3^2+E_1^2-E_2^2) & -E_2E_3 \\ -E_3E_1 & -E_3E_2 & \frac{1}{2}(E_1^2+E_2^2-E_3^2) \end{vmatrix} \qquad (53)$$

We observe that the condition of symmetry $S_{ki} = S_{ik}$ is fulfilled. It is, above all, important to notice that the components of the stress tensor at a point depend only on the electrical intensity of field at this point. (They, moreover, depend only on the field, and not on the charge.) Whenever a force p can be retraced by (47) to stresses S, which form a symmetrical tensor of the second order only dependent on the values of the phase-quantities describing the physical state at the point in question, we shall have to regard these stresses as the primary factors and the actions of the forces as their consequent. The mathematical justification for this point of view is brought to light by the fact that the force p results from differentiating the stress. Compared with forces, stresses are thus, so to speak, situated on the next lower plane of differentiation, and yet do not depend on the whole series of values traversed by the phase-quantities, as would be the case for an arbitrary integral, but only on its value at the point under consideration. It further follows from the fact that the electrostatic forces which charged bodies exert on one another can be retraced to a symmetrical stress tensor, that the resulting total force as well as the resulting couple vanishes (because the integral taken over the whole space has a divergence $= 0$). This means that an isolated system of charged masses which is initially at rest cannot of itself acquire a translational or rotational motion as a whole.

The tensor (53) is, of course, independent of the choice of co-ordinate system. If we introduce the square of the value of the field intensity

$$|E|^2 = E_i E^i$$

then we have

$$S_{ik} = \tfrac{1}{2} g_{ik} |E|^2 - E_i E_k.$$

These are the co-variant stress components not only in a Cartesian but also in any arbitrary affine co-ordinate system, if E_i are the co-variant components of the field intensity. The physical significance of these stresses is extremely simple. If, for a certain point, we use rectangular co-ordinates, the x_1 axis of which points in the direction **E**: then

$$E_1 = |E|, \qquad E_2 = 0, \qquad E_3 = 0;$$

we thus find them to be composed of a tension having the intensity $\frac{1}{2}|E|^2$ in the direction of the lines of force, and of a pressure of the same intensity acting perpendicularly to them.

The fundamental laws of electrostatics may now be summarised in the following invariant tensor form:—

$$\left.\begin{array}{ll} \text{(I)} & \dfrac{\partial E_i}{\partial x_k} - \dfrac{\partial E_k}{\partial x_i} = 0, \text{ or } E_i = \dfrac{\partial \phi}{\partial x_i} \text{ respectively;} \\ \text{(II)} & \dfrac{\partial E^i}{\partial x_i} = \rho; \\ \text{(III)} & S_{ik} = \tfrac{1}{2} g_{ik} |E|^2 - E_i E_k. \end{array}\right\} \qquad (54)$$

A system of discrete point-charges $e_1, e_2, e_3, \ldots$ has potential energy

$$U = \frac{1}{8\pi} \sum_{i \neq k} \frac{e_i e^k}{r_{ik}}$$

in which r_{ik} denotes the distance between the two charges e_i and e_k. This signifies that the virtual work which is performed by the forces acting at the separate points (owing to the charges at the remaining points) for an infinitesimal displacement of the points is a total differential, viz. δU. For continuously distributed charges this formula resolves into

$$U = \iint \frac{\rho(P)\rho(P')}{8\pi r_{PP'}}\, dV\, dV'$$

in which both volume integrations with respect to P and P' are to be taken over the whole space, and $r_{PP'}$ denotes the distance between these two points. Using the potential ϕ we may write

$$U = -\tfrac{1}{2}\int \rho\phi\, dV.$$

The integrand is $\phi \cdot \operatorname{div} \mathbf{E}$. In consequence of the equation

$$\operatorname{div}(\phi\mathbf{E}) = \phi \cdot \operatorname{div} \mathbf{E} + \mathbf{E} \operatorname{grad} \phi$$

and of Gauss's theorem, according to which the integral of $\operatorname{div}(\phi\mathbf{E})$ taken over the whole space is equal to 0, we have

$$-\int \rho\phi\, dV = \int (\mathbf{E} \operatorname{grad} \phi)\, dV = \int |E|^2\, dV;$$

i.e.

$$U = \int \tfrac{1}{2}|\mathbf{E}|^2\, dV. \tag{55}$$

This representation of the energy makes it directly evident that the energy is a **positive** quantity. If we trace the forces back to stresses, we must picture these stresses (like those in an elastic body) as being everywhere associated with positive potential energy of strain. The seat of the energy must hence be sought in the field. Formula (55) gives a fully satisfactory account of this point. It tells us that the energy associated with the strain amounts to $\frac{1}{2}|E|^2$ per unit volume, and is thus exactly equal to the tension and the pressure which are exerted along and perpendicularly to the lines of force. The deciding factor which makes this view permissible is again the circumstance that the value obtained for the energy-density depends solely on the value, **at the point in question**, of the phrase-quantity $\mathbf{E}$ which characterises the field. Not only the field as a whole, but every portion of the field has a definite amount of potential energy $= \int \frac{1}{2}|E|^2\, dV$. In statics, it is only the total energy which comes into consideration. Only later, when we pass on to consider variable fields, shall we arrive at irrefutable confirmation of the correctness of this view.

In the case of conductors in a statical field the charges collect on the outer surface and there is no field in the interior. The equations (51) then suffice to determine the electrical field in free space in the "æther". If, however, there are non-conductors, dielectrics in the field, the phenomenon of **dielectric polarisation** (displacement) must be taken into consideration. Two charges $+e$ and $-e$ at the points P_1 and P_2 respectively, "source and sink" as we shall call them, produce a field, which arises from the potential

$$\frac{e}{4\pi}\left(\frac{1}{r_1} - \frac{1}{r_2}\right)$$

in which r_1 and r_2 denote the distances of the points P_1, P_2 from the origin, O. Let the product of e and the vector $\overrightarrow{P_1P_2}$ be called the moment $\mathbf{m}$ of the "source and sink" pair. If we now suppose the two charges to approach one another in a definite direction at a point P, the charge increasing simultaneously in such a way that the moment $\mathbf{m}$ remains constant, we get, in the limit, a "doublet" of moment $\mathbf{m}$, the potential of which is given by

$$\frac{\mathbf{m}}{4\pi} \operatorname{grad}_P \frac{1}{r}.$$

The result of an electric field in a dielectric is to give rise to these doublets in the separate elements of volume: this effect is known as **polarisation**. If $\mathbf{m}$ is the electric moment of the doublets per unit volume, then, instead of (50), the following formula holds for the potential

$$-4\pi\phi = \int \frac{\rho}{r} dV + \int \mathbf{m} \cdot \operatorname{grad}_P \frac{1}{r} dv. \tag{56}$$

From the point of view of the theory of electrons this circumstance becomes immediately intelligible. Let us, for example, imagine an atom to consist of a positively charged "nucleus" at rest, around which an oppositely charged electron rotates in a circular path. The mean position of the electron for the mean time of a complete revolution of the electron round the nucleus will then coincide with the position of the nucleus, and the atom will appear perfectly neutral from without. But if an electric field acts, it exerts a force on the negative electron, as a result of which its path will lie excentrically with respect to the atomic nucleus, e.g. will become an ellipse with the nucleus at one of its foci. In the mean, for times which are great compared with the time of revolution of the electron, the atom will act like a doublet; or if we treat matter as being continuous we shall have to assume continuously distributed doublets in it. Even before entering upon an exact atomistic treatment of this idea we can say that, at least to a first approximation, the moment $\mathbf{m}$ per unit volume will be proportional to the intensity $\mathbf{E}$ of the electric field: i.e. $\mathbf{m} = k\mathbf{E}$, in which k denotes a constant characteristic of the matter, which is dependent on its chemical constitution, viz. on the structure of its atoms and molecules.

Since

$$\operatorname{div}\left(\frac{\mathbf{m}}{r}\right) = \mathbf{m}\operatorname{grad}\frac{1}{r} + \frac{\operatorname{div}\mathbf{m}}{r}$$

we may replace equation (56) by

$$-4\pi\phi = \int \frac{\rho - \operatorname{div}\mathbf{m}}{r}\, dV.$$

From this we get for the field intensity $\mathbf{E} = \operatorname{grad}\phi$

$$\operatorname{div}\mathbf{E} = \rho - \operatorname{div}\mathbf{m}.$$

If we now introduce the "electric displacement"

$$\mathbf{D} = \mathbf{E} + \mathbf{m}$$

the fundamental equations become:

$$\operatorname{curl}\mathbf{E} = \mathbf{0}, \qquad \operatorname{div}\mathbf{D} = \rho. \tag{57}$$

They correspond to equations (51); in one of them the intensity $\mathbf{E}$ of field now occurs, in the other $\mathbf{D}$ the electric displacement. With the above assumption $\mathbf{m} = k\mathbf{E}$ we get the law of matter

$$\mathbf{D} = \epsilon\mathbf{E} \tag{58}$$

if we insert the constant $\epsilon = 1 + k$, characteristic of the matter, called the **dielectric constant**.

These laws are excellently confirmed by observation. The influence of the intervening medium which was experimentally proved by Faraday, and which expresses itself in them, has been of great importance in the development of the theory of action by contact. We may here pass over the corresponding extension of the formulæ for stress, energy, and force.

It is clear from the mode of derivation that (57) and (58) are not rigorously valid laws, since they relate only to mean values and are deduced for spaces containing a great number of atoms and for times which are great compared with the times of revolution of the electrons round the atom. **We still look upon** (51) **as expressing the physical laws exactly.** Our objective here and in the sequel is above all to derive the strict physical laws. But if we start from phenomena, such "phenomenological laws" as (57) and (58) are necessary stages in passing from the results of direct observation to the exact theory. In general, it is possible to work out such a theory only by starting in this way. The validity of the theory is then established if, with the aid of definite ideas about the atomic structure of matter, we can again arrive at the phenomenological laws by using mean value arguments. If the atomic structure is known, this process must, in addition, yield the values of the constants occurring in these laws and characteristic of the matter in question (such constants do not occur in exact physical laws). Since laws of matter such as (58), which only take the influence of massed matter into account, certainly fail for events in which the fine structure of matter cannot be neglected, the range of validity of the phenomenological theory must be furnished by an atomistic theory of this kind, as must also those laws which have to be substituted in its place for the region beyond this range. In all this the electron theory has met with great success, although, in view of the difficulty of the task, it is far from giving a complete statement of the more detailed structure of the atom and its inner mechanism.

In the first experiments with permanent magnets, magnetism appears to be a mere repetition of electricity: here Coulomb's Law holds likewise! A characteristic difference, however, immediately asserts itself in the fact that positive and negative magnetism cannot be dissociated from one another. There are no sources, but only doublets in the magnetic field. Magnets consist of infinitely small elementary magnets, each of which itself contains positive and negative magnetism. The amount of magnetism in every portion of matter is *de facto* nil; this would appear to mean that there is really no such thing as magnetism. The explanation of this was furnished by Oersted's discovery of the magnetic action of electric currents. The exact quantitative formulation of this action as expressed by Biot and Savart's Law leads, just like Coulomb's Law, to two simple laws of action by contact. If **s** denotes the density of the electric current, and **H** the intensity of the magnetic field, then

$$\operatorname{curl}\mathbf{H} = \mathbf{s}, \qquad \operatorname{div}\mathbf{H} = 0. \tag{59}$$

The second equation asserts the non-existence of sources in the magnetic field. Equations (59) are exactly analogous to (51) if div and curl be interchanged. These two operations of vector analysis correspond to one another in exactly the same way as do scalar and vectorial multiplication in vector algebra (div denotes scalar, curl vectorial, multiplication by the symbolic vector "differentiation"). The solution of the equations (59) vanishes for infinite distances; for a given distribution of current it is given by

$$\mathbf{H} = \int \frac{[\mathbf{s}, \mathbf{r}]}{4\pi r^3}\, dV, \tag{60}$$

which is exactly analogous to (49) and is, indeed, the expression of Biot and Savart's Law. This solution may be derived from a "vector potential" $-\mathbf{f}$ in accordance with the formulæ

$$\mathbf{H} = -\operatorname{curl}\mathbf{f}, \qquad -4\pi\mathbf{f} = \int \frac{\mathbf{s}}{r}\,dV.$$

Finally the formula for the density of force in the magnetic field is

$$\mathbf{p} = [\mathbf{s}, \mathbf{H}] \tag{61}$$

corresponding exactly with (52).

There is no doubt that these laws give us a true statement of magnetism. They are not a repetition but an exact counterpart of electrical laws, and bear the same relation to the latter as vectorial products to scalar products. From them it may be proved mathematically that a small circular current acts exactly like a small elementary magnet thrust through it perpendicularly to its plane. Following Ampère we have thus to imagine the magnetic action of magnetised bodies to depend on **molecular currents**; according to the electron theory these are straightway given by the electrons circulating in the atom.

The force $\mathbf{p}$ in the magnetic field may also be traced back to stresses, and we find, indeed, that we get the same values for the stress components as in the electrostatic field: we need only replace $\mathbf{E}$ by $\mathbf{H}$. Consequently we shall use the corresponding value $\frac{1}{2}\mathbf{H}^2$ for the density of the potential energy contained in the field. This step will only be properly justified when we come to the theory of fields varying with the time.

It follows from (59) that the current distribution is free of sources: $\operatorname{div}\mathbf{s} = 0$. The current field can therefore be entirely divided into current tubes all of which again merge into themselves, i.e. are continuous. The same total current flows through every cross-section of each tube. In no wise does it follow from the laws holding in a stationary field, nor does it come into consideration for such a field, that this current is an electric current in the ordinary sense, i.e. that it is composed of electricity in motion; this is, however, without doubt the case. In view of this fact the law $\operatorname{div} \mathbf{s} = 0$ asserts that electricity is neither created nor destroyed. It is only because the flux of the current vector through a closed surface is nil that the density of electricity remains everywhere unchanged—so that electricity is neither created nor destroyed. (We are, of course, dealing with stationary fields exclusively.) The expression **vector potential f**, introduced above, also satisfies the equation $\operatorname{div} \mathbf{f} = 0$.

Being an electric current, **s** is without doubt a vector in the true sense of the word. It then follows, however, from the Law of Biot and Savart that **H is not a vector but a linear tensor of the second order**. Let its components in any co-ordinate system (Cartesian or even merely affine) be H_{ik}. The vector potential **f** is a true vector. If ϕ_i are its co-variant components and s^i the contra-variant components of the current-density (the current is like velocity fundamentally a contra-variant vector), the following table gives us the final form (independent of the dimensional number) of **the laws which hold in the magnetic field produced by a stationary electric current.**

$$\frac{\partial H_{kl}}{\partial x_i} + \frac{\partial H_{li}}{\partial x_k} + \frac{\partial H_{ik}}{\partial x_l} = 0, \tag{62$_1$}$$

$$H_{ik} = \frac{\partial \phi_i}{\partial x_k} - \frac{\partial \phi_k}{\partial x_i} \quad \text{respectively}$$

$$\frac{\partial H_{ik}}{\partial x_k} = s^i. \tag{62$_2$}$$

The stresses are determined by:

$$S_i^k = H_{ir}H^{kr} - \tfrac{1}{2}\delta_i^k |H|^2 \tag{62$_3$}$$

in which $|H|$ signifies the strength of the magnetic field:

$$|H|^2 = H_{ik}H^{ik}.$$

The stress tensor is symmetrical, since

$$H_{ir}H_k^r = H_i^r H_{kr} = g^{rs} H_{ir} H_{ks}.$$

The components of the force-density are

$$p_i = H_k^i s^k. \tag{62$_4$}$$

The energy-density $= \frac{1}{2}|H|^2$.

These are the laws that hold for the field in empty space. We regard them as being exact physical laws which are generally valid, as in the case of electricity. For a phenomenological theory it is, however, necessary to take into consideration the **magnetisation**, a phenomenon analogous to dielectric polarisation. Just as **D** occurred in conjunction with **E**, so the "magnetic induction" **B** associates itself with the intensity of field **H**. The laws

$$\operatorname{curl} \mathbf{H} = \mathbf{s}, \qquad \operatorname{div} \mathbf{B} = 0$$

hold in the field, as does the law which takes account of the magnetic character of the matter

$$\mathbf{B} = \mu \mathbf{H}. \tag{63}$$

The constant μ is called magnetic permeability. But whereas the single atom only becomes polarised by the action of the intensity of the electrical field (i.e. becomes a doublet), (this takes place in the direction of the field intensity), the atom is from the outset an elementary magnet owing to the presence of rotating electrons in it (at least, in the case of para- and ferro-magnetic substances). All these elementary magnets, however, neutralise one another's effects, as long as they are irregularly arranged and all positions of the electronic orbits occur equally frequently on the average. The imposed magnetic force merely fulfils the function of **directing** the existing doublets. It evidently is due to this fact that the range within which (63) holds is much less than the corresponding range of (63). Permanent magnets and ferro-magnetic bodies (iron, cobalt, nickel) are, above all, not subject to it.

In the phenomenological theory there must be added to the laws already mentioned that of **Ohm**:

$$\mathbf{s} = \sigma \mathbf{E} \qquad (\sigma = \text{conductivity}).$$

It asserts that the current follows the fall of potential and is proportional to it for a given conductor. Corresponding to Ohm's Law we have in the atomic theory the fundamental law of mechanics, according to which the motion of the "free" electrons is determined by the electric and magnetic forces acting on them which thus produce an electric current. Owing to collisions with the molecules no permanent acceleration can come about, but (just as in the case of a heavy body which is falling and experiences the resistance of the air) a mean limiting velocity is reached, which may, to a first approximation at least, be put proportional to the driving electric force $\mathbf{E}$. In this way Ohm's Law acquires a meaning.

If the current is produced by a voltaic cell or an accumulator, the chemical action which takes place maintains a constant difference of potential, the "electromotive force," between the two ends of the conducting wire. Since the events which occur in the contrivance producing the current can obviously be understood only in the light of an atomic theory, it leads to the simplest result phenomenologically to represent it by means of a cross-section taken through the conducting circuit at each end, beyond which the potential makes a sudden jump equal to the electromotive force.

This brief survey of Maxwell's theory of stationary fields will suffice for what follows. We have not the space here to enlarge upon details and concrete applications.

CHAPTER II

THE METRICAL CONTINUUM

§ 10. Note on Non-Euclidean Geometry*

DOUBTS as to the validity of Euclidean geometry seem to have been raised even at the time of its origin, and are not, as our philosophers usually assume, outgrowths of the hypercritical tendency of modern mathematicians. These doubts have from the outset hovered round the fifth postulate. The substance of the latter is that in a plane containing a given straight line g and a point P external to the latter (but in the plane) there is only one straight line through P which does not intersect g: it is called the straight line parallel to P. Whereas the remaining axioms of Euclid are accepted as being self-evident, even the earliest exponents of Euclid have endeavoured to prove this theorem from the remaining axioms. Nowadays, knowing that this object is unattainable, we must look upon these reflections and efforts as the beginning of "non-Euclidean" geometry, i.e. of the construction of a geometrical system which can be developed logically by accepting all the axioms of Euclid, except the postulate of parallels. A report of Proclus (A.D. 5) about these attempts has been handed down to posterity. Proclus utters an emphatic warning against the abuse that may be practised by calling propositions self-evident. This warning cannot be repeated too often; on the other hand, we must not fail to emphasise the fact that, in spite of the frequency with which this property is wrongfully used, the "self-evident" property is the final root of all knowledge, including empirical knowledge. Proclus insists that "asymptotic lines" may exist.

* *Vide* note 1.

We may picture this as follows. Suppose a straight line g be given in a plane, also a point P outside it in the plane, and a straight line s passing through P and which may be rotated about P. Let s be perpendicular to g initially. If we now ro-

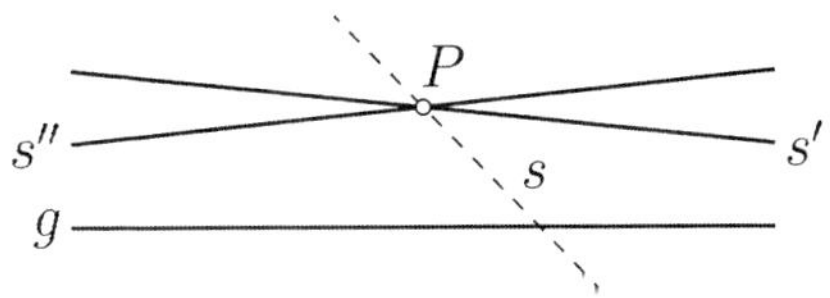

FIG. 2.

tate s, the point of intersection of s and g glides along g, e.g. to the right, and if we continue turning, a definite moment arrives at which this point of intersection just vanishes to infinity; s then occupies the position of an "asymptotic" straight line. If we continue turning, Euclid assumes that, at even this same moment, a point of intersection already appears on the left. Proclus, on the other hand, points out the possibility that one may perhaps have to turn s through a further definite angle before a point of intersection arises to the left. We should then have two "asymptotic" straight lines, one to the right, viz. s', and the other to the left, viz. s''. If the straight line s through P were then situated in the angular space between s'' and s' (during the rotation just described) it would cut g; if it lay between s' and s'', it would **not** intersect g. There must be at least **one** non-intersecting straight line; this follows from the other axioms of Euclid. I shall recall a familiar figure of our early studies in plane geometry, consisting of the straight line h and two straight lines g and g' which intersect h at A and A' and make equal angles with it, g and g' are each divided into a right and a left half by their point of intersection with h. Now, if g and g' had a common point s to the right of h, then, since $BAA'B'$ is congruent with $C'A'AC$ (*vide* Fig. 3),

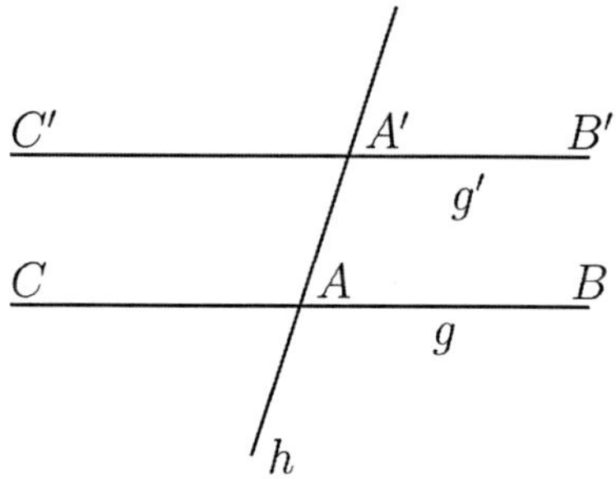

FIG. 3.

there would also be a point of intersection S^* to the left of h. But this is impossible since there is only one straight line that passes through two given points S and S^*.

Attempts to prove Euclid's postulate were continued by Arabian and western mathematicians of the Middle Ages. Passing straight to a more recent period we shall mention the names of only the last eminent forerunners of non-Euclidean geometry, viz. the Jesuit father Saccheri (beginning of the eighteenth century) and the mathematicians Lambert and Legendre. Saccheri was aware that the question whether the postulate of parallels is valid is equivalent to the question whether the sum of the angles of a triangle are equal to or less than 180°. If they amount to 180° in **one** triangle, then they must do so in every triangle and Euclidean geometry holds. If the sum is $< 180°$ in one triangle then it is $< 180°$ in every triangle. That they cannot be $> 180°$ is excluded for the same reason for which we just now concluded that not all the straight lines through P can cut the fixed straight line g. Lambert discovered that if we assume the sum of the three angles to be $< 180°$ there must be a unique length in geometry. This is closely related to an observation which Wallis had previously made that there can be no similar figures of different sizes in non-Euclidean geometry (just as in the case of the geometry of the surface of a rigid sphere). Hence if there is such a thing as "form" independent of size, Euclidean geometry is justified in its claims. Lambert, moreover, deduced a formula for the area of a triangle, from which it is clear that, in the case of non-Euclidean geometry, this area cannot increase beyond all limits. It appears that the researches of these men has gradually spread the belief in wide circles that the postulate of parallels cannot be proved. At that time this problem occupied many minds. D'Alembert pronounced it a scandal of geometry that it had not yet been decisively settled. Even the authority of Kant, whose philosophic system claims Euclidean geometry as *a priori* knowledge representing the content of pure space-intuition in adequate judgments, did not succeed in settling these doubts permanently.

Gauss also set out originally to prove the axiom of parallels, but he early gained the conviction that this was impossible and thereupon developed the principles of a non-Euclidean geometry, for which the axioms of parallels does not hold, to such an extent that, from it, the further development could be carried out with the same ease as for Euclidean geometry. He did not make his investigations known for, as he later wrote in a private letter, he feared "the outcry of the Bœotians"; for, he said, there were only a few people who understood what was the true essence of these questions. Independently of Gauss, Schweikart, a professor of jurisprudence, gained a full insight into the conditions of non-Euclidean geometry, as is evident from a concise note addressed to

Gauss. Like the latter he considered it in no wise self-evident, and established that Euclidean geometry is valid in our actual space. His nephew Taurinus whom he encouraged to study these questions was, in contrast to him, a believer of Euclidean geometry, but we are nevertheless indebted to Taurinus for the discovery of the fact that the formulæ of spherical trigonometry are real on a sphere which has an imaginary radius $= \sqrt{-1}$, and that through them a geometrical system is constructed along analytical lines which satisfies all the axioms of Euclid except the fifth postulate.

For the general public the honour of discovering and elaborating non-Euclidean geometry must be shared between Nikolaj Iwanowitsch Lobatschefskij (1793–1856), a Russian professor of mathematics at Kasan, and Johann Bolyai (1802–1860), a Hungarian officer in the Austrian army. The ideas of both assumed a tangible form in 1826. The chief manuscript of both, by which the public were informed of their discovery and which offered an argument of the new geometry in the manner of Euclid, had its origin in 1830–1831. The discussion by Bolyai is particularly clear, inasmuch as he carries the argument as far as possible without making an assumption as to the validity or non-validity of the fifth postulate, and only afterwards derives the theorems of Euclidean and non-Euclidean geometry from the theorems of his "absolute" geometry according to whether one decides in favour of or against Euclid.

Although the structure was thus erected, it was by no means definitely decided whether, in absolute geometry, the axiom of parallels would not after all be shown to be a dependent theorem. The strict proof that **non-Euclidean geometry is absolutely consistent in itself** had yet to follow. This resulted almost of itself in the further development of non-Euclidean geometry. As often happens, the simplest way of proving this was not discovered at once. It was discovered by Klein as late as 1870 and depends on the construction of a **Euclidean model** for non-Euclidean geometry (*vide* note 2). Let us confine our attention to the plane! In a Euclidean plane with rectangular co-ordinates x and y we shall draw a circle U of radius unity with the origin as centre. Introducing homogeneous co-ordinates

$$x = \frac{x_1}{x_3}, \qquad y = \frac{x_2}{x_3}$$

(so that the position of a point is defined by the ratio of three numbers, i.e. $x_1 : x_2 : x_3$), the equation to the circle becomes

$$-x_1^2 - x_2^2 + x_3^2 = 0.$$

Let us denote the quadratic form on the left by $\Omega(x)$ and the corresponding symmetrical bilinear form of two systems of value, x_i, x_i' by $\Omega(x, x')$. A transformation which assigns to every point x a transformed point x' according to the linear formulæ

$$x_i' = \sum_{k=1}^{3} \alpha_i^k x_k \qquad (|\alpha_i^k| \neq 0)$$

is called, as we know, a collineation (affine transformations are a special class of collineations). It transforms every straight line, point for point, into another straight line and leaves the cross ratio of four points on a straight line unaltered. We shall now set up a little dictionary by which we translate the conceptions of Euclidean geometry into a new language, that of non-Euclidean geometry; we use inverted commas to distinguish its words. The vocabulary of this dictionary is composed of only three words.

The word "point" is applied to any point on the inside of U (Fig. 4).

A "straight line" signifies the portion of a straight line lying wholly in U. The collineations which transform the circle U into itself are of two kinds; the first leaves the sense in which U is

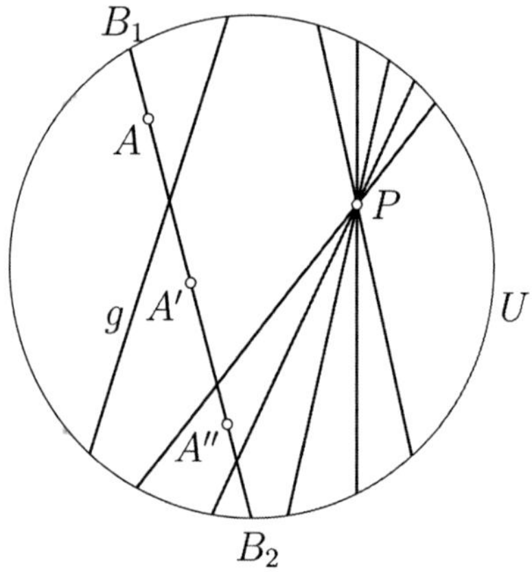

FIG. 4.

described unaltered, whereas the second reverses it. The former are called "congruent" transformations; two figures composed of points are called "congruent" if they can be transformed into one another by such a transformation. All the axioms of Euclid except the postulate of parallels hold for these "points," "straight lines," and the conception "congruence". A whole sheaf of "straight lines" passing through the "point" P which do not cut the one "straight line" g is shown in Fig. 4. This suffices to prove the consistency of non-Euclidean geometry, for things and relations are shown for which all the theorems of Euclidean geometry are valid provided that the appropriate nomenclature be adopted. It is evident, without further explanation, that Klein's model is also applicable to spatial geometry.

We now determine the non-Euclidean distance between two "points" in this model, viz. between

$$A = (x_1 : x_2 : x_3) \text{ and } A' = (x'_1 : x'_2 : x'_3).$$

Let the straight line AA' cut the circle U in the two points, B_1, B_2. The homogeneous co-ordinates y_i of these two points are of the form

$$y_i = \lambda x_i + \lambda' x'_i$$

and the corresponding ratio of the parameters, $\lambda : \lambda'$, is given by the equation $\Omega(y) = 0$, viz.

$$\frac{\lambda}{\lambda'} = \frac{-\Omega(x, x') \pm \sqrt{\Omega^2(x, x') - \Omega(x)\Omega(x')}}{\Omega(x)}.$$

Hence the cross-ratio of the four points, A, A', B_1, B_2 is

$$[AA'] = \frac{\Omega(x, x') + \sqrt{\Omega^2(x, x') - \Omega(x)\Omega(x')}}{\Omega(x, x') - \sqrt{\Omega^2(x, x') - \Omega(x)\Omega(x')}}.$$

This quantity which depends on the two arbitrary "points," A, A', is not altered by a "congruent" transformation. If A, A', A'' are any three "points" lying on a "straight line" in the order written, then

$$[AA''] = [AA'] \cdot [A'A''].$$

The quantity

$$\tfrac{1}{2} \log[AA'] = \overline{AA'} = r$$

has thus the functional property

$$\overline{AA'} + \overline{A'A''} = \overline{AA''}.$$

As it has the same value for "congruent" distances AA' too, we must regard it as the non-Euclidean distance between the two points, A, A'. Assuming the logs to be taken to the base e, we get an absolute determination for the unit of measure, as was recognised by Lambert. The definition may be written in the shorter form:

$$\cosh r = \frac{\Omega(x, x')}{\sqrt{\Omega(x) \cdot \Omega(x')}} \tag{1}$$

(cosh denotes the hyperbolic cosine).

This measure-determination had already been enunciated before Klein by Cayley* who referred it to an arbitrary real or imaginary conic section $\Omega(x) = 0$: he called it the "projective measure-determination". But it was reserved for Klein to recognise that in the case of a real conic it leads to non-Euclidean geometry.

It must not be thought that Klein's model shows that the non-Euclidean plane is finite. On the contrary, using non-Euclidean measures I can mark off the same distance on a "straight line" an infinite number of times in succession. It is only by using **Euclidean** measures in the **Euclidean** model that the distances of these "equidistant" points becomes smaller and smaller. For non-Euclidean geometry the bounding circle U represents unattainable, infinitely distant, regions.

If we use an imaginary conic, Cayley's measure-determination leads to ordinary spherical geometry, such as holds on the surface

* *Vide* note 3.

of a sphere in Euclidean space. Great circles take the place of straight lines in it, but every pair of points at the end of the same diameter must be regarded as a single "point," in order that two "straight lines" may only intersect at one "point". Let us project the points on the sphere by means of (straight) rays from the centre on to the tangential plane at a point on the surface of the sphere, e.g. the south pole. Two diametrically opposite points will then coincide on the tangential plane as a result of the transformation. We must, in addition, as in projective geometry, furnish this plane with an infinitely distant straight line; this is given by the projection of the equator. We shall now call two figures in this plane "congruent" if their projections (through the centre) on to the surface of the sphere are congruent in the ordinary Euclidean sense. Provided this conception of "congruence" is used, a non-Euclidean geometry, in which all the axioms of Euclid except the fifth postulate are fulfilled, holds in this plane. Instead of this postulate we have the fact that each pair of straight lines, without exception, intersects, and, in accordance with this, the sum of the angles in a triangle $> 180°$. This seems to conflict with the Euclidean proof quoted above. The apparent contradiction is explained by the circumstance that in the present "spherical" geometry the straight line is closed, whereas Euclid, although he does not explicitly state it in his axioms, tacitly assumes that it is an open line, i.e. that each of its points divides it into two parts. The deduction that the hypothetical point of intersection S on the "right-hand" side is different from that S^* on the "left-hand" side is rigorously true only if this "openness" be assumed.

Let us mark out in space a Cartesian co-ordinate system x_1, x_2, x_3, having its origin at the centre of the sphere and the line connecting the north and south poles as its x_3 axis, the radius of the sphere being the unit of length. If x_1, x_2, x_3 are the co-ordinates of any point on the sphere, i.e.

$$\Omega(x) \equiv x_1^2 + x_2^2 + x_3^2 = 1$$

then $\frac{x_1}{x_3}$ and $\frac{x_2}{x_3}$ are respectively the first and second co-ordinate of the transformed point in our plane $x_3 = 1$, i.e. $x_1 : x_2 : x3$ is the ratio of the homogeneous co-ordinates of the transformed point. Congruent transformations of the sphere are linear transformations which leave the quadratic form $\Omega(x)$ invariant. The "congruent" transformations of the plane in terms of our "spherical" geometry are thus given by such linear transformations of the homogeneous co-ordinates as convert the equation $\Omega(x) = 0$, which signifies an imaginary conic, into itself. This proves the statement made above concerning the relationship between spherical geometry and Cayley's measure-relation. This agreement is expressed in the formula for the distance r between two points A, A', which is here

$$\cos r = \frac{\Omega(x, x')}{\sqrt{\Omega(x)\Omega(x')}}. \tag{2}$$

At the same time we have confirmed the discovery of Taurinus that Euclidean geometry is identical with non-Euclidean geometry on a sphere of radius $\sqrt{-1}$.

Euclidean geometry occupies an intermediate position between that of Bolyai-Lobatschefsky and spherical geometry. For if we make a real conic section change to a degenerate one, and thence to an imaginary one, we find that the plane with its corresponding Cayley measure-relation is at first Bolyai-Lobatschefskyan, then Euclidean, and finally spherical.

§ 11. The Geometry of Riemann

The next stage in the development of non-Euclidean geometry that concerns us chiefly is that due to Riemann. It links up with the foundations of Differential Geometry, in particular with that of the theory of surfaces as set out by Gauss in his *Disquisitiones circa superficies curvas.*

The most fundamental property of space is that its points form a three-dimensional manifold. What does this convey to us? We say, for example, that ellipses form a two-dimensional manifold (as regards their size and form, i.e. considering congruent ellipses similar, non-congruent ellipses as dissimilar), because each separate ellipse may be distinguished in the manifold by two given numbers, the lengths of the semi-major and semi-minor axis. The difference in the conditions of equilibrium of an ideal gas which is given by two independent variables, such as pressure and temperature, form a two-dimensional manifold, likewise the points on a sphere, or the system of pure tones (in terms of intensity and pitch). According to the physiological theory which states that the sensation of colour is determined by the combination of three chemical processes taking place on the retina (the black-white, red-green, and the yellow-blue process, each of which can take place in a definite direction with a definite intensity), colours form a three-dimensional manifold with respect to quality and intensity, but colour qualities form only a two-dimensional manifold. This is confirmed by Maxwell's familiar construction of the colour triangle. The possible positions of a rigid body form a six-dimensional manifold, the possible positions of a mechanical system having n degrees of freedom constitute, in general, an n-dimensional manifold. **The characteristic of an**

n-dimensional manifold is that each of the elements composing it (in our examples, single points, conditions of a gas, colours, tones) **may be specified by the giving of n quantities, the "co-ordinates," which are continuous functions within the manifold.** This does not mean that the whole manifold with all its elements must be represented in a single and reversible manner by value systems of n co-ordinates (e.g. this is impossible in the case of the sphere, for which $n = 2$); it signifies only that if P is an arbitrary element of the manifold, then in every case a certain domain surrounding the point P must be representable singly and reversibly by the value system of n co-ordinates. If x_i is a system of n co-ordinates, x'_i another system of n co-ordinates, then the co-ordinate values x_i, x'_i of the same element will in general be connected with one another by relations

$$x_i = f_i(x'_1, x'_2, \ldots, x'_n) \qquad (i = 1, 2, \ldots, n) \tag{3}$$

which can be resolved into terms of x'_i and in which the f_i's are continuous functions of their arguments. As long as nothing more is known about the manifold, we cannot distinguish any one co-ordinate system from the others. For an analytical treatment of arbitrary continuous manifolds we thus require a theory of invariance with regard to arbitrary transformation of co-ordinates, such as (3), whereas for the development of affine geometry in the preceding chapter we used only the much more special theory of invariance for the case of **linear** transformations.

Differential geometry deals with curves and surfaces in three-dimensional Euclidean space; we shall here consider them mapped out in Cartesian co-ordinates x, y, z. A **curve** is in general a one-dimensional point-manifold; its separate points can be distinguished from one another by the values of a parameter u. If the point u on the curve happens to be at the point x, y, z in space, then x, y, z will be certain continuous functions of u:

$$x = x(u), \qquad y = y(u), \qquad z = z(u) \tag{4}$$

and (4) is called the "parametric" representation of the curve. If we interpret u as the time, then (4) is the law of motion of a point which traverses the given curve. The curve itself does not, however, determine singly the parametric representation (4) of the curve; the parameter u may, indeed, be subjected to any arbitrary continuous transformation.

A two-dimensional point-manifold is called a **surface**. Its points can be distinguished from one another by the values of two parameters u_1, u_2. It may therefore be represented parametrically in the form

$$x = x(u_1, u_2), \qquad y = y(u_1, u_2), \qquad z = z(u_1, u_2). \tag{5}$$

The parameters u_1, u_2 may likewise undergo any arbitrary continuous transformation without affecting the represented curve. We shall assume that the functions (5) are not only continuous but have also continuous differential co-efficients. Gauss, in his general theory, starts from the form (5) of representing any surface; the parameters u_1, u_2 are hence called the Gaussian (or curvilinear) co-ordinates on the surface. For example, if, as in the preceding section, we project the points of the surface of the unit sphere in a small region encircling the origin of the co-ordinate system on to the tangent plane $z = 1$ at the south pole, and if we make x, y, z the co-ordinates of any arbitrary point on the sphere, u_1 and u_2 being respectively the x and y co-ordinates of the point of projection in this plane, then

$$x = \frac{u_1}{\sqrt{1 + u_1^2 + u_2^2}}, \quad y = \frac{u_2}{\sqrt{1 + u_1^2 + u_2^2}}, \quad z = \frac{1}{\sqrt{1 + u_1^2 + u_2^2}}. \tag{6}$$

This is a parametric representation of the sphere. It does not, however, embrace the whole sphere, but only a certain region round the south pole, viz. the part from the south pole to the equator, excluding the latter. Another illustration of a parametric representation is given by the geographical co-ordinates, latitude and longitude.

In thermodynamics we use a graphical representation consisting of a plane on which two rectangular co-ordinate axes are drawn, and in which the state of a gas as denoted by its pressure p and temperature θ is represented by a point having the rectangular co-ordinates p, θ. The same procedure may be adopted here. With the point u_1, u_2 on the surface, we associate a point in the "representative" plane having the rectangular co-ordinates u_1, u_2. The formulæ (5) do not then represent only the surface, but also at the same time a definite continuous **representation** of this surface on the u_1, u_2 plane. Geographical maps are familiar instances of such representations of curved portions of surface by means of planes. A curve on a surface is given mathematically by a parametric representation

$$u_1 = u_1(t), \qquad u_2 = u_2(t), \tag{7}$$

whereas a portion of a surface is given by a "mathematical region" expressed in the variables u_1, u_2, and which must be characterised by inequalities involving u_1 and u_2; i.e. graphically by means of the representative curve or the representative region in the u_1-u_2-plane. If the representative plane be marked out with a network of co-ordinates in the manner of squared paper, then this becomes transposed, through the representation, to the curved surface as a net consisting of meshes having the form of little parallelograms, and composed of the two families of "co-ordinate lines" u_1 = const., u_2 = const., respectively. If the meshes be made sufficiently fine it becomes possible to map out any given figure of the representative plane on the curved surface.

The distance ds between two infinitely near points of the surface, namely,

$$(u_1, u_2) \quad \text{and} \quad (u_1 + du_1, u_2 + du_2)$$

is determined by the expression

$$ds^2 = dx^2 + dy^2 + dz^2$$

if we set

$$dx = \frac{\partial x}{\partial u_1} du_1 + \frac{\partial x}{\partial u_2} du_2 \tag{8}$$

in it, with corresponding expressions for dy and dz. We then get a quadratic differential form for ds^2 thus:

$$ds^2 = \sum_{i,k=1}^{2} g_{ik}\, du_i\, du_k \qquad (g_{ki} = g_{ik}) \tag{9}$$

in which the co-efficients are

$$g_{ik} = \frac{\partial x}{\partial u_i}\frac{\partial x}{\partial u_k} + \frac{\partial y}{\partial u_i}\frac{\partial y}{\partial u_k} + \frac{\partial z}{\partial u_i}\frac{\partial z}{\partial u_k}$$

and are not, in general, functions of u_1 and u_2.

In the case of the parametric representation of the sphere (6) we have

$$ds^2 = \frac{(1 + u_1^2 + u_2^2)(du_1^2 + du_2^2) - (u_1\,du_1 + u_2\,du_2)^2}{(1 + u_1^2 + u_2^2)^2}. \tag{10}$$

Gauss was the first to recognise that the metrical groundform is the determining factor for **geometry on surfaces**. The lengths of curves, angles, and the size of given regions on the surface depend on it alone. The geometries on two different surfaces is accordingly identical if, for a representation in appropriate parameters, the co-efficients g_{ik} of the metrical groundform coincide in value.

Proof.—The length of any arbitrary curve, given by (7), on the surface is furnished by the integral

$$\int ds = \int \sqrt{\sum_{i,k} g_{ik} \frac{du_i}{dt}\frac{du_k}{dt}} \cdot dt.$$

If we fix our attention on a definite point $P^0 = (u_1^0, u_2^0)$ on the surface and use the relative co-ordinates

$$u_i - u_i^0 = du_i, \qquad x - x^0 = dx, \qquad y - y^0 = dy, \qquad z - z^0 = dz$$

for its immediate neighbourhood, then equation (8), in which the derivatives are to be taken for the point P^0, will hold more exactly the smaller du_1, du_2, are taken; we say that it holds for "infinitely small" values du_1 and du_2. If we add to these the analogous equations for dy and dz, then they express that the immediate neighbourhood of P^0 is a plane, and that du_1, du_2 are affine co-ordinates on it.* Accordingly we may apply the formulæ of affine geometry

*We here assume that the determinants of the second order which can be

to the region immediately adjacent to P^0. For the angle θ between two line-elements or infinitesimal displacements having the components du_1, du_2 and δu_1, δu_2 respectively, we get

$$\cos\theta = \frac{Q(d,\delta)}{\sqrt{Q(d,d)Q(\delta,\delta)}}$$

in which $Q(d,\delta)$ stands for the symmetrical bilinear form

$$\sum_{i,k} g_{ik}\, du_i\, \delta u_k \text{ corresponding to (9).}$$

The area of the infinitesimal parallelogram marked out by these two displacements is found to be

$$\sqrt{g}\begin{vmatrix} du_1 & du_2 \\ \delta u_1 & \delta u_2 \end{vmatrix}$$

in which g denotes the determinant of the g_{ik}'s. The area of a curved portion of surface is accordingly given by the integral

$$\iint \sqrt{g}\, du_1\, du_2$$

---

formed from the table of co-efficients of these equations,

$$\begin{vmatrix} \frac{\partial x}{\partial u_1} & \frac{\partial y}{\partial u_1} & \frac{\partial z}{\partial u_1} \\ \frac{\partial x}{\partial u_2} & \frac{\partial y}{\partial u_2} & \frac{\partial z}{\partial u_2} \end{vmatrix},$$

do not all vanish. This condition is fulfilled for the regular points of the surface, at which there is a tangent plane. The three determinants are identically equal to 0, if, and only if, the surface degenerates to a curve, i.e. the functions x, y, z of u_1 and u_2 actually depend only on one parameter, a function of u_1 and u_2.

taken over the corresponding part of the representative plane. This proves Gauss' statement. The values of the expressions obtained are of course independent of the choice of parametric representation. This invariance with respect to arbitrary transformations of the parameters can easily be confirmed analytically. All the geometric relations holding on the surface can be studied on the representative plane. The geometry of this plane is the same as that of the curved surface if we agree to accept the distance ds of two infinitely near points as expressed by (9) and **not** by Pythagoras' formula

$$ds^2 = du_1^2 + du_2^2.$$

The geometry of the surface deals with the inner measure relations of the surface that belong to it independently of the manner in which it is embedded in space. They are the relations that can be determined by **measurements carried out on the surface itself**. Gauss in his investigation of the theory of surfaces started from the practical task of surveying Hanover geodetically. The fact that the earth is not a plane can be ascertained by measuring a sufficiently large portion of the earth's surface. Even if each single triangle of the network is taken too small for the deviation from a plane to come into consideration, they cannot be put together to form a closed net on a plane in the way they do on the earth's surface. To show this a little more clearly let us draw a circle C on a sphere of radius unity (the earth), having its centre P on the surface of the sphere. Let us further draw radii of this circle, i.e. arcs of great circles of the sphere radiating from P and ending at the circumference of C (let these arcs be $< \frac{\pi}{2}$). By carrying out measurements on the sphere's surface we can now ascertain that these radii starting out in all directions are the shortest lines connecting P to the circle C, and that they are all of the same length r; by measurement we find the closed curve C to be of length s. If we were dealing with a plane we should infer from this that the "radii" are straight lines and hence the curve C would be a circle and we should expect s to be equal to $2\pi r$. Instead of this, however, we find that s is less than the value given by the above formula, for in the actual case $s = 2\pi \sin r$. We thus discover by measurements carried out on the surface of the sphere that this surface is not a plane. If, on the other hand, we draw figures on a sheet of paper and then roll it up, we shall find the same values for measurements of these figures in their new condition as before, provided that no distortion has occurred through rolling up the paper. The same geometry will hold on it now as on the plane. It is impossible for me to ascertain that it is curved by carrying out geodetic measurements. Thus, in general, the same geometry holds for two surfaces that can be transformed into one another without distortion or tearing.

The fact that plane geometry does not hold on the sphere means analytically that it is impossible to convert the quadratic differential form (10) by means of a transformation

$$u_1 = u_1(u'_1, u'_2) \qquad u'_1 = u'_1(u_1, u_2)$$
$$u_2 = u_2(u'_1, u'_2) \qquad u'_2 = u'_2(u_1, u_2)$$

into the form

$$(du'_1)^2 + (du'_2)^2.$$

We know, indeed, that it is possible to do this for each point by a linear transformation of the differentials, viz. by

$$du'_i = \alpha_{i1}\, du_1 + \alpha_{i2}\, du_2 \qquad (i = 1, 2), \tag{11}$$

but it is impossible to choose the transformation of the differentials at each point so that the expressions (11) become **total** differentials for du'_1, du'_2.

Curvilinear co-ordinates are used not only in the theory of surfaces but also in the treatment of space problems, particularly in mathematical physics in which it is often necessary to adapt the co-ordinate system to the bodies presented, as is instanced in the case of cylindrical, spherical, and elliptic co-ordinates. The square of the distance, ds^2, between two infinitely near points in space, is always expressed by a quadratic form

$$\sum_{i,k=1}^{3} g_{ik}\, dx_i\, dx_k \tag{12}$$

in which x_1, x_2, x_3 are any arbitrary co-ordinates. If we uphold Euclidean geometry, we express the belief that this quadratic form can be brought by means of some transformation into one which has constant co-efficients.

These introductory remarks enable us to grasp the full meaning of the ideas developed fully by Riemann in his inaugural address, "Concerning the Hypotheses which lie at the Base of Geometry".* It is evident from Chapter I that Euclidean geometry holds for a three-dimensional **linear** point-configuration in a four-dimensional Euclidean space; but curved three-dimensional spaces, which exist in four-dimensional space just as much as curved surfaces occur in three-dimensional space, are of a different type. Is it not possible that our three-dimensional space of ordinary experience is curved? Certainly. It is not embedded in a four-dimensional space; but it is conceivable that its inner

* *Vide* note 4.

measure-relations are such as cannot occur in a "plane" space; it is conceivable that a very careful geodetic survey of our space carried out in the same way as the above-mentioned survey of the earth's surface might disclose that it is not plane. We shall continue to regard it as a three-dimensional manifold, and to suppose that infinitesimal line elements may be compared with one another in respect to length independently of their position and direction, and that the square of their lengths, the distance between two infinitely near points, may be expressed by a quadratic form (12), any arbitrary co-ordinates x_i being used. (There is a very good reason for this assumption; for, since every transformation from one co-ordinate system to another entails **linear** transformation-formulæ for the co-ordinate differentials, a quadratic form must always again pass into a quadratic form as a result of the transformation.) We no longer assume, however, that these co-ordinates may in particular be chosen as affine co-ordinates such that they make the co-efficients g_{ik} of the groundform become constant.

The transition from Euclidean geometry to that of Riemann is founded in principle on the same idea as that which led from physics based on action at a distance to physics based on infinitely near action. We find by observation, for example, that the current flowing along a conducting wire is proportional to the difference of potential between the ends of the wire (Ohm's Law). But we are firmly convinced that this result of measurement applied to a long wire does not represent a physical law in its most general form; we accordingly deduce this law by reducing the measurements obtained to an infinitely small portion of wire. By this means we arrive at the expression (Chap. I, page 111) on which Maxwell's theory is founded. Proceeding in the reverse direction, we derive from this differential law by mathematical processes the integral law, which we observe directly, on the supposition **that conditions are everywhere similar** (homogeneity). We have the same circumstances here. The fundamental fact of Euclidean geometry is that the square of the distance between two points is a quadratic form of the relative co-ordinates of the two points (*Pythagoras' Theorem*). *But if we look upon this law as being strictly valid only for the case when these two points are infinitely near, we enter the domain of Riemann's geometry.* This at the same time allows us to dispense with defining the co-ordinates more exactly since Pythagoras' Law expressed in this form (i.e. for infinitesimal distances) is invariant for arbitrary transformations. We pass from Euclidean "finite" geometry to Riemann's "infinitesimal" geometry in a manner exactly analogous to that by which we pass from "finite" physics to "infinitesimal" (or "contact") physics. Riemann's geometry is Euclidean geometry formulated to meet the requirements of continuity, and in virtue of this formulation it assumes a much more general character. Euclidean finite geometry is the appropriate instrument for investigating the straight line and the plane, and the treatment of these problems directed its development. As soon as we pass over to differential geometry, it becomes natural and reasonable to start from the property of infinitesimals set out by Riemann. This gives rise to no complications, and excludes all speculative considerations tending to overstep the boundaries of geometry. In Riemann's space, too, a surface, being a two-dimensional manifold, may be represented parametrically in the form $x_i = x_i(u_1, u_2)$. If we substitute the resulting differentials,

$$dx_i = \frac{\partial x_i}{\partial u_1} \cdot du_1 + \frac{\partial x_i}{\partial u_2} \cdot du_2$$

in the metrical groundform (12) of Riemann's space, we get for the square of the distance between two infinitely near surface-points a quadratic differential form in du_1, du_2 (as in Euclidean space). The measure-relations of three-dimensional Riemann space may be applied directly to any surface existing in it, and thus converts it into a two-dimensional Riemann space. Whereas from the Euclidean standpoint space is assumed at the very outset to be of a much simpler character than the surfaces possible in it, viz. to be rectangular, Riemann has generalised the conception of space just sufficiently far to overcome this discrepancy. **The principle of gaining knowledge of the external world from the behaviour of its infinitesimal parts** is the mainspring of the theory of knowledge in infinitesimal physics as in Riemann's geometry, and, indeed, the mainspring of all the eminent work of Riemann, in particular, that dealing with the theory of complex functions. The question of the validity of the "fifth postulate," on which historical development started its attack on Euclid, seems to us nowadays to be a somewhat accidental point of departure. The knowledge that was necessary to take us beyond the Euclidean view was, in our opinion, revealed by Riemann.

We have yet to convince ourselves that the geometry of Bolyai and Lobatschefsky as well as that of Euclid and also spherical geometry (Riemann was the first to point out that the latter was a possible case of non-Euclidean geometry) are all included as particular cases in Riemann's geometry. We find, in fact, that if we denote a point in the Bolyai-Lobatschefsky plane by the rectangular co-ordinates u_1, u_2 of its corresponding point in Klein's model the distance ds between two infinitely near points is by (1)

$$ds^2 = \frac{(1 - u_1^2 - u_2^2)(du_1^2 + du_2^2) + (u_1\,du_1 + u_2\,du_2)^2}{(1 - u_1^2 - u_2^2)^2}. \tag{13}$$

By comparing this with (10) we see that the Theorem of Taurinus is again confirmed. The metrical groundform of three-dimensional non-Euclidean space corresponds exactly to this expression.

If we can find a curved surface in Euclidean space for which formula (13) holds, provided appropriate Gaussian co-ordinates u_1, u_2 be chosen, then the geometry of Bolyai and Lobatschefsky is valid on it. Such surfaces can actually be constructed; the simplest is the surface of revolution derived from the tractrix. The tractrix is a plane curve of the shape shown in Fig. 5, with one vertex and one asymptote. It is characterised geometrically by the property that any tangent measured from the point of contact to the point of intersection with the asymptote is of constant length. Suppose the curve to revolve about its asymptote as axis. Non-Euclidean geometry holds on the surface generated. This Euclidean model of striking simplicity was first mentioned by Beltrami (*vide* note 5). There are certain shortcomings in it; in the first place the form in which it is presented confines it to two-dimensional geometry; secondly, each of the two halves of the surface of revolution into which the sharp edge divides it represents only a part of the non-Euclidean plane. Hilbert proved rigorously that there cannot be a surface free from singularities in Euclidean space which pictures the whole of Lobatschefsky's plane (*vide* note 6). Both of these weaknesses are absent in the elementary geometrical model of Klein.

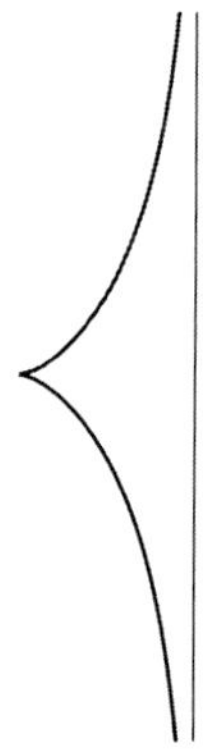

FIG. 5.

So far we have pursued a speculative train of thought and have kept within the boundaries of mathematics. There is, however, a difference in demonstrating the consistency of non-Euclidean geometry and **inquiring whether it or Euclidean geometry holds in actual space**. To decide this question Gauss long ago measured the triangle having for its vertices Inselsberg, Brocken, and Hoher Hagen (near Göttingen), using methods of the greatest refinement, but the deviation of the sum of the angles from 180° was found to lie within the limits of errors of observation. Lobatschefsky concluded from the very small value of the parallaxes of the stars that actual space could differ from Euclidean space only by an extraordinarily small amount. Philosophers have put forward the thesis that the validity or non-validity of Euclidean geometry cannot be proved by empirical observations. It must in fact be granted that in all such observations essentially physical assumptions, such as the statement that the path of a ray of light is a straight line and other similar statements, play a prominent part. This merely bears out the remark already made above that it is only the whole composed of geometry and physics that may be tested empirically. Conclusive experiments are thus possible only if physics in addition to geometry is worked out for Euclidean space **and** generalised Riemann space. We shall soon see that without making artificial limitations we can easily translate the laws of the electromagnetic field, which were originally set up on the basis of Euclidean geometry, into terms of Riemann's space. Once this has been done there is no reason why experience should not decide whether the special view of Euclidean geometry or the more general one of Riemann geometry is to be upheld. It is clear that at the present stage this question is not yet ripe for discussion.

In this concluding paragraph we shall once again present the foundations of Riemann's geometry in the form of a résumé, in which we do not restrict ourselves to the dimensional number $n = 3$.

An n-dimensional Riemann space is an n-dimensional manifold, not of an arbitrary nature, but one which derives its measure-relations from a definitely positive quadratic differential form. The two principal laws according to which this form determines the metrical quantities are expressed in (1) and (2) in which the x_i's denote any co-ordinates whatsoever.

1. If g is the determinant of the co-efficients of the groundform, then the size of any portion of space is given by the integral

$$\int \sqrt{g}\, dx_1\, dx_2 \ldots dx_n \tag{14}$$

which is to be taken over the mathematical region of the variables x_i, which corresponds to the portion of space in question.

2. If $Q(d, \delta)$ denote the symmetrical bilinear form, corresponding to the quadratic groundform, of two line elements d and δ situated at the same point, then the angle θ between them is given by

$$\cos\theta = \frac{Q(d, \delta)}{\sqrt{Q(d, d) \cdot Q(\delta, \delta)}}. \tag{15}$$

An m-dimensional manifold existing in n-dimensional space $(1 \leq m \leq n)$ is given in parametric terms by

$$x_i = x_i(u_1, u_2, \ldots, u_m) \qquad (i = 1, 2, \ldots, n).$$

By substituting the differentials

$$dx_i = \frac{\partial x_i}{\partial u_1} \cdot du_1 + \frac{\partial x_i}{\partial u_2} \cdot du_2 + \cdots + \frac{\partial x_i}{\partial u_m} \cdot du_m$$

in the metrical groundform of space we get the metrical groundform of this m-dimensional manifold. The latter is thus itself an

m-dimensional Riemann space, and the size of any portion of it may be calculated from formula (14) in the case $m = n$. In this way the lengths of segments of lines and the areas of portions of surfaces may be determined.

§ 12. Continuation. Dynamical View of Metrical Properties

We shall now revert to the theory of surfaces in Euclidean space. The **curvature** of a plane curve may be defined in the following way as the measure of the rate at which the normals to the curve diverge. From a fixed point O we trace out the vector Op, the "normal" to the curve at an arbitrary point P, and make it of unit length. This gives us a point P, corresponding to P, on the circle of radius unity. If P traverses a small arc Δs of the curve, the corresponding point p will traverse an arc $\Delta\sigma$ of the circle; $\Delta\sigma$ is the plane angle which is the sum of the angles that the normals erected at all points of the arc of the curve make with their respective neighbours. The limiting value of the quotient $\dfrac{\Delta\sigma}{\Delta s}$ for an element of arc Δs which contracts to a point P is the curvature

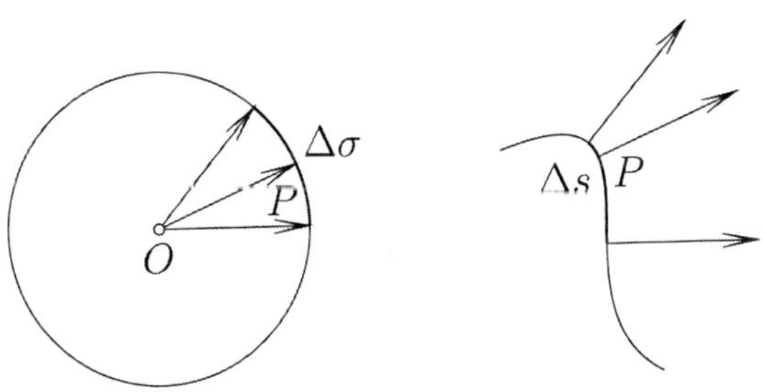

FIG. 6.

at P. Gauss defined the curvature of a surface as the measure of the rate at which its normals diverge in an exactly analogous manner. In place of the unit circle about O, he uses the unit sphere. Applying the same method of representation he makes a small portion $d\omega$ of this sphere correspond to a small area do of the surface; $d\omega$ is equal to the solid angle formed by the normals erected at the points of do. The ratio $\frac{d\omega}{do}$ for the limiting case when do becomes vanishingly small is the *Gaussian measure of curvature. Gauss made the important discovery that this curvature is determined by the inner measure-relations of the surface alone, and that it can be calculated from the co-efficients of the metrical groundform as a differential expression of the second order.* The curvature accordingly remains unaltered if the surface be bent without being distorted by stretching. By this geometrical means a **differential invariant of the quadratic differential forms** of two variables was discovered, that is to say, a quantity was found, formed of the co-efficients of the differential form in such a way that its value was the same for two differential forms that arise from each other by a transformation (and also for parametric pairs which correspond to one another in the transformation).

Riemann succeeded in extending the conception of curvature to quadratic forms of three and more variables. He then found that it was no longer a scalar but a tensor (we shall discuss this in § 15 of the present chapter). More precisely it may be stated that Riemann's space has a definite curvature at every point in the normal direction of every surface. The characteristic of Euclidean space is that its curvature is nil at every point and in every direction. Both in the case of Bolyai-Lobatschefsky's geometry and spherical geometry the curvature has a value a independent of the place and of the surface passing through it: this value is positive in the case of spherical geometry, negative in that of Bolyai-Lobatschefsky. (It may therefore be put $= \pm 1$ if a suitable unit of length be chosen.) If an n-dimensional space has a constant curvature a, then if we choose appropriate co-ordinates x_i, its metrical groundform must be of the form

$$\frac{(1 + a\sum_i x_i^2) \cdot \sum_i dx_i^2 - a\left(\sum_i x_i\, dx_i\right)^2}{(1 + a\sum_i x_i^2)^2}.$$

It is thus completely defined in a single-valued manner. If space is everywhere homogeneous in all directions, its curvature must be constant, and consequently its metrical groundform must be of the form just given. Such a space is necessarily either Euclidean, spherical, or Lobatschefskyan. Under these circumstances not only have the line elements an existence which is independent of place and direction, but any arbitrary finitely extended figure may be transferred to any arbitrary place and put in any arbitrary direction without altering its metrical conditions, i.e. its displacements are congruent. This brings us back to congruent transformations which we used as a starting-point for our reflections on space in § 1. Of these three possible cases the Euclidean one is characterised by the circumstance that the group of translations having the special properties set out in § 1 are unique in the group of congruent transformations. The facts which are summarised in this paragraph are mentioned briefly in Riemann's essay; they have been discussed in greater detail by Christoffel, Lipschitz, Helmholtz, and Sophus Lie (*vide* note 7).

Space is a form of phenomena, and, by being so, is necessarily homogeneous. It would appear from this that out of the rich abundance of possible geometries included in Riemann's conception only the three special cases mentioned come into consideration from the outset, and that all the others must be rejected without further examination as being of no account: *parturiunt montes, nascetur ridiculus mus!* Riemann held a different opinion, as is evidenced by the concluding remarks of his essay. Their full purport was not grasped by his contemporaries, and his words died away almost unheard (with the exception of a solitary echo in the writings of W. K. Clifford). Only now that Einstein has removed the scales from our eyes by the magic light of his theory of gravitation do we see what these words actually mean. To make them quite clear I must begin by remarking that Riemann contrasts **discrete** manifolds, i.e. those composed of single isolated elements, with **continuous** manifolds. The measure of every part of such a discrete manifold is determined by the **number** of elements belonging to it. Hence, as Riemann expresses it, a discrete manifold has the principle of its metrical relations in itself, *a priori*, as a consequence of the concept of number. In Riemann's own words:—

"The question of the validity of the hypotheses of geometry in the infinitely small is bound up with the question of the ground of the metrical relations of space. In this question, which we may still regard as belonging to the doctrine of space, is found the application of the remark made above; that in a discrete manifold, the principle or character of its metric relations is already given in the notion of the manifold, whereas in a continuous manifold this ground has to be found elsewhere, i.e. has to come from outside. Either, therefore, the reality which underlies space must form a discrete manifold, or we must seek the ground of its metric relations (measure-conditions) outside it, in binding forces which act upon it.

"A decisive answer to these questions can be obtained only by starting from the conception of phenomena which has hitherto been justified by experience, to which Newton laid the foundation, and then making in this conception the successive changes required by facts which admit of no explanation on the old theory; researches of this kind, which commence with general notions, cannot be other than useful in preventing the work from being hampered by too narrow views, and in keeping progress in the knowledge of the inter-connections of things from being checked by traditional prejudices.

"This carries us over into the sphere of another science, that of physics, into which the character and purpose of the present discussion will not allow us to enter."

If we discard the first possibility, "that the reality which underlies space forms a discrete manifold"—although we do not by this in any way mean to deny finally, particularly nowadays in view of the results of the quantum-theory, that the ultimate solution of the problem of space may after all be found in just this possibility—we see that Riemann rejects the opinion that had prevailed up to his own time, namely, that the metrical structure of space is fixed and inherently independent of the physical phenomena for which it serves as a background, and that the real content takes possession of it as of residential flats. *He asserts, on the contrary, that space in itself is nothing more than a three-dimensional manifold devoid of all form; it acquires a definite form only through the advent of the material content filling it and determining its metric relations.* There remains the problem of ascertaining the laws in accordance with which this is brought about. In any case, however, the metrical groundform will alter in the course of time just as the disposition of matter in the world changes. We recover the possibility of displacing a body without altering its metric relations by making the body carry along with it the "metrical field" which it has produced (and which is represented by the metrical groundform); just as a mass, having assumed a definite shape in equilibrium under the influence of the field of force which it has itself produced, would become deformed if one could keep the field of force fixed while displacing the mass to another position in it; whereas, in reality, it retains its shape during motion (supposed to be sufficiently slow), since it carries the field of force, which it has produced, along with itself. We shall illustrate in greater detail this bold idea of Riemann concerning the metrical field produced by matter, and we shall show that if his opinion is correct, any two portions of space which can be transformed into one another by a continuous deformation, must be recognised as being congruent in the sense we have adopted, and that the same material content can fill one portion of space just as well as the other.

To simplify this examination of the underlying principles we assume that the material content can be described fully by scalar phase quantities such as mass-density, density of charge, and so forth. We fix our attention on a definite moment of time. During this moment the density ρ of charge, for example, will, if we choose a certain co-ordinate system in space, be a definite function $f(x_1, x_2, x_3)$ of the co-ordinates x_1 but will be represented by a different function $f^*(x_1^*, x_2^*, x_3^*)$ if we use another co-ordinate system in x_i^*. *A parenthetical note.* Beginners are often confused by failing to notice that in mathematical literature symbols are used throughout to designate **functions**, whereas in physical literature (including the mathematical treatment of physics) they are used exclusively to denote "**magnitudes**" (quantities). For example, in thermodynamics the energy of a gas is denoted by a definite letter, say E, irrespective of whether it is a function of the pressure p and the temperature θ or a function of the volume v and the temperature θ. The mathematician, however, uses two different symbols to express this:—

$$E = \phi(p, \theta) = \psi(v, \theta).$$

The partial derivatives $\frac{\partial\phi}{\partial\theta}$, $\frac{\partial\psi}{\partial\theta}$, which are totally different in meaning, consequently occur in physics books under the common expression $\frac{\partial E}{\partial\theta}$. A suffix must be added (as was done by Boltzmann), or it must be made clear in the text that in one case p, in the other case v, is kept constant. The symbolism of the mathematician is clear without any such addition.*

Although the true state of things is really more complex we shall assume the most simple system of geometrical optics, the fundamental law of which states that the ray of light from a point M emitting light to an observer at P is a "geodetic" line, which is the shortest of all the lines connecting M with P: we take no account of the finite velocity with which light is propagated. We ascribe to the receiving consciousness merely an optical faculty of perception and simplify this to a "point-eye" that immediately observes the differences of direction of the impinging rays, these directions being the values of θ given by (15); the "point-eye" thus obtains a picture of the directions in which the surrounding objects lie (colour factors are ignored). The Law of Continuity governs not only the action of physical things on one another but also psychophysical interactions. The direction in which we observe objects

*This is not to be taken as a criticism of the physicist's nomenclature which is fully adequate to the purposes of physics, which deals with **magnitudes**.

is determined not by their places of occupation alone, but also by the direction of the ray from them that strikes the retina, that is, by the state of the optical field directly in contact with that elusive body of reality whose essence it is to have an objective world presented to it in the form of experiences of consciousness. To say that a material content G is the same as the material content G' can obviously mean no more than saying that to every point of view P with respect to G there corresponds a point of view P' with respect to G' (and conversely) in such a way that an observer at P' in G' receives the same "direction-picture" as an observer in G receives at P.

Let us take as a basis a definite co-ordinate system x_i. The scalar phase-quantities, such as density of electrification ρ, are then represented by definite functions

$$\rho = f(x_1, x_2, x_3).$$

Let the metrical groundform be

$$\sum_{i,k=1}^{3} g_{ik}\, dx_i\, dx_k$$

in which the g_{ik}'s likewise (in "mathematical" terminology) denote definite functions of x_1, x_2, x_3. Furthermore, suppose any continuous transformation of space into itself to be given, by which a point P' corresponds to each point P respectively. Using this co-ordinate system and the modes of expression

$$P = (x_1, x_2, x_3), \qquad P' = (x'_1, x'_2, x'_3),$$

suppose the transformation to be represented by

$$x'_i = \phi(x_1, x_2, x_3). \tag{16}$$

Suppose this transformation convert the portion **S** of space into **S**$'$, I shall show that if Riemann's view is correct **S**$'$ is congruent with **S** in the sense defined.

I make use of a second co-ordinate system by taking as co-ordinates of the point P the values of x'_i given by (16); the expressions (16) then become the formulæ of transformation. The mathematical region in three variables represented by **S** in the co-ordinates x' is identical with that represented by **S**$'$ in the co-ordinates x. An arbitrary point P has the same co-ordinates in x' as P' has in x. I now imagine space to be filled by matter in some other way, namely, that represented by the formulæ

$$\rho = f(x'_1, x'_2, x'_3)$$

at the point P, with similar formulæ for the other scalar quantities. If the metric relations of space are taken to be independent of the contained matter, the metrical groundform will, as in the case of the first content, be of the form

$$\sum_{i,k} g_{ik}\,dx_i\,dx_k = \sum_{i,k} g'_{ik}(x'_1, x'_2, x'_3)\,dx'_i\,dx'_k,$$

the right-hand member of which denotes the expression after transformation to the new co-ordinate system. If, however, the metric relations of space are determined by the matter filling it—we assume, with Riemann, that this is actually so—then, since the second occupation by matter expresses itself in the co-ordinates x' in exactly the same way as does the first in x, the metrical groundform for the second occupation will be

$$\sum_{i,k} g_{ik}(x'_1, x'_2, x'_3)\,dx'_i\,dx'_k.$$

In consequence of our underlying principle of geometrical optics assumed above, the content in the portion $\mathbf{S}'$ of space during the first occupation will present exactly the same appearance to an observer at P' as the material content in $\mathbf{S}$ during the second occupation presents to an observer at P. If the older view of "residential flats" is correct, this would of course not be the case.

The simple fact that I can squeeze a ball of modelling clay with my hands into any irregular shape totally different from a sphere would seem to reduce Riemann's view to an absurdity. This, however, proves nothing. For if Riemann is right, a deformation of the inner atomic structure of the clay is entirely different from that which I can effect with my hands, and a rearrangement of the masses in the universe, would be necessary to make the distorted ball of clay appear spherical to an observer from all points of view. The essential point is that a piece of space has no visual form at all, but that this form depends on the material content occupying the world, and, indeed, occupying it in such a way that by means of an appropriate rearrangement of the mode of occupation I can give it any visual form. By this I can also metamorphose any two **different** pieces of space into the **same** visual form by choosing an appropriate disposition of the matter. Einstein helped to lead Riemann's ideas to victory (although he was not directly influenced by Riemann). Looking back from the stage to which Einstein has brought us, we now recognise that these ideas could give rise to a valid theory only after **time** had been added as a fourth dimension to the three-space dimensions in the manner set forth in the so-called special theory of relativity. As, according to Riemann, the conception "congruence" leads to no metrical system at all, not even to the general metrical system of Riemann, which is governed by a quadratic differential form, we see that "the inner ground of the metric relations" must indeed be sought elsewhere. Einstein affirms that it is to be found in the "binding forces" of **Gravitation**. In Einstein's theory (Chapter IV) the co-efficients g_{ik} of the metrical groundform play the same part as does gravitational potential in Newton's theory of gravitation. The laws according to which space-filling matter determines the metrical structure are the laws of gravitation. The gravitational field affects light rays and "rigid" bodies used as measuring rods in such a way that when we use these rods and rays in the usual manner to take measurements of objects, a geometry of measurement is found to hold which deviates very little from that of Euclid in the regions accessible to observation. These metric relations are not the outcome of space being a form of phenomena, but of the physical behaviour of measuring rods and light rays as determined by the gravitational field.

After Riemann had made known his discoveries, mathematicians busied themselves with working out his system of geometrical ideas formally; chief among these were Christoffel, Ricci, and Levi-Civita (*vide* note 8). Riemann, in the last words of the above quotation, clearly left the real development of his ideas in the hands of some subsequent scientist whose genius as a physicist could rise to equal flights with his own as a mathematician. After a lapse of seventy years this mission has been fulfilled by Einstein.

Inspired by the weighty inferences of Einstein's theory to examine the mathematical foundations anew the present writer made the discovery that Riemann's geometry goes only half-way towards attaining the ideal of a pure infinitesimal geometry. It still remains to eradicate the last element of geometry "at a distance," a remnant of its Euclidean past. Riemann assumes that it is possible to compare the lengths of two line elements at **different** points of space, too; **it is not permissible to use comparisons at a distance in an "infinitely near" geometry**. One principle alone is allowable; by this a division of length is transferable from one point to that infinitely adjacent to it.

After these introductory remarks we now pass on to the systematic development of pure infinitesimal geometry (*vide* note 9), which will be traced through three stages; from the **continuum**, which eludes closer definition, by way of **affinely connected manifolds**, to **metrical space**. This theory which, in my opinion, is the climax of a wonderful sequence of logically-connected ideas, and in which the result of these ideas has found its ultimate shape, is a true *geometry*, a doctrine of *space itself* and not merely like Euclid, and almost everything else that has been done under the name of geometry, a doctrine of the configurations that are possible in space.

§ 13. Tensors and Tensor-densities in any Arbitrary Manifold

An n-dimensional Manifold.—Following the scheme outlined above we shall make the sole assumption about space that it is an n-dimensional continuum. It may accordingly be referred to n-co-ordinates $x_1, x_2, \ldots, x_n$, of which each has a definite numerical value at each point of the manifold; different value-systems of the co-ordinates correspond to different points. If $\bar{x}_1, \bar{x}_2, \ldots \bar{x}_n$ is a second system of co-ordinates, then there are certain relations

$$x_i = f_i(\bar{x}_1, \bar{x}_2, \ldots \bar{x}_n) \text{ where } (i = 1, 2, \ldots, n) \tag{17}$$

between the x-co-ordinates and the $\bar{x}$-co-ordinates; these relations are conveyed by certain functions f_i. We do not only assume that they are continuous, but also that they have continuous derivatives

$$\alpha^i_k = \frac{\partial f_i}{\partial \bar{x}_k}$$

whose determinant is non-vanishing. The latter condition is necessary and sufficient to make affine geometry hold in infinitely small regions, that is, so that reversible linear relations exist between the differentials of the co-ordinates in both systems, i.e.

$$dx_i = \sum_k \alpha^i_k \, d\bar{x}_k. \tag{18}$$

We assume the existence and continuity of higher derivatives wherever we find it necessary to use them in the course of our investigation. In every case, then, a meaning which is invariant and independent of the co-ordinate system has been assigned to the conception of continuous functions of a point which have continuous first, second, third, or higher derivatives as required; the co-ordinates themselves are such functions.

Conception of a Tensor.—The relative co-ordinates dx of a point $P' = (x_i + dx_i)$ infinitely near to the point $P = (x_i)$ are the components of a **line element** at P or of an **infinitesimal displacement** $\overrightarrow{PP'}$ of P. The transformation to another co-ordinate system is effected for these components by formulæ (18), in which α^i_k denote the values of the respective derivatives at the point P. The infinitesimal displacements play the same part in the development of Tensor Calculus as do displacements in Chapter I. It must, however, be noticed that, here, **a displacement is essentially bound to a point**, and that there is no meaning in saying that the infinitesimal displacements of two different points are the equal or unequal. It might occur to us to adopt the convention of calling the infinitesimal displacements of two points equal if they have the same components; but it is obvious from the fact that the α^i_k's in (18) are not constants, that if this were the case for one co-ordinate system it need in no wise be true for another. Consequently we may only speak of the infinitesimal displacement of a **point** and not, as in Chapter I, of the whole of space; hence we cannot talk of a vector or tensor simply, but must talk of a **vector** or **tensor** as being **at a point** P. A tensor at a point P is a linear form, in several series of variables, which is dependent on a co-ordinate system to which the immediate neighbourhood of P is referred in the following way: the expressions of the linear form in any two co-ordinate systems x and $\bar{x}$ pass into one another if certain of the series of variables (with upper indices) are transformed co-grediently, the remainder (with lower indices) contra-grediently, to the differentials dx_i, according to the scheme

$$\xi^i = \sum_k \alpha^i_k \bar{\xi}^k \text{ and } \bar{\xi}_i = \sum_k \alpha^k_i \xi_k \text{ respectively.} \tag{19}$$

By α^i_k we mean the values of these derivatives **at the point** P. The co-efficients of the linear form are called the components of the tensor in the co-ordinate system under consideration; they are co-variant in those indices that belong to the variables with an upper index, contra-variant in the remaining ones. The conception of tensors is possible owing to the circumstance that the transition from one co-ordinate system to another expresses itself as a **linear** transformation in the differentials. One here uses the exceedingly fruitful mathematical device of making a problem "linear" by reverting to infinitely small quantities. The whole of **Tensor Algebra**, by whose operations only tensors **at the same point** are associated, **can now be taken over from Chapter I**. Here, again, we shall call tensors of the first order **vectors**. There are contra-variant and co-variant vectors. Whenever the word vector is used without being defined more exactly we shall understand it as meaning a contra-variant vector. Infinitesimal quantities of this type are the line elements in P. Associated with every co-ordinate system there are n "unit vectors" $\mathbf{e}_i$ at P, namely, those which have components

$$\begin{array}{c|ccccc} \mathbf{e}_1 & 1, & 0, & 0, & \ldots & 0 \\ \mathbf{e}_2 & 0, & 1, & 0, & \ldots & 0 \\ \ldots & \ldots & \ldots & \ldots & \ldots & \ldots \\ \mathbf{e}_n & 0, & 0, & 0, & \ldots & 1 \end{array}$$

in the co-ordinate system. Every vector $\mathbf{x}$ at P may be expressed in linear terms of these unit vectors. For if ξ^i are its components, then

$$\mathbf{x} = \xi^1\mathbf{e}_1 + \xi^2\mathbf{e}_2 + \cdots + \xi^n\mathbf{e}_n \text{ holds.}$$

The unit vectors $\bar{\mathbf{e}}_i$ of another co-ordinate system $\bar{x}$ are derived from the $\mathbf{e}_i$'s according to the equations

$$\bar{\mathbf{e}}_i = \sum_k \alpha_i^k \mathbf{e}_k.$$

The possibility of passing from co-variant to contra-variant components of a tensor does not, of course, come into question here. Each two linearly independent line elements having components dx_i, δx_i map out a **surface element** whose components are

$$dx_i\,\delta x_k - dx_k\,\delta x_i = \Delta x_{ik}.$$

Each three such line elements map out a three-dimensional space element and so forth. Invariant differential forms that assign a number linearly to each arbitrary line element, surface element, etc., respectively are **linear tensors** (= co-variant skew-symmetrical tensors, *vide* § 7). The above convention about omitting signs of summation will be retained.

Conception of a Curve.—If to every value of a parameter s a point $P = P(s)$ is assigned in a continuous manner, then if we interpret s as time, a "**motion**" is given. In default of a better expression we shall apply this name in a purely mathematical sense, even when we do not interpret s in this way. If we use a definite co-ordinate system we may represent the motion in the form

$$x_i = x_i(s) \tag{20}$$

by means of n continuous functions $x_i(s)$, which we assume not only to be continuous, but also continuously differentiable.* In passing from the parametric value s to $s + ds$, the corresponding point P suffers an infinitesimal displacement having components dx_i. If we divide this vector at P by ds, we get the "**velocity**," a vector at P having components $\frac{dx_i}{ds} = u^i$. The formulæ (20) is at the same time a parametric representation of the **trajectory** of the motion. Two motions describe the same **curve** if, and only if, the one motion arises from the other when the parameter s is subjected to a transformation $s = \omega(\bar{s})$, in which ω is a continuous and continuously differentiable uniform function ω. Not the components of velocity at a point are determinate for a curve, but only their ratios (which characterise the **direction** of the curve).

Tensor Analysis.—A **tensor field** of a certain kind is defined in a region of space if to every point P of this region a tensor

*I.e. have continuous differential co-efficients.

of this kind at P is assigned. Relatively to a co-ordinate system the components of the tensor field appear as definite functions of the co-ordinates of the variable "point of emergence" P: we assume them to be continuous and to have continuous derivatives. The Tensor Analysis worked out in Chapter I, § 8, cannot, without alteration, be applied to any arbitrary continuum. For in defining the general process of differentiation we earlier used arbitrary co-variant and contra-variant vectors, whose components were **independent of the point in question.** This condition is indeed invariable for linear transformations, but not for any arbitrary ones since, in these, the α^i_k's are not constants. For an arbitrary manifold we may, therefore, set up only the analysis of **linear** tensor fields: this we proceed to show. Here, too, there is derived from a scalar field f by means of differentiation, independently of the co-ordinate system, a linear tensor field of the first order having components

$$f_i = \frac{\partial f}{\partial x_i}. \tag{21}$$

From a linear tensor field f_i of the first order we get one of the second order

$$f_{ik} = \frac{\partial f_i}{\partial x_k} - \frac{\partial f_k}{\partial x_i}. \tag{22}$$

From one of the second order, f_{ik}, we get a linear tensor field of the third order

$$f_{ikl} = \frac{\partial f_{kl}}{\partial x_i} + \frac{\partial f_{li}}{\partial x_k} + \frac{\partial f_{ik}}{\partial x_l}, \tag{23}$$

and so forth.

If ϕ is a given scalar field in space and if x_i, $\bar{x}_i$ denote any two co-ordinate systems, then the scalar field will be expressed in each in turn as a function of the x_i's or $\bar{x}_i$'s respectively, i.e.

$$\phi = f(x_1, x_2, \ldots, x_n) = \bar{f}(\bar{x}_1, \bar{x}_2, \ldots, \bar{x}_n).$$

If we form the increase of ϕ for an infinitesimal displacement of the current point, we get

$$d\phi = \sum_i \frac{\partial f}{\partial x_i}\, dx_i = \sum_i \frac{\partial \bar{f}}{\partial \bar{x}_i}\, d\bar{x}_i.$$

From this we see that the $\dfrac{\partial f}{\partial x_i}$'s are components of a co-variant tensor field of the first order, which is derived from the scalar field ϕ in a manner independent of all co-ordinate systems. We have here a simple illustration of the conception of vector fields. At the same time we see that the operation "grad" is invariant not only for linear transformations, but also for any arbitrary transformations of the co-ordinates whatsoever, and this is what we enunciated.

To arrive at (22) we perform the following construction. From the point $P = P_{00}$ we draw the two line elements with components dx_i and δx_i, which lead to the two infinitely near points P_{10} and P_{01}. We displace (by "variation") the line element dx in some way so that its point of emergence describes the distance $P_{00}P_{01}$; suppose it to have got to $\overrightarrow{P_{01}P_{11}}$ finally. We shall call this process the displacement δ. Let the components dx_i have increased by δdx_i, so that

$$\delta dx_i = \big\{x_i(P_{11}) - x_i(P_{01})\big\} - \big\{x_i(P_{10}) - x_i(P_{00})\big\}.$$

We now interchange d and δ. By an analogous displacement d of the line element δx along $P_{00}P_{10}$, by which it finally takes up the position $\overrightarrow{P_{10}P'_{11}}$, its components are increased by

$$d\delta x_i = \big\{x_i(P'_{11}) - x_i(P_{10})\big\} - \big\{x_i(P_{01}) - x_i(P_{00})\big\}.$$

Hence it follows that

$$\delta dx_i - d\delta x_i = x_i(P_{11}) - x_i(P'_{11}). \tag{24}$$

If, and only if, the two points P_{11} and P'_{11} coincide, i.e. if the two line elements dx and δx sweep out the same infinitesimal "parallelogram" during their displacements δ and d respectively—that is how we shall view it—then we shall have

$$\delta dx_i - d\delta x_i = 0. \tag{25}$$

If, now, a co-variant vector field with components f_i is given, then we form the change in the invariant $df = f_i\,dx_i$ owing to the displacement δ thus:

$$\delta df = \delta f_i\,dx_i + f_i\,\delta dx_i.$$

Interchanging d and δ, and then subtracting, we get

$$\Delta f = (\delta d - d\delta)f = (\delta f_i\,dx_i - df_i\,\delta x_i) + f_i(\delta dx_i - d\delta x_i)$$

and if both displacements pass over the same infinitesimal parallelogram we get, in particular,

$$\Delta f = \delta f_i\,dx_i - df_i\,\delta x_i = \left(\frac{\partial f_i}{\partial x_k} - \frac{\partial f_k}{\partial x_i}\right) dx_i\,\delta x_k. \tag{26}$$

If one is inclined to distrust these perhaps too venturesome operations with infinitesimal quantities the differentials may be replaced by differential co-efficients. Since an infinitesimal element of surface is only a part (or more correctly, the limiting value of the part) of an arbitrarily small but finitely extended surface, the argument will run as follows. Let a point (s, t) of our manifold be assigned to every pair of values of two parameters s, t (in a certain region encircling $s = 0$, $t = 0$). Let the functions $x_i = x_i(s, t)$, which represents this "two-dimensional motion" (extending over a surface) in any co-ordinate system x_i, have continuous first and second differential co-efficients. For every point (s, t) there are two velocity vectors with components $\frac{dx_i}{ds}$ and $\frac{dx_i}{dt}$. We may assign our parameters so that a prescribed point $P = (0, 0)$ corresponds to $s = 0$, $t = 0$, and that the two velocity vectors at it coincide with two arbitrarily given vectors u^i, v^i (for this it is merely necessary to make the x_i's linear functions of s and t). Let d denote the differentiation $\frac{d}{ds}$, and δ denote $\frac{d}{dt}$. Then

$$df = f_i \frac{dx_i}{ds}, \qquad \delta df = \frac{\partial f_i}{\partial x_k} \frac{dx_i}{ds} \frac{dx_k}{dt} + f_i \frac{d^2 x_i}{dt\, ds}.$$

By interchanging d and δ, and then subtracting, we get

$$\Delta f = \delta df - d\delta f = \left(\frac{\partial f_i}{\partial x_k} - \frac{\partial f_k}{\partial x_i} \right) \frac{dx_i}{ds} \frac{dx_k}{dt}. \tag{27}$$

By setting $s = 0$ and $t = 0$, we get the invariant at the point P

$$\left(\frac{\partial f_i}{\partial x_k} - \frac{\partial f_k}{\partial x_i} \right) u^i v^k$$

which depends on two arbitrary vectors u, v at that point. The connection between this view and that which uses infinitesimals consists in the fact that the latter is applied in rigorous form to the infinitesimal parallelograms into which the surface $x_i = x_i(s, t)$ is divided by the co-ordinate lines $s =$ const. and $t =$ const.

Stokes' Theorem may be recalled in this connection. The invariant linear differential $f_i\, dx_i$ is called **integrable** if its integral along every closed curve (its "curl") $= 0$. (This is true, as we know, only for a total differential.) Let any arbitrary surface given in a parametric form $x_i = x_i(s, t)$ be spread out within the closed curve, and be divided into infinitesimal parallelograms by the co-ordinate lines. The curl taken around the perimeter of the whole surface may then be traced back to the single curls around these little surface meshes, and their values are given for every mesh by our expression (27), after it has been multiplied by $ds\, dt$. A differential division of the curl is produced in this way, and the tensor (22) is a measure of the "intensity of the curl" at every point.

In the same way we pass on to the next higher stage (23). In place of the infinitesimal parallelogram we now use the three-dimensional parallelepiped mapped out by the three line elements d, δ, and ∂. We shall just indicate the steps of the argument briefly.

$$\partial(f_{ik}\, dx_i\, \delta x_k) = \frac{\partial f_{ik}}{\partial x_l}\, dx_i\, \delta x_k\, \partial x_l + f_{ik}(\partial dx_i \cdot \delta x_k + \partial \delta x_k \cdot dx_i). \quad (28)$$

Sincc $f_{ki} = -f_{ik}$, the second term on the right is

$$= f_{ik}(\partial dx_i \cdot \delta x_k - \partial \delta x_i \cdot dx_k). \quad (29)$$

If we interchange d, δ, and ∂ cyclically, and then sum up, the six members arising out of (29) will destroy each other in pairs on account of the conditions of symmetry (25).

Conception of Tensor-density.—If $\int \mathbf{W}\, dx$, in which dx represents briefly the element of integration $dx_1\, dx_2 \ldots dx_n$, is an invariant integral, then **W** is a quantity dependent on the co-ordinate system in such a way that, when transformed to another co-ordinate system, its value become multiplied by the absolute (numerical) value of the functional determinant. If we regard this integral as a measure of the quantity of substance occupying the region of integration, then **W** is its density. We may, therefore, call a quantity of the kind described a **scalar-density**.

This is an important conception, equally as valuable as the conception of scalars; it cannot be reduced to the latter. In an analogous sense we may speak of **tensor-densities** as well as scalar-densities. A linear form of several series of variables which is dependent on the co-ordinate system, some of the variables carrying upper indices, others lower ones, is a **tensor-density** at a point P, if, when the expression for this linear form is known for a given co-ordinate system, its expression for any other arbitrary co-ordinate system, distinguished by bars, is obtained by multiplying it with the absolute or numerical value of the functional determinant

$$\Delta = \text{abs.}\, |\alpha_i^k| \quad \text{i.e. the absolute value of } |\alpha_i^k|,$$

and by transforming the variable according to the old scheme (19). The words, components, co-variant, contra-variant, symmetrical, skew-symmetrical, field, and so forth, are used exactly as in the case of tensors. By contrasting tensors and tensor-densities, it seems to me that we have grasped rigorously the difference between **quantity** and **intensity**, so far as this difference has a physical meaning: **tensors are the magnitudes of intensity, tensor-densities those of quantity**. The same unique part that co-variant skew-symmetrical tensors play among tensors is taken among tensor-densities by contra-variant symmetrical tensor-densities, which we shall term briefly **linear tensor-densities**.

Algebra of Tensor-densities.—As in the realm of tensors so have here the following operations:—

1. Addition of tensor-densities of the same type; multiplication of a tensor-density by a number.
2. Contraction.
3. Multiplication of a tensor by a tensor-density (**not** multiplication of two tensor-densities by each other). For, if two scalar-densities, for example, were to be multiplied together, the result would not again be a scalar-density but a quantity which, to be transformed to another co-ordinate system, would have to be multiplied by the square of the functional determinant. Multiplying a tensor by a tensor-density, however, always leads to a tensor-density (whose order is equal to the sum of the orders of both factors). Thus, for example, if a contra-variant vector with components f^i and a co-variant tensor-density with components $\mathbf{w}_{ik}$ be multiplied together, we get a mixed tensor-density of the third order with components $f^i\mathbf{w}_{kl}$ produced in a manner independent of the co-ordinate system.

The analysis of tensor-densities can be established only for **linear** fields in the case of an arbitrary manifold. It leads to the following **processes resembling the operation of divergence**:—

$$\frac{\partial \mathbf{w}^i}{\partial x_i} = \mathbf{w}, \tag{30}$$

$$\frac{\partial \mathbf{w}^{ik}}{\partial x_k} = \mathbf{w}^i, \tag{31}$$

....

As a result of (30) a linear tensor-density field $\mathbf{w}^i$ of the first order gives rise to a scalar-density field $\mathbf{w}$, whereas (31) produces from a linear field of the second order ($\mathbf{w}^{ki} = -\mathbf{w}^{ik}$) a linear field of the first order, and so forth. These operations are independent of the co-ordinate system. The divergence (30) of a field $\mathbf{w}^i$ of the first order which has been produced from one, $\mathbf{w}^{ik}$, of the second order by means of (31) is $= 0$; an analogous result holds for the higher orders. To prove that (30) is invariant, we use the following known result of the theory of the motion of continuously extended masses.

If ξ^i is a given vector field, then

$$\bar{x}_i = x_i + \xi^i \cdot \delta t \tag{32}$$

expresses an **infinitesimal displacement** of the points of the continuum, by which the point with the co-ordinates x_i is transferred to the point with the co-ordinates $\bar{x}_i$. Let the constant infinitesimal factor δt be defined as the element of time during which the deformation takes place. The determinant of transformation $A = \left|\frac{\partial x^i}{\partial x_k}\right|$ differs from unity by $\delta t \frac{\partial \xi^i}{\partial x_i}$. The displacement causes portion $\mathbf{G}$ of the continuum, to which, if x^i's are used to denote its co-ordinates, the mathematical region $\mathfrak{X}$ in the variables x_i corresponds, to pass into the region $\overline{\mathbf{G}}$, from which $\mathbf{G}$ differs by an infinitesimal amount. If $\mathbf{s}$ is a scalar-density field, which we regard as the density of a substance occupying the medium, then the quantity of substance present in **G**

$$= \int_{\mathfrak{X}} \mathbf{s}(x)\,dx$$

whereas that which occupies $\overline{\mathbf{G}}$

$$= \int \mathbf{s}(\bar{x})\,d\bar{x} = \int_{\Upsilon} \mathbf{s}(\bar{x}) A\,dx,$$

whereby the values (32) are to be inserted in the last expression for the arguments $\bar{x}_i$ of $\mathbf{s}$. (I am here displacing the volume with respect to the substance; instead of this, we can of course make the substance flow through the volume; $\mathbf{s}\xi^i$ then represents the intensity of the current.) The increase in the amount of substance that the region **G** gains by the displacement is given by the integral $\mathbf{s}(\bar{x})A - \mathbf{s}(x)$ taken with respect to the variables x_i over $\mathfrak{X}$. We, however, get for the integrand

$$\mathbf{s}(\bar{x})(A-1) + \left\{\mathbf{s}(\bar{x}) - \mathbf{s}(x)\right\} = \delta t \left(\mathbf{s}\,\frac{\partial \xi^i}{\partial x_i} + \frac{\partial \mathbf{s}}{\partial x_i}\,\xi^i\right) = \delta t \cdot \frac{\partial(\mathbf{s}\xi^i)}{\partial x_i}.$$

Consequently the formula

$$\frac{\partial(\mathbf{s}\xi^i)}{\partial x_i} = \mathbf{w}$$

establishes an invariant connection between the two scalar-density fields $\mathbf{s}$ and $\mathbf{w}$ and the contra-variant vector field with the components ξ^i. Now, since every vector-density $\mathbf{w}^i$ is representable in the form $\mathbf{s}\xi^i$—for if in a **definite** co-ordinate system a scalar-density $\mathbf{s}$ and a vector field ξ be defined by $\mathbf{s} = 1$, $\xi^i = \mathbf{w}^i$, then the equation $\mathbf{w}^i = \mathbf{s}\xi^i$ holds for **every** co-ordinate system—the required proof is complete.

In connection with this discussion we shall enunciate the **Principle of Partial Integration** which will be of frequent use below. If the functions $\mathbf{w}^i$ vanish at the boundary of a region $\mathbf{G}$, then the integral

$$\int_{\mathbf{G}} \frac{\partial \mathbf{w}^i}{\partial x_i}\, dx = 0.$$

For this integral, multiplied by δt, signifies the change that the "volume" $\int dx$ of this region suffers through an infinitesimal deformation whose components $= \delta t \cdot \mathbf{w}^i$.

The invariance of the process of divergence (30) enables us easily to advance to further stages, the next being (31). We enlist the help of a co-variant vector field f_i, which has been derived from a potential f; i.e.

$$f_i = \frac{\partial f}{\partial x_i}.$$

We then form the linear tensor-density $\mathbf{w}^{ik} f_i$ of the first order and also its divergence

$$\frac{\partial(\mathbf{w}^{ik} f_i)}{\partial x_k} = f_i \frac{\partial \mathbf{w}^{ik}}{\partial x_k}.$$

The observation that the f_i's may assume any arbitrarily assigned values at a point P concludes the proof. In a similar way we proceed to the third and higher orders.

§ 14. Affinely Related Manifolds

The Conception of Affine Relationship.—We shall call a point P of a manifold affinely related to its neighbourhood if we are given the vector P' into which every vector at P is transformed by a parallel displacement from P to P'; P' is here an arbitrary point infinitely near P (*vide* note 10). No more and no less is required of this conception than that it is endowed with all the properties that were ascribed to it in the affine geometry of Chapter I. That is, we postulate: *There is a co-ordinate system (for the immediate neighbourhood of P) such that, in it, the components of any vector at P are not altered by an infinitesimal parallel displacement.* This postulate characterises parallel displacements as being such that they may rightly be regarded as leaving vectors **unchanged**. Such co-ordinate systems are called **geodetic** at P. What is the effect of this in an arbitrary co-ordinate system x_i? Let us suppose that, in it, the point P has the co-ordinate x_i^0, P' the co-ordinates $x_i^0 + dx_i$; let ξ^i be the components of an arbitrary vector at P, $\xi^i + d\xi^i$ the components of the vector resulting from it by parallel displacement towards P'. Firstly, since the parallel displacement from P to P' causes all the vectors at P to be mapped out linearly or affinely by all the vectors at P', $d\xi^i$ must be linearly dependent on ξ^i, i.e.

$$d\xi^i = -d\gamma_r^i \xi^r. \tag{33}$$

Secondly, as a consequence of the postulate with which we started, the $d\gamma_r^i$'s must be linear forms of the differentials dx_i, i.e.

$$d\gamma_r^i = \Gamma_{rs}^i \, dx_s \tag{33'}$$

in which the number co-efficients Γ, the "components of the affine relationship," satisfy the condition of symmetry

$$\Gamma_{sr}^i = \Gamma_{rs}^i. \tag{33''}$$

To prove this, let $\bar{x}_i$ be a geodetic co-ordinate system at P; the formulæ of transformation (17) and (18) then hold. It follows from the geodetic character of the co-ordinate system $\bar{x}_i$ that, for a parallel displacement,

$$d\xi^i = d(\alpha^i_r \bar{\xi}^r) = d\alpha^i_r \bar{\xi}^r.$$

If we regard the ξ^i's as components δx_i of a line element at P we must have

$$-d\gamma^i_r \, \delta x_r = \frac{\partial^2 f_i}{\partial \bar{x}_r \, \partial \bar{x}_s} \, \delta \bar{x}_r \, d\bar{x}_s$$

(in the case of the second derivatives we must of course insert their values at P). The statement contained in our enunciation follows directly from this. Moreover, the symmetrical bilinear form

$$-\Gamma^i_{rs} \, \delta \bar{x}_r \, d\bar{x}_s \quad \text{is derived from} \quad \frac{\partial^2 f_i}{\partial \bar{x}_r \, \partial \bar{x}_s} \, \delta \bar{x}_r \, d\bar{x}_s \qquad (34)$$

by transformation according to (18). This exhausts all the aspects of the question. Now, if Γ^i_{rs} are arbitrarily given numbers that satisfy the condition of symmetry (33″), and if we define the affine relationship by (33) and (33′), the transformation formulæ lead to

$$x_i - x_i^0 = \bar{x}_i - \tfrac{1}{2}\Gamma^i_{rs}\bar{x}_r\bar{x}_s,$$

that is, to a geodetic co-ordinate system $\bar{x}_i$ at P, since the equations (34) are fulfilled for them at P. In fact this transformation at P gives us

$$\bar{x}_i = 0, \qquad d\bar{x}_i = dx_i \quad (\alpha^i_k = \delta^i_k), \qquad \frac{\partial^2 f_i}{\partial \bar{x}_r \, \partial \bar{x}_s} = -\Gamma^i_{rs}.$$

The formulæ according to which the components Γ^i_{rs} of the affine relationship are transformed in passing from one co-ordinate system to another may easily be obtained from the above discussion; we do not, however, require them for subsequent work. The Γ's are certainly **not** components of a tensor (contra-variant in i, co-variant in r and s) at the point P; they have this character with regard to *linear* transformations, but lose it when subjected to *arbitrary* transformations. For they all vanish in a geodetic co-ordinate system. Yet every virtual change of the affine relationship $[\Gamma^i_{rs}]$, whether it be finite or "infinitesimal," is a tensor. For

$$[d\xi^i] = [\Gamma^i_{rs}]\xi^r \, dx_s$$

is the difference of the two vectors that arise as a result of the two parallel displacements of the vector ξ from P to P'.

The meaning of the **parallel displacement of a co-variant vector** ξ_i at the point P to the infinitely near point P' is defined uniquely by the postulate that the invariant product $\xi_i\eta^i$ of the vector ξ_i and any arbitrary contra-variant vector η^i remain unchanged after the simultaneous parallel displacements, i.e.

$$d(\xi_i\eta^i) = (d\xi_i \cdot \eta^i) + (\xi_r \, d\eta^r) = (d\xi_i - d\gamma^r_i \, \xi_r)\eta^i = 0,$$

whence

$$d\xi_i = \sum_r d\gamma^r_i \, \xi_r. \tag{35}$$

We shall call a contra-variant **vector field** ξ^i *stationary* at the point P, if the vectors at the points P' infinitely near P arise from the vector at P by parallel displacement, that is, if the total differential equations

$$d\xi^i + d\gamma^i_r \, \xi^r = 0 \quad \left(\text{or } \frac{\partial \xi^i}{\partial x_s} + \Gamma^i_{rs}\xi^r = 0\right)$$

are satisfied at P. A vector field can obviously always be found such that it has arbitrary given components at a point P (this remark will be used in a construction which is to be carried out in the sequel). The same conception may be set up for a co-variant vector field.

From now onwards we shall occupy ourselves with **affine manifolds; they are such that every point of them is affinely related to its neighbourhood**. For a definite co-ordinate system the components Γ^i_{rs} of the affine relationship are continuous functions of the co-ordinates x_i. By selecting the appropriate co-ordinate system the Γ^i_{rs}'s may, of course, be made to vanish at a single point P, but it is, in general, not possible to achieve this simultaneously for all points of the manifold. There is no difference in the nature of any of the affine relationships holding between the various points of the manifold and their immediate neighbourhood. The manifold is homogeneous in this sense. There are not various types of manifolds capable of being distinguished by the nature of the affine relationships governing each kind. The postulate with which we set out admits of only one definite kind of affine relationship.

Geodetic Lines.—If a point which is in motion carries a vector (which is arbitrarily variable) with it, we get for every value of the time parameter s not only a point

$$P = (s) : \ x = x_i(s)$$

of the manifold, but also a vector at this point with components $v^i = v^i(s)$ dependent on s. The vector remains stationary at the moment s if

$$\frac{dv^i}{ds} + \Gamma^i_{\alpha\beta}\, v^\alpha \frac{dx_\beta}{ds} = 0. \tag{36}$$

(This will relieve the minds of those who disapprove of operations with differentials; they have here been converted into differential co-efficients.) In the case of a vector being carried along according to any arbitrary rule, the left-hand side V^i of (36) consists of the components of a vector in (s) connected invariantly with the motion and indicating how much the vector v^i changes per unit of time at this point. For in passing from the point $P = (s)$ to $P' = (s + ds)$, the vector v^i at P becomes the vector

$$v^i + \frac{dv^i}{ds}\,ds$$

at P'. If, however, we displace v^i from P to P' leaving it unchanged, we there get

$$v^i + \delta v^i = v^i - \Gamma^i_{\alpha\beta}\, v^\alpha\, dx_\beta.$$

Accordingly, the difference between these two vectors at P', the change in v during the time ds has components

$$\frac{dv^i}{ds}\,ds - \delta v^i = V^i\, ds.$$

In analytical language the invariant character of the vector V may be recognised most readily as follows. Let us take an arbitrary auxiliary co-variant vector $\xi_i = (s)$ at P, and let us form the change in the invariant $\xi_i v^i$ in its passage from (s) to $(s + ds)$, whereby the vector ξ_i is taken along unchanged. We get

$$\frac{d(\xi_i v^i)}{ds} = \xi_i V^i.$$

If V vanishes for every value of s, the vector v glides with the point P along the trajectory during the motion *without becoming changed.*

Every motion is accompanied by the vector $u^i = \dfrac{dx_i}{ds}$ of its velocity; for this particular case, V is the vector

$$U^i = \frac{du^i}{ds} + \Gamma^i_{\alpha\beta}\, u^\alpha u^\beta = \frac{d^2x}{ds^2} + \Gamma^i_{\alpha\beta}\, \frac{dx_\alpha}{ds}\, \frac{dx_\beta}{ds}:$$

namely, the **acceleration**, which is a measure of the change of velocity per unit of time. A motion, in the course of which the velocity remains unchanged throughout, is called a **translation**. The trajectory of a translation, being a curve which preserves its direction unchanged, is a **straight** or **geodetic line**. According to the translational view (cf. Chapter I, § 1) this is the inherent property of the straight line.

The analysis of tensors and tensor-densities may be developed for an affine manifold just as simply and completely as for the linear geometry of Chapter I. For example, if f_i^k are the components (co-variant in i, contra-variant in k) of a tensor field of the second order, we take two auxiliary arbitrary vectors at the point P, of which the one, ξ, is contra-variant and the other, η, is co-variant, and form the invariant

$$f_i^k \xi^i \eta_k$$

and its change for an infinitesimal displacement d of the current point P, by which ξ and η are displaced parallel to themselves. Now

$$d(f_i^k \xi^i \eta_k) = \frac{\partial f_i^k}{\partial x_l}\, \xi^i \eta_k\, dx_l - f_r^k \eta_k\, d\gamma_i^r\, \xi^i + f_i^r \xi^i\, d\gamma_r^k\, \eta_k,$$

hence

$$f_{il}^k = \frac{\partial f_i^k}{\partial x_l} - \Gamma^r_{il} f_r^k + \Gamma^k_{rl} f_i^r$$

are the components of a tensor field of the third order, co-variant in i, l and contra-variant in k: this tensor field is derived from the given one of the second order by a process independent of the co-ordinate system. The additional terms, which the components of the affine relationship contain, are characteristic quantities in which, following Einstein, we shall later recognise the influence of the gravitational field. The method outlined enables us to differentiate a tensor in every conceivable case.

Just as the operation "grad" plays the fundamental part in tensor analysis and all other operations are derivable from it, so the operation "div" defined by (30) is the basis of the analysis of tensor-densities. The latter leads to processes of a similar character for tensor-densities of any order. For instance, if we wish to find an expression for the divergence of a mixed tensor-density $\mathbf{w}_i^k$ of the second order, we make use of an auxiliary stationary vector field $\xi^i \mathbf{w}_i^k$ at P and find the divergence of the tensor-density $\xi^i \mathbf{w}_i^k$:

$$\frac{\partial(\xi^i \mathbf{w}_i^k)}{\partial x_k} = \frac{\partial \xi^r}{\partial x_k} \mathbf{w}_r^k + \xi^i \frac{\partial \mathbf{w}_i^k}{\partial x_k} = \xi^i \left(-\Gamma_{ik}^r \, \mathbf{w}_r^k + \frac{\partial \mathbf{w}_i^k}{\partial x_k} \right).$$

This quantity is a scalar-density, and since the components of a vector field which is stationary at P may assume any values at this point (P), namely,

$$\frac{\partial \mathbf{w}_i^k}{\partial x_k} - \Gamma_{is}^r \mathbf{w}_r^s, \tag{37}$$

it is a co-variant tensor-density of the first order which has been derived from $\mathbf{w}_i^k$ in a manner independent of every co-ordinate system.

Moreover, not only can we reduce a tensor-density to one of the next lower order by carrying out the process of **divergence**, but we can also transpose a tensor-density into one of the next higher order by **differentiation**. Let $\mathbf{s}$ denote a scalar-density, and let us again use a stationary vector field ξ^i at P: we then form the divergence of current-density, $\mathbf{s}\xi^i$:

$$\frac{\partial(\mathbf{s}\xi^i)}{\partial x_i} = \frac{\partial \mathbf{s}}{\partial x_i}\xi^i + \mathbf{s}\frac{\partial \xi^i}{\partial x_i} = \left(\frac{\partial \mathbf{s}}{\partial x_i} - \Gamma^r_{ir}\mathbf{s}\right)\xi^i.$$

We thus get

$$\frac{\partial \mathbf{s}}{\partial x_i} - \Gamma^r_{ir}\mathbf{s}$$

as the components of a co-variant vector-density. To extend differentiation beyond scalar tensor-densities to any tensor-densities whatsoever, for example, to the mixed tensor-density $\mathbf{w}_i^k$ of the second order, we again proceed, as has been done repeatedly above, to make use of two stationary vector fields at P, namely, ξ^i and η^i, the latter being co-variant and the former contra-variant. We differentiate the scalar-density $\mathbf{w}_i^k\xi^i\eta_k$. If the tensor-density that has been derived by differentiation be contracted with respect to the symbol of differentiation and one of the contra-variant indices, the divergence is again obtained.

§ 15. Curvature

If P and P^* are two points connected by a curve, and if a vector is given at P, then this vector may be moved parallel to itself along the curve from P to P^*. Equations (36), giving the unknown components v^i of the vector which is being subjected to a continuous parallel displacement, have, for given initial values of v^i, one and only one solution. The **vector transference** that comes about in this way is in general **non-integrable**, that is, the vector which we get at P^* is dependent on the path of the displacement along which the transference is effected. Only in the particular case, in which integrability occurs, is it allowable to speak of the *same* vector at two different points P and P^*; this comprises those vectors that are generated from one another by parallel displacement. Let such a manifold be called **Euclidean-affine**. If we subject all points of such a manifold to an infinitesimal displacement, which is in each case representable by an "**equal**" infinitesimal vector, then the space is said to have undergone an infinitesimal **total translation**. With the help of this conception, and following the line of reasoning of Chapter I. (without entering on a rigorous proof), we may construct "linear" co-ordinate systems which are characterised by the fact that, in them, the same vectors have the same components at different points of the systems. In a linear co-ordinate system the components of the affine relationship vanish identically. Any two such systems are connected by **linear** formulæ of transformation. The manifold is then an affine space in the sense of Chapter I.: *The integrability of the vector transference is the infinitesimal geometrical property which distinguishes "linear" spaces among affinely related spaces.*

We must now turn our attention to the **general case**; it must not be expected in this that a vector that has been taken round a closed curve by parallel displacement finally returns to its initial position. Just as in the proof of Stokes's Theorem, so here we stretch a surface over the closed curve and divide it into infinitely small parallelograms by parametric lines. The change in any arbitrary vector after it has traversed the periphery of the surface is reduced to the change effected after it has traversed each of the infinitesimal parallelograms marked out by two line elements dx_i and δx_i at a point P. This change has now to be determined.

We shall adopt the convention that the amount $\Delta\mathbf{x} = (\Delta\xi^i)$, by which a vector $\mathbf{x} = \xi^i$ increases, is derived from $\mathbf{x}$ by a linear transformation, a matrix $\Delta\mathbf{F}$, i.e.

$$\Delta\mathbf{x} = \Delta\mathbf{F}(\mathbf{x}); \qquad \Delta\xi^\alpha = \Delta F^\alpha_\beta \cdot \xi^\beta. \tag{38}$$

If $\Delta\mathbf{F} = 0$, then the manifold is "**plane**" at the point P in the surface direction assumed by the surface element; if this is true for all elements of a finitely extended portion of surface, then every vector that is subjected to parallel displacement along the edge of the surface returns finally to its initial position. $\Delta\mathbf{F}$ is linearly dependent on the element of surface:

$$\Delta\mathbf{F} = \mathbf{F}_{ik}\, dx_i\, \delta x_k = \tfrac{1}{2}\mathbf{F}_{ik}\Delta x_{ik} \qquad (\Delta x_{ik} = dx_i\, \delta x_k - dx_k\, \delta x_i,$$

and

$$\mathbf{F}_{ki} = -\mathbf{F}_{ik}). \tag{39}$$

The differential form that occurs here characterises the **curvature**, that is, the deviation of the manifold from plane-ness at the point P for all possible directions of the surface; since its co-efficients are not numbers, but matrices, we might well speak of a "linear matrix-tensor of the second order," and this would undoubtedly best characterise the quantitative nature of curvature. If, however, we revert from the matrices back to their components—supposing $F^\alpha_{\beta ik}$ to be the components of $\mathbf{F}_{ik}$ or else the co-efficients of the form

$$\Delta F^\alpha_\beta = F^\alpha_{\beta ik}\, dx_i\, \delta x_k \tag{40}$$

—then we arrive at the formula

$$\Delta\mathbf{x}\, F^\alpha_{\beta ik}\mathbf{e}_\alpha\xi^\beta\, dx_i\, \delta x_k. \tag{41}$$

From this we see that the $F^\alpha_{\beta ik}$'s are the components of a tensor of the fourth order which is contra-variant in α and co-variant in β, i and k. Expressed in terms of the components Γ^i_{rs} of the affine relationship, it is

$$F^\alpha_{\beta ik} = \left(\frac{\partial \Gamma^\alpha_{\beta k}}{\partial x_i} - \frac{\partial \Gamma^\alpha_{\beta i}}{\partial x_k}\right) + (\Gamma^\alpha_{ri}\Gamma^r_{\beta k} - \Gamma^\alpha_{rk}\Gamma^r_{\beta i}). \tag{42}$$

According to this they fulfil the conditions of "skew" and "cyclical" symmetry, namely:—

$$F^\alpha_{\beta ki} = -F^\alpha_{\beta ik}; \qquad F^\alpha_{\beta ik} + f^\alpha_{ik\beta} + F^\alpha_{k\beta i} = 0. \tag{43}$$

The vanishing of the curvature is the invariant differential law which distinguishes Euclidean spaces among affine spaces in terms of general infinitesimal geometry.

To prove the statements above enunciated we use the same process of sweeping twice over an infinitesimal parallelogram as we used on page 157 to derive the curl tensor; we use the same notation as on that occasion. Let a vector $\mathbf{x} = \mathbf{x}(P_{00})$ with components ξ^i be given at the point P_{00}. The vector $\mathbf{x}(P_{10})$ that is derived from $\mathbf{x}(P_{00})$ by parallel displacement along the line element dx is attached to the end point P_{10} of the same line element. If the components of $\mathbf{x}(P_{10})$ are $\xi^i + d\xi^i$ then

$$d\xi^\alpha = -d\gamma^\alpha_\beta\,\xi^\beta = -\Gamma^\alpha_{\beta i}\,\xi^\beta\,dx_i.$$

Throughout the displacement δ to which the line element dx is to be subjected (and which need by no means be a parallel displacement) let the vector at the end point be bound always by the specified condition to the vector at the initial point. The $d\xi^\alpha$'s are then increased, owing to the displacement, by an amount

$$\delta d\xi^\alpha = -\delta\Gamma^\alpha_{\beta i}\,dx_i\,\xi^\beta - \Gamma^\alpha_{\beta i}\,\delta dx_i\,\xi^\beta - d\gamma^\alpha_r\,\delta\xi^r.$$

If, in particular, the vector at the initial point of the line element remains parallel to itself during the displacement, then $\delta\xi^r$ must be replaced in this formula by $-\delta\gamma^r_\beta\,\xi^\beta$. In the final position $\overrightarrow{P_{01}P_{11}}$ of the line element we then get, at the point P_{01}, the vector $\mathbf{x}(P_{01})$, which is derived from $\mathbf{x}(P_{00})$ by parallel displacement along $\overrightarrow{P_{00}P_{01}}$; at P_{11} we get the vector $\mathbf{x}(P_{11})$, into which $\mathbf{x}(P_{01})$ is converted by parallel displacement along $\overrightarrow{P_{01}P_{11}}$, and we have

$$\delta d\xi^\alpha = \big\{\xi^\alpha(P_{11}) - \xi^\alpha(P_{01})\big\} - \big\{\xi^\alpha(P_{10}) - \xi^\alpha(P_{00})\big\}.$$

If the vector that is derived from $\mathbf{x}(P_{10})$ by parallel displacement along $\overrightarrow{P_{10}P_{11}}$ is denoted by $\mathbf{x}_*P_{11}$, then, by interchanging d and δ, we get an analogous expression for

$$d\delta\xi^\alpha = \big\{\xi^\alpha_*(P_{11}) - \xi^\alpha(P_{10})\big\} - \big\{\xi^\alpha(P_{01}) - \xi^\alpha(P_{00})\big\}.$$

By subtraction we get

$$\begin{aligned}\Delta\xi^\alpha &= \delta d\xi^\alpha - d\delta\xi^\alpha\\ &= \left\{\begin{matrix} -\,\delta\Gamma^\alpha_{\beta i}\,dx_i + d\gamma^\alpha_r\,\delta\gamma^r_\beta - \Gamma^\alpha_{\beta i}\,\delta dx_i\\ +\,d\Gamma^\alpha_{\beta k}\,\delta x_k - d\gamma^\alpha_r\,d\gamma^r_\beta + \Gamma^\alpha_{\beta i}\,d\delta x_i\end{matrix}\right\}\xi^\beta.\end{aligned}$$

Since $\delta dx_i = d\delta x_i$ the two last terms on the right destroy one another, and we are left with

$$\Delta\xi^\alpha = \Delta F^\alpha_\beta\cdot\xi^\beta$$

in which the $\Delta\xi^\alpha$'s are the components of a vector $\Delta\mathbf{x}$ at P_{11}, which is the difference of the two vectors $\mathbf{x}$ and $\mathbf{x}_*$ **at the same point**, i.e.

$$-\Delta\xi^\alpha = \xi^\alpha(P_{11}) - \xi^\alpha_*(P_{11}).$$

Since, when we proceed to the limit, P_{11} coincides with $P = P_{00}$, this proves the statements enunciated above.

The foregoing argument, based on infinitesimals, become rigorous as soon as we interpret d and δ in terms of the differentiations $\frac{d}{ds}$ and $\frac{d}{dt}$, as was done earlier. To trace the various stages of the vector $\mathbf{x}$ during the sequence of infinitesimal displacements, we may well adopt the following plan. Let us ascribe to every pair of values s, t, not only a point $P = (s, t)$, but also a co-variant vector at P with components $f_i(s, t)$. If ξ^i is an arbitrary vector at P, then $d(f_i\xi^i)$ signifies the value that $\frac{d(f_i\xi^i)}{ds}$ assumes if ξ^i is taken along unchanged from the point (s, t) to the point $(s+ds, t)$. And $d(f_i\xi^i)$ is itself again an expression of the form $f_i\xi^i$ excepting that instead of f_i there are now other functions f_i' of s and t. We may, therefore, again subject it to the same process, or to the analogous one δ. If we do the latter, and repeat the whole operation in the reverse order, and then subtract, we get

$$\delta d(f_i\xi^i) = \delta df_i\,\xi^i + df_i\,\delta\xi^i + \delta f_i\,d\xi^i + f_i\,\delta d\xi^i,$$

and then, since

$$\delta df_i = \frac{d^2 f_i}{dt\,ds} = \frac{d^2 f_i}{ds\,dt} = d\delta f_i,$$

we have

$$\Delta(f_i\xi^i) = (\delta d - d\delta)(f_i\xi^i) = f_i\,\Delta\xi^i.$$

In the last expression $\Delta\xi^i$ is precisely the expression found above. The invariant obtained is, for the point $P = (0, 0)$,

$$F^\alpha_{\beta ik} f_\alpha \xi^\beta u^i v^k.$$

It depends on an arbitrary co-variant vector with components f_i at this point, and on three contra-variant vectors ξ, u, v; the $F^{\alpha}_{\beta ik}$'s are accordingly the components of a tensor of the fourth order.

§ 16. Metrical Space

The Conception of Metrical Manifolds.—A manifold **has a measure-determination at the point** P, if the line elements at P may be compared with respect to length; we herein assume that the Pythagorean law (of Euclidean geometry) is valid for infinitesimal regions. *Every vector* $\mathbf{x}$ *then defines a distance at* P*; and there is a non-degenerate quadratic form* $\mathbf{x}^2$*, such that* $\mathbf{x}$ *and* $\mathbf{y}$ *define the same distance if, and only if,* $\mathbf{x}^2 = \mathbf{y}^2$. This postulate determines the quadratic form fully, if a factor of proportionality differing from zero be prefixed. The fixing of the latter serves to **calibrate** the manifold at the point P. We shall then call $\mathbf{x}^2$ the measure of the vector $\mathbf{x}$, or since it depends only on the distance defined by $\mathbf{x}$, we may call it the **measure** l **of this distance**. Unequal distances have different measures; the distances at a point P therefore constitute a one-dimensional totality. If we replace this calibration by another, the new measure $\bar{l}$ is derived from the old one l by multiplying it by a constant factor $\lambda \neq 0$, independent of the distance; that is, $\bar{l} = \lambda l$. The relations between the measures of the distances are independent of the calibration. So we see that just as the characterisation of a vector at P by a system of numbers (its components) depends on the choice of the co-ordinate system, so the fixing of a distance by a number depends on the calibration; and just as the components of a vector undergo a homogeneous linear transformation in passing to another co-ordinate system, so also the measure of an arbitrary distance when the calibration is altered. We shall call two vectors $\mathbf{x}$ and $\mathbf{y}$ (at P), for which the symmetrical bilinear form $\mathbf{x} \cdot \mathbf{y}$ corresponding to $\mathbf{x}^2$ vanishes, **perpendicular** to one another; this reciprocal relation is not affected by the calibration factor. The fact that the form $\mathbf{x}^2$ is definite is of no account in our subsequent mathematical propositions, but, nevertheless, we wish to keep this case uppermost in our minds in the sequel. If this form has p positive and q negative dimensions ($p+q=n$), we say that the manifold is $(p+q)$-dimensional at the point in question. If $p \neq q$ we fix the sign of the metrical fundamental form $\mathbf{x}^2$ once and for all by the postulate that $p > q$; the calibration ratio λ is then always positive. After choosing a definite co-ordinate system and a certain calibration factor, suppose that, for every vector $\mathbf{x}$ with components ξ^i, we have

$$\mathbf{x}^2 = \sum_{i,k} g_{ik}\xi^i\xi^k \qquad (g_{ki} = g_{ik}). \tag{44}$$

We now assume that our manifold has a measure-determination at every point. Let us calibrate it everywhere, and insert in the manifold a system of n co-ordinates x_i—we must do this so as to be able to express in numbers all quantities that occur—then the g_{ik}'s in (44) are perfectly definite functions of the co-ordinates x_i; we assume that these functions are continuous and differentiable. Since the determinant of the g_{ik}'s vanishes at no point, the integral numbers p and q will remain the same in the whole domain of the manifold; we assume that $p > q$.

For a manifold to be a metrical space, it is not sufficient for it to have a measure-determination at every point; in addition, every point must be **metrically related** to the domain surrounding it. The conception of metrical relationship is analogous to that of affine relationship; just as the latter treats of **vectors**, so the former deals with distances. A point is thus metrically related to the domain in its immediate neighbourhood, if the distance is known to which every distance at P gives rise when it passes by a congruent displacement from P to any point P' infinitely near P. The immediate vicinity of P may be calibrated in such a way that the measure of any distance at P has undergone no change after congruent displacements to infinitely near points. Such a calibration is called *geodetic* at P. If, however, the manifold is calibrated in any way, and if l is the measure of any arbitrary distance at P, and $l+dl$ the measure of the distance at P' resulting from a congruent displacement to the infinitely near point P', there is necessarily an equation

$$dl = -l\,d\phi \tag{45}$$

in which the infinitesimal factor $d\phi$ is independent of the displaced distance, for the displacement effects a representation of the distances at P similar to that at P'. In (45), $d\phi$ corresponds to the $d\gamma^i_r$'s in the formula for vector displacements (33). If the calibration is altered at P and its neighbouring points according to the formula $\bar{l} = l\lambda$ (the calibration ratio λ is a positive function of the position), we get in place of (45)

$$d\bar{l} = -\bar{l}\,d\bar{\phi} \text{ in which } d\bar{\phi} = d\phi - \frac{d\lambda}{\lambda}. \tag{46}$$

The necessary and sufficient condition that an appropriate value of λ make $d\bar{\phi}$ vanish identically at P with respect to the infinitesimal displacement $\overrightarrow{PP'} = (dx_i)$ is clearly that $d\phi$ must be a differential form, that is,

$$d\phi = \phi_i\,dx_i. \tag{45'}$$

The inferences that may be drawn from the postulate enunciated at the outset are exhausted in (45) and (45′). (In short, the ϕ_i's are definite numbers at the point P. If P has co-ordinates $x_i = 0$, we need only assume $\log\lambda$ equal to the linear function $\sum \phi_i x_i$ to get $d\phi = 0$ there.) All points of the manifold are identical as regards the measure-determinations governing each and as regards their metrical relationship with their neighbouring points. Yet, according as n is even or odd, there are respectively $\frac{n}{2}+1$ or $\frac{n+1}{2}$ different types of metrical manifolds which are distinguishable from one another by the inertial index of the metrical groundform. One kind, with which we shall occupy ourselves particularly, is given by the case in which $p = n$, $q = 0$ (or $p = 0$, $q = n$); other cases are $p = n-1$, $q = 1$ (or $p = 1$, $q = n-1$), or $p = n-2$, $q = 2$ (or $p = 2$, $q = n-2$), and so forth.

We may summarise our results thus. *The metrical character of a manifold is characterised relatively to a system of reference (= co-ordinate system + calibration) by two fundamental forms, namely, a quadratic differential form* $Q = \sum_{i,k} g_{ik}\, dx_i\, dx_k$ *and a linear one* $d\phi = \sum_i \phi_i\, dx_i$. *They remain invariant during transformations to new co-ordinate systems. If the calibration is changed, the first form receives a factor* λ, *which is a positive function of position with continuous derivatives, whereas the second function becomes diminished by the differential of* $\log\lambda$. Accordingly all quantities or relations that represent metrical conditions analytically must contain the functions $g_{ik}\phi_i$ in such a way that invariance holds (1) for any transformation of co-ordinate (*co-ordinate invariance*), (2) for the substitution which replaces g_{ik} and ϕ_i respectively by

$$\lambda \cdot g_{ik} \quad \text{and} \quad \phi_i - \frac{1}{\lambda} \cdot \frac{\partial \lambda}{\partial x_i}$$

no matter, in (2), what function of the co-ordinates λ may be. (This may be termed *calibration invariance.*)

In the same way as in § 15, in which we determined the change in a vector which, remaining parallel to itself, traverses the periphery of an infinitesimal parallelogram bounded by dx_i, δx_i, so here we calculate the change Δl in the measure l of a distance subjected to an analogous process. Making use of $dl = -l\, d\phi$ we get

$$\delta dl = -\delta l\, d\phi - l\, \delta d\phi = l\, \delta\phi\, d\phi - l\, \delta d\phi,$$

i.e.

$$\Delta l = \delta dl - d\delta l = -l\, \Delta\phi$$

where

$$\Delta\phi = (\delta d - d\delta)\phi = f_{ik}\, dx_i\, \delta x_k \quad \text{and} \quad f_{ik} = \frac{\partial \phi_i}{\partial x_k} - \frac{\partial \phi_k}{\partial x_i}. \qquad (47)$$

Hence we may call the linear tensor of the second order with components f_{ik} the *distance curvature* of metrical space as an analogy to the *vector curvature* of affine space, which was derived in § 15. Equation (46) confirms analytically that the distance curvature is independent of the calibration; it satisfies the equations of invariance

$$\frac{\partial f_{kl}}{\partial x_i} + \frac{\partial f_{li}}{\partial x_k} + \frac{\partial f_{ik}}{\partial x_l} = 0.$$

Its vanishing is the necessary and sufficient condition that every distance may be transferred from its initial position, in a manner independent of the path, to all points of the space. This is the only case that Riemann considered. If metrical space is a **Riemann space**, there is meaning in speaking of the **same** distance at different points of space; the manifold may then be calibrated (*normal calibration*) so that $d\phi$ vanishes identically. (Indeed, it follows from $f_{ik} = 0$, that $d\phi$ is a total differential, namely, the differential of a function $\log \lambda$; by re-calibrating in the calibration ratio λ, $d\phi$ may then be made equal to zero everywhere.) In normal calibration the metrical groundform Q of Riemann's space is determined except for an arbitrary **constant** factor, which may be fixed by choosing once and for all a unit distance (no matter at which point; the normal meter may be transported to any place).

The Affine Relationship of a Metrical Space.—We now arrive at a fact, which may almost be called the **key-note of infinitesimal geometry**, inasmuch as it leads the logic of geometry to a wonderfully harmonious conclusion. In a metrical space the conception of infinitesimal parallel displacements may be given in only one way if, in addition to our previous postulate, it is also to satisfy the almost self-evident one: *parallel displacement of a vector must leave unchanged the distance which it determines. Thus, the principle of transference of distances or lengths which is the basis of metrical geometry, carries with it a principle of transference of direction*; in other words, **an affine relationship is inherent in metrical space**.

Proof.—We take a definite system of reference. In the case of all quantities a^i which carry an upper index i (not necessarily excluding others) we shall define the lowering of the index by equations

$$a_i = \sum_j g_{ij} a^j$$

and the reverse process of raising the index by the corresponding inverse equations. If the vector ξ^i at the point $P = (x_i)$ is to be transformed into the vector $\xi^i + d\xi^i$ at $P' = (x_i + dx_i)$ by the parallel displacement to P' which we are about to explain, then

$$d\xi^i = -d\gamma^i_k\,\xi^k, \qquad d\gamma^i_k = \Gamma^i_{kr}\,dx_r,$$

and the equation

$$dl = -l\,d\phi$$

must hold for the measure

$$l = g_{ik}\xi^i\xi^k$$

according to the postulate enunciated, and this gives

$$2\xi_i\,d\xi^i + \xi^i\xi^k\,dg_{ik} = -(g_{ik}\xi^i\xi^k)\,d\phi.$$

The first term on the left

$$= -2\xi_i\xi^k\,d\gamma^i_k = -2\xi^i\xi^k\,d\gamma_{ik} = -\xi^i\xi^k(d\gamma_{ik} + d\gamma_{ki}).$$

Hence we get

$$d\gamma_{ik} + d\gamma_{ki} = dg_{ik} + g_{ik}\,d\phi,$$

or

$$\Gamma_{i,kr} + \Gamma_{k,ir} = \frac{\partial g_{ik}}{\partial x_r} + g_{ik}\phi_r. \tag{48}$$

By interchanging the indices i, k, r cyclically, then adding the last two and subtracting the first from the resultant sum, we get, bearing in mind that the Γ's must be symmetrical in their last two indices,

$$\Gamma_{r,ik} = \tfrac{1}{2}\left(\frac{\partial g_{ir}}{\partial x_k} + \frac{\partial g_{kr}}{\partial x_i} - \frac{\partial g_{ik}}{\partial x_r}\right) + \tfrac{1}{2}(g_{ir}\phi_k + g_{kr}\phi_i - g_{ik}\phi_r). \tag{49}$$

From this the Γ^r_{ik} are determined according to the equation

$$\Gamma_{r,ik} = g_{rs}\Gamma^s_{ik} \quad \text{or, explicitly,} \quad \Gamma^r_{ik} = g^{rs}\Gamma_{s,ik}. \tag{50}$$

These components of the affine relationship fulfil all the postulates that have been enunciated. It is in the nature of metrical space to be furnished with this affine relationship; in virtue of it the whole analysis of tensors and tensor-densities with all the conceptions worked out above, such as geodetic line, curvature, etc., may be applied to metrical space. If the curvature vanishes identically, the space is metrical and Euclidean in the sense of Chapter I.

In the case of **vector curvature** we have still to derive an important decomposition into components, by means of which we prove that distance curvature is an inherent constituent of the former. This is quite to be expected since vector transference is automatically accompanied by distance transference. If we use the symbol $\Delta = \delta d - d\delta$ relating to parallel displacement as before, then the measure l of a vector ξ^i satisfies

$$\Delta l = -l\,\Delta\phi, \qquad \Delta \xi_i \xi^i = -(\xi_i \xi^i)\,\Delta\phi. \tag{47}$$

Just as we found for the case in which f_i are any functions of position that

$$\Delta(f_i \xi^i) = f_i\,\Delta\xi^i$$

so we see that

$$\Delta(\xi_i \xi^i) = \Delta(g_{ik}\xi^i\xi^k) - g_{ik}\,\Delta\xi^i \cdot \xi^k + g_{ik}\xi^i \cdot \Delta\xi^k = 2\xi_i\,\Delta\xi^i,$$

and equation (47) then leads to the following result. If for the vector $\mathbf{x} = (\xi^i)$ we set

$$\Delta\mathbf{x} = {*\Delta\mathbf{x}} - \mathbf{x} \cdot \tfrac{1}{2}\Delta\phi,$$

then $\Delta\mathbf{x}$ appears split up into a component at right angles to $\mathbf{x}$ and another parallel to $\mathbf{x}$, namely, $*\Delta\mathbf{x}$ and $-\mathbf{x}\cdot\frac{1}{2}\Delta\phi$ respectively. This is accompanied by an analogous resolution of the curvature tensor, i.e.

$$F^{\alpha}_{\beta ik} = *F^{\alpha}_{\beta ik} - \tfrac{1}{2}\delta^{\alpha}_{\beta} f_{ik}. \tag{51}$$

The first component $*F$ will be called "**direction curvature**"; it is defined by

$$*\Delta\mathbf{x} = *F^{\alpha}_{\beta ik}\mathbf{e}_{\alpha}\xi^{\beta}\,dx_i\,\delta x_k.$$

The perpendicularity of $*\Delta\mathbf{x}$ to $\mathbf{x}$ is expressed by the formula

$$*F^{\alpha}_{\beta ik}\xi_{\alpha}\xi^{\beta}\,dx_i\,\delta x_k = *F_{\alpha\beta ik}\xi^{\alpha}\xi^{\beta}\,dx_i\,\delta x_k = 0.$$

The system of numbers $*F_{\alpha\beta ik}$ is skew-symmetrical not only with respect to i and k but also with respect to the index pair α and β. In consequence we have also, in particular,

$$*F^{\alpha}_{\alpha ik} = 0.$$

Corollaries.—If the co-ordinate system and calibration around a point P is chosen so that they are geodetic at P, then we have, at P, $\phi_i = 0$, $\Gamma^{r}_{ik} = 0$, or, according to (48) and (49), the equivalent

$$\phi_i = 0, \qquad \frac{\partial g_{ik}}{\partial x_r} = 0.$$

The linear form $d\phi$ vanishes at P and the co-efficients of the quadratic groundform become stationary; in other words, those conditions come about at P, which are obtained in Euclidean space simultaneously for all points by a single system of reference. This results in the following explicit definition of the parallel displacement of a vector in metrical space. A geodetic system of reference at P may be recognised by the property that the ϕ_i's at P vanish relatively to it and the g_{ik}'s assume stationary values. A vector is displaced from P parallel to itself to the infinitely near point P' by leaving its components in **a system of reference belonging to** P unaltered. (There are always geodetic systems of reference; the choice of them does not affect the conception of parallel displacements.)

Since, in a **translation** $x_i = x_i(s)$, the velocity vector $u_i = \dfrac{dx_i}{ds}$ moves so that it remains parallel to itself, it satisfies

$$\frac{d(u_i u^i)}{ds} + (u_i u^i)(\phi_i u^i) = 0 \quad \text{in metrical geometry.} \tag{52}$$

If at a certain moment the u^i's have such values that $u_i u^i = 0$ (a case that may occur if the quadratic groundform Q is indefinite), then this equation persists throughout the whole translation: we shall call the trajectory of such a translation a **geodetic null-line**. An easy calculation shows that the geodetic null-lines do not alter if the metric relationship of the manifold is changed in any way, as long as the measure-determination is kept fixed at every point.

Tensor Calculus.—It is an essential characteristic of a tensor that its components depend only on the co-ordinate system and not on the calibration. In a generalised sense we shall, however, also call a linear form which depends on the co-ordinate system and the **calibration** a tensor, if it is transformed in the usual way when the co-ordinate system is changed, but becomes multiplied by the factor λ^e (where λ = the calibration ratio) when the calibration is changed; we say that it is of **weight** e. Thus the g_{ik}'s are components of a symmetrical co-variant tensor of the second order and of weight 1. Whenever tensors are mentioned without their weight being specified, we shall take this to mean that those of weight 0 are being considered. The relations which were discussed in tensor analysis are relations, which are independent of calibration and co-ordinate system, between tensors and tensor-densities **in this special sense**. We regard the extended conception of a tensor, and also the analogous one of tensor-density of weight e, merely as an auxiliary conception, which is introduced to simplify calculations. They are convenient for two reasons: (1) They make it possible to "juggle with indices" in this extended region. By lowering a contra-variant index in the components of a tensor of weight e we get the components of a tensor of weight $e + 1$, the components being co-variant with respect to this index. The process may also be carried out in the reverse direction. (2) Let g denote the determinant of the g_{ik}'s, furnished with a plus or minus sign according as the number q of the negative dimensions is even or uneven, and let $\sqrt{g}$ be the positive root of this positive number g. Then, **by multiplying any tensor by $\sqrt{g}$ we get a tensor-density whose weight is $\frac{n}{2}$ more than that of the tensor**; from a tensor of weight $-\frac{n}{2}$ we get, in particular, a tensor-density in the true sense. The proof is based on the evident fact that $\sqrt{g}$ is itself a scalar-density of weight $\frac{n}{2}$. We shall always indicate when a quantity is multiplied by $\sqrt{g}$ by changing the ordinary letter which designates the quantity into the corresponding one printed in Clarendon type. Since, in Riemann's geometry, the quadratic groundform Q is fully determined by normal calibration (we need not consider the arbitrary **constant** factor), the difference in the weights of tensors disappears here: since, in this case, every quantity that may be represented by a tensor may also be represented by the tensor-density that is derived from it by multiplying it by $\sqrt{g}$, the difference between tensors and tensor-densities (as well as between co-variant and contra-variant) is effaced. This makes it clear why for a long time tensor-densities did not come into their right as compared with tensors. The main use of tensor calculus in geometry is an **internal** one, that is, to construct fields that are derived invariantly from the metrical structures. We shall give two examples that are of importance for later work. Let the metrical manifold be $(3+1)$-dimensional, so that $-g$ will be the determinant of the g_{ik}'s. In this space, as in every other, the distance curvature with components f_{ik} is a true linear tensor field of the second order. From it is derived the contra-variant tensor f^{ik} of weight -2, which, on account of its weight differing from zero, is of no actual importance; multiplication by $\sqrt{g}$ leads to $\mathbf{f}^{ik}$, a true linear tensor-density of the second order.

$$\mathbf{l} = \tfrac{1}{4} f_{ik}\mathbf{f}^{ik} \tag{53}$$

is the simplest scalar-density that can be formed; consequently $\int \mathbf{l}\, dx$ is the simplest invariant integral associated with the metrical basis of a $(3+1)$-dimensional manifold. On the other hand, the integral $\int \sqrt{g}\, dx$, which occurs in Riemann's geometry as "volume," is meaningless in general geometry. We can derive the intensity of current (vector-density) from $\mathbf{f}^{ik}$ by means of the operation divergence thus:

$$\frac{\partial \mathbf{f}^{ik}}{\partial x_k} = \mathbf{s}^i.$$

In physics, however, we use the tensor calculus not to describe the metrical condition but to describe fields expressing physical states in metrical space—as, for example, the electromagnetic field—and to set up the laws that hold in them. Now, we shall find at the close of our investigations that this distinction between physics and geometry is false, and that physics does not extend beyond geometry. The world is a $(3+1)$-dimensional metrical manifold, and all physical phenomena that occur in it are only modes of expression of the metrical field. In particular, the affine relationship of the world is nothing more than the gravitational field, but its metrical character is an expression of the state of the "æther" that fills the world; even matter itself is reduced to this kind of geometry and loses its character as a permanent substance. Clifford's prediction, in an article of the *Fortnightly Review* of 1875, becomes confirmed here with remarkable accuracy; in this he says that "the theory of space curvature hints at a possibility of describing matter and motion in terms of extension only".

These are, however, as yet dreams of the future. For the present, we shall maintain our view that physical states are foreign states in space. Now that the principles of infinitesimal geometry have been worked out to their conclusion, we shall set out, in the next paragraph, a number of observations about the special case of Riemann's space and shall give a number of formulæ which will be of use later.

§ 17. Observations about Riemann's Geometry as a Special Case

General tensor analysis is of great utility even for Euclidean geometry whenever one is obliged to make calculations, not in a Cartesian or affine co-ordinate system, but in a curvilinear co-ordinate system, as often happens in mathematical physics. To illustrate this application of the tensor calculus we shall here write out the fundamental equations of the electrostatic and the magnetic field due to stationary currents in terms of general curvilinear co-ordinates.

Firstly, let E_i be the components of the electric intensity of field in a Cartesian co-ordinate system. By transforming the quadratic and the linear differential forms

$$ds^2 = dx_1^2 + dx_2^2 + dx_3^2 \qquad E_1\,dx_1 + E_2\,dx_2 + E_3\,dx_3$$

respectively, into terms of arbitrary curvilinear co-ordinates (again denoted by x_i), each form being independent of the Cartesian co-ordinate system, suppose we get

$$ds^2 = g_{ik}\,dx_i\,dx_k \quad \text{and} \quad E_i\,dx_i.$$

Then the E_i's are in every co-ordinate system the components of the same co-variant vector field. From them we form a vector-density with components

$$\mathbf{E}^i = \sqrt{g}\cdot g^{ik}E_k \qquad (g = |g_{ik}|).$$

We transform the potential $-\phi$ as a scalar into terms of the new co-ordinates, but we define the density ρ of electricity as being the electric charge given by $\int \rho\,dx_1\,dx_2\,dx_3$ contained in any portion of space; ρ is not then a scalar but a scalar density. The laws are expressed by

$$\left.\begin{array}{c} E_i = \dfrac{\partial \phi}{\partial x_i} \qquad \dfrac{\partial E_i}{\partial x_k} - \dfrac{\partial E_k}{\partial x_i} = 0 \\ \dfrac{\partial \mathbf{E}^i}{\partial x_i} = \rho \\ \mathbf{S}_i^k = E_i \mathbf{E}^k - \frac{1}{2}\delta_i^k \mathbf{S}, \end{array}\right\} \tag{54}$$

in which $\mathbf{S}, = E_i\mathbf{E}^i$, are the components of a mixed tensor-density of the second order, namely, the potential difference. The proof is sufficiently indicated by the remark that these equations, in the form we have written them, are absolutely invariant in character, but pass into the fundamental equations, which were set up earlier, for a Cartesian co-ordinate system.

The magnetic field produced by stationary currents was characterised in Cartesian co-ordinate systems by an invariant skew-symmetrical bilinear form $H_{ik}\, dx_i\, \delta x_k$. By transforming the latter into terms of arbitrary curvilinear co-ordinates, we get H_{ik}, the components of a linear tensor of the second order, namely, of the *magnetic field*, these components being co-variant with respect to arbitrary transformations of the co-ordinates. Similarly, we may deduce the components ϕ_i of the vector potential as components of a co-variant vector field in any curvilinear co-ordinate system. We now introduce a linear tensor-density of the second order by means of the equations

$$\mathbf{H}^{ik} = \sqrt{g} \cdot g^{i\alpha} g^{k\beta} H_{\alpha\beta}.$$

The laws are then expressed by

$$\left.\begin{array}{c} H_{ik} = \dfrac{\partial\phi_i}{\partial x_k} - \dfrac{\partial\phi_k}{\partial x_i} \quad \text{or} \quad \dfrac{\partial H_{kl}}{\partial x_i} + \dfrac{\partial H_{li}}{\partial x_k} + \dfrac{\partial H_{ik}}{\partial x_l} = 0 \\ \text{respectively,} \\ \dfrac{\partial \mathbf{H}^{ik}}{\partial x_k} = \mathbf{s}^i, \\ \mathbf{S}_i^k = H_{ir}\mathbf{H}^{kr} - \tfrac{1}{2}\delta_i^k\mathbf{S}, \qquad \mathbf{S} = \tfrac{1}{2}H_{ik}\mathbf{H}^{ik}. \end{array}\right\} \tag{55}$$

The $\mathbf{s}^i$'s are the components of a vector-density, the *electric intensity of current*; the potential differences $\mathbf{S}_i^k$ have the same invariant character as in the electric field. These formulæ may be specialised for the case of, for example, spherical and cylindrical co-ordinates. No further calculations are required to do this, if we have an expression for ds^2, the distance between two adjacent points, expressed in these co-ordinates; this expression is easily obtained from considerations of infinitesimal geometry.

It is a matter of greater fundamental importance that (54) and (55) furnish us with the underlying laws of stationary electromagnetic fields if unforeseen reasons should compel us to give up the use of Euclidean geometry for physical space and replace it by **Riemann's geometry** with a new groundform. For even in the case of such generalised geometric conditions our equations, in virtue of their invariant character, represent statements that are independent of all co-ordinate systems, and that express formal relationships between charge, current, and field. In no wise can it be doubted that they are the direct transcription of the laws of the stationary electric field that hold in Euclidean space; it is indeed astonishing how simply and naturally this transcription is effected by means of the tensor calculus. The question whether space is Euclidean or not is quite irrelevant for the laws of the electromagnetic field. The property of being Euclidean is expressed in a universally invariant form by differential equations of the second order in the g_{ik}'s (denoting the vanishing of the curvature) but only the g_{ik}'s and their first derivatives appear in these laws. It must be emphasised that a transcription of such a simple kind is possible only for laws dealing with **action at infinitesimal distances**. To derive the laws of action at a distance corresponding to Coulomb's, and Biot and Savart's Law from these laws of contiguous action is a purely mathematical problem that amounts in essence to the following. In place of the usual potential equation $\Delta\phi = 0$ we get as its invariant generalisation (*vide* (54)) in Riemann's geometry the equation

$$\frac{\partial}{\partial x_i}\left(\sqrt{g}\cdot g^{ik}\frac{\partial\phi}{\partial x_k}\right) = 0$$

that is, a linear differential equation of the second order whose co-efficients are, however, no longer constants. From this we are to get the "standard solution," tending to infinity, at any arbitrary given point; this solution corresponds to the "standard solution" $\frac{1}{r}$ of the potential equation. It presents a difficult mathematical problem that is treated in the theory of linear partial differential equations of the second order. The same problem is presented when we are limited to Euclidean space if, instead of investigating events in empty space, we have to consider those taking place in a non-homogeneous medium (for example, in a medium whose dielectric constant varies at different places with the time). Conditions are not so favourable for transcribing electromagnetic laws, if real space should become disclosed as a metrical space of a still more general character than Riemann assumed. In that case it would be just as inadmissible to assume the possibility of a calibration that is independent of position in the case of currents and charges as in the case of distances. Nothing is gained by pursuing this idea. The true solution of the problem lies, as was indicated in the concluding words of the previous paragraph, in quite another direction.

Let us rather add a few observations about **Riemann's space as a special case**. Let the unit measure (1 centimetre) be chosen once and for all; it must, of course, be the same at all points. The metrical structure of the Riemann space is then described by an invariant quadratic differential form $g_{ik}\, dx_i\, dx_k$ or, what amounts to the same thing, by a co-variant symmetrical tensor field of the second order. The quantities ϕ_i, that are now equal to zero, must be struck out everywhere in the formulæ of general metrical geometry. Thus, the components of the affine relationship, which here bear the name "Christoffel three-indices symbols" and are usually denoted by $\begin{Bmatrix} ik \\ r \end{Bmatrix}$, are determined from

$$\begin{bmatrix} ik \\ r \end{bmatrix} = \tfrac{1}{2}\left(\frac{\partial g_{ir}}{\partial x_k} + \frac{\partial g_{kr}}{\partial x_i} - \frac{\partial g_{ik}}{\partial x_r}\right), \qquad \begin{Bmatrix} ik \\ r \end{Bmatrix} = g^{rs}\begin{bmatrix} ik \\ s \end{bmatrix}. \tag{56}$$

(We give way to the usual nomenclature—although it disagrees flagrantly with our own convention regarding rules about the position of indices—so as to conform to the usage of the text-books.)

The following formulæ are now tabulated for future reference:—

$$\frac{1}{\sqrt{g}}\frac{\partial\sqrt{g}}{\partial x_i} - \begin{Bmatrix} ir \\ r \end{Bmatrix} = 0, \tag{57}$$

$$\frac{1}{\sqrt{g}}\frac{\partial(\sqrt{g}\cdot g^{ik})}{\partial x_k} + \begin{Bmatrix} rs \\ i \end{Bmatrix} g^{rs} = 0, \tag{57'}$$

$$\frac{1}{\sqrt{g}}\frac{\partial(\sqrt{g}\cdot g^{ik})}{\partial x_l}+\begin{Bmatrix} lr \\ i \end{Bmatrix}g^{rk}+\begin{Bmatrix} lr \\ k \end{Bmatrix}g^{ri}-\begin{Bmatrix} lr \\ r \end{Bmatrix}g^{ik}=0. \tag{57''}$$

These equations hold because $\sqrt{g}$ is a scalar and $\sqrt{g}\cdot g^{ik}$ is a tensor-density; hence, according to the rules given by the analysis of tensor-densities, the left-hand members of these equations, multiplied by $\sqrt{g}$, are likewise tensor-densities. If, however, we use a co-ordinate system $\left(\frac{\partial g^{ik}}{\partial x_r}\right)=0$, which is geodetic at P, then all terms vanish. Hence, in virtue of the invariant nature of these equations, they also hold in every other co-ordinate system. Moreover,

$$\frac{dg}{g}=g^{ik}\,dg_{ik}, \qquad \frac{d\sqrt{g}}{\sqrt{g}}=\tfrac{1}{2}g^{ik}\,dg_{ik}. \tag{58}$$

For the total differential of a determinant with n^2 (independent and variable) elements g_{ik} is equal to $G^{ik}\,dg_{ik}$, where G^{ik} denotes the minor of g_{ik}. If $\mathbf{t}^{ik}$ $(=\mathbf{t}^{ki})$ is any symmetrical system of numbers, then we always have

$$\mathbf{t}^{ik}\,dg_{ik}=-\mathbf{t}_{ik}\,dg^{ik}. \tag{59}$$

From

$$g_{ij}g^{jk}=\delta_i^k$$

it follows that

$$g_{ij}\,dg^{jk}=-g^{jk}\,dg_{ij}.$$

If these equations are multiplied by $\mathbf{t}_k^i$ (this symbol cannot be misinterpreted since $\mathbf{t}_k^i=g_{kl}\mathbf{t}^{il}=g_{kl}\mathbf{t}^{li}=\mathbf{t}_k^i$) the required result follows. In particular, in place of (58) we may also write

$$\frac{dg}{g}=-g_{ik}\,dg^{ik}. \tag{58'}$$

The co-variant **components** $R_{\alpha\beta ik}$ **of curvature** in Riemann's space, which we denote by R instead of F, satisfy the conditions of symmetry

$$R_{\alpha\beta ki} = -R_{\alpha\beta ik}, \qquad R_{\beta\alpha ki} = -R_{\alpha\beta ik},$$
$$R_{\alpha\beta ki} + R_{\alpha ik\beta} + R_{\alpha k\beta i} = 0,$$

(for the "distance curvature" vanishes). It is easy to show that, from them, it follows that (*vide* note 11)

$$R_{ik\alpha\beta} = R_{\alpha\beta ik}.$$

As the result of an observation on page 83, it follows that all those conditions taken together enable us to characterise the curvature tensor completely by means of a quadratic form that is dependent on an arbitrary element of surface, namely,

$$\tfrac{1}{4} R_{\alpha\beta ik}\, \Delta x_{\alpha\beta}\, \Delta x_{ik} \qquad (\Delta x_{ik} = dx_i\, \delta x_k - dx_k\, \delta x_i).$$

If this quadratic form is divided by the square of the magnitude of the surface element, the quotient depends only on the ratio of the Δx_{ik}'s, i.e. on the position of the surface element; Riemann calls this number the curvature of the space at the point P in the surface direction in question. In two-dimensional Riemann space (on a surface) there is only one surface direction and the tensor degenerates into a scalar (Gaussian curvature). In Einstein's theory of gravitation the contracted tensor of the second order

$$R^{\alpha}_{i\alpha k} = R_{ik}$$

which is symmetrical in Riemann's space, becomes of importance: its components are

$$R_{ik} = \frac{\partial}{\partial x_r}\begin{Bmatrix} ik \\ r \end{Bmatrix} - \frac{\partial}{\partial x_k}\begin{Bmatrix} ir \\ r \end{Bmatrix} + \begin{Bmatrix} ik \\ r \end{Bmatrix}\begin{Bmatrix} rs \\ s \end{Bmatrix} - \begin{Bmatrix} ir \\ s \end{Bmatrix}\begin{Bmatrix} ks \\ r \end{Bmatrix}. \qquad (60)$$

Only in the case of the second term on the right, the symmetry with respect to i and k is not immediately evident; according to (57), however, it is equal to

$$\frac{1}{2}\frac{\partial^2(\log g)}{\partial x_i\,\partial x_k}.$$

Finally, by applying contraction once more we may form the **scalar of curvature**

$$R = g^{ik}R_{ik}.$$

In general metrical space the analogously formed scalar of curvature F is expressed in the following way (as is easily shown) by the Riemann expression R, which is dependent only on the g_{ik}'s and which has no distinct meaning in that space:—

$$F = R - (n-1)\frac{1}{\sqrt{g}}\frac{\partial(\sqrt{g}\phi^i)}{\partial x_i} - \frac{(n-1)(n-2)}{4}(\phi_i\phi^i). \qquad (61)$$

F is a scalar of weight -1. Hence, in a region in which $F \neq 0$ we may define a unit of length by means of the equation $F =$ constant. This is a remarkable result inasmuch as it contradicts in a certain sense the original view concerning the transference of lengths in general metrical space, according to which a direct comparison of lengths at a distance is not possible; it must be noticed, however, that the unit of length which arises in this way is dependent on the conditions of curvature of the manifold. (The existence of a unique uniform calibration of this kind is no more extraordinary than the possibility of introducing into Riemann's space certain unique co-ordinate systems arising out of the metrical structure.) The "volume" that is measured by using this unit of length is represented by the invariant integral

$$\int \sqrt{g\cdot F^n}\,dx. \qquad (62)$$

For two vectors ξ^i, η^i that undergo parallel displacement we have, in metrical space,

$$d(\xi_i\eta^i) + (\xi_i\eta^i)\,d\phi = 0.$$

In Riemann's space, the second term is absent. From this it follows that in Riemann's space the parallel displacement of a contravariant vector ξ is expressed in exactly the same way in terms of the quantities $\xi_i = g_{ik}\xi^k$ as the parallel displacement of a covariant vector is expressed in terms of its components ξ_i:

$$d\xi_i - \begin{Bmatrix} i\alpha \\ \beta \end{Bmatrix} dx_\alpha\, \xi_\beta = 0 \quad \text{or} \quad d\xi_i - \begin{bmatrix} i\alpha \\ \beta \end{bmatrix} dx_\alpha\, \xi^\beta = 0.$$

Accordingly, for a translation we have

$$\frac{du_i}{ds} - \tfrac{1}{2}\frac{\partial g_{\alpha\beta}}{\partial x_i} u^\alpha u^\beta = 0 \qquad \left(u^i = \frac{dx_i}{ds},\; u_i = g_{ik}u^k\right) \tag{63}$$

for, by equation (48),

$$\begin{bmatrix} i\alpha \\ \beta \end{bmatrix} + \begin{bmatrix} i\beta \\ \alpha \end{bmatrix} = \frac{\partial g_{\alpha\beta}}{\partial x_i}$$

and hence for any symmetrical system of numbers $\mathbf{t}^{\alpha\beta}$:—

$$\tfrac{1}{2}\frac{\partial g_{\alpha\beta}}{\partial x_i} \cdot \mathbf{t}_{\alpha\beta} = \begin{bmatrix} i\alpha \\ \beta \end{bmatrix} \mathbf{t}^{\alpha\beta} = \begin{Bmatrix} i\alpha \\ \beta \end{Bmatrix} \mathbf{t}^{\alpha}_{\beta}. \tag{64}$$

Since the numerical value of the velocity vector remains unchanged during translations, we get

$$g_{ik}\frac{dx_i}{ds}\frac{dx_k}{ds} = u_i u^i = \text{const.} \tag{65}$$

If, for the sake of simplicity, we assume the metrical groundform to be definitely positive, then every curve $x_i = x_i(s)$ $[a \leq s \leq b]$

has a **length**, which is independent of the mode of parametric representation. This length is

$$\int_a^b \sqrt{Q}\, ds \qquad \left(Q = g_{ik}\,\frac{dx_i}{ds}\,\frac{dx_k}{ds}\right).$$

If we use the length of arc itself as the parameter, Q becomes equal to 1. Equation (65) states that a body in translation traverses its path, the geodetic line, with constant speed, namely, that the time-parameter is proportional to s, the length of arc. In Riemann's space the geodetic line possesses not only the differential property of preserving its direction unaltered, but also **the integral property that every portion of it is the shortest line connecting its initial and its final point**. This statement must not, however, be taken literally, but must be understood in the same sense as the statement in mechanics that, in a position of equilibrium, the potential energy is a minimum, or when it is said of a function $f(x, y)$ in two variables that it has a minimum at points where its differential

$$df = \frac{\partial f}{\partial x}\,dx + \frac{\partial f}{\partial y}\,dy$$

vanishes identically in dx and dy; whereas the true expression is that it assumes a "stationary" value at that point, which may be a minimum, a maximum, or a "point of inflexion". The geodetic line is not necessarily a curve of least length but is a curve of stationary length. On the surface of a sphere, for instance, the great circles are geodetic lines. If we take any two points, A and B, on such a great circle, the shorter of the two arcs AB is indeed the shortest line connecting A and B, but the other arc AB is also a geodetic line connecting A and B; it is not of least but of stationary length.

We shall seize this opportunity of expressing in a rigorous form the principle of infinitesimal variation.

Let any arbitrary curve be represented parametrically by

$$x_i = x_i(s), \qquad (a \leq s \leq b).$$

We shall call it the "initial" curve. To compare it with neighbouring curves we consider an arbitrary family of curves involving one parameter:

$$x_i = x_i(s; \epsilon), \qquad (a \leq s \leq b).$$

The parameter ϵ varies within an interval about $\epsilon = 0$; $x_i(s; \epsilon)$ are to denote functions that resolve into $x_i(s)$ when $\epsilon = 0$. Since all curves of the family are to connect the same initial point with the same final point, $x_i(a; \epsilon)$ and $x_i(b; \epsilon)$ are independent of ϵ. The length of such a curve is given by

$$L(\epsilon) = \int_a^b \sqrt{Q}\, ds.$$

Further, we assume that s denotes the length of an arc of the initial curve, so that $Q = 1$ for $\epsilon = 0$. Let the direction components $\frac{dx_i}{ds}$ of the initial curve $\epsilon = 0$ be denoted by u^i. We also set

$$\epsilon \cdot \left(\frac{dx_i}{d\epsilon}\right)_{\epsilon=0} = \xi^i(s) = \delta x_i.$$

These are the components of the "infinitesimal" displacement which makes the initial curve change into the neighbouring curve due to the "variation" corresponding to an infinitely small value of ϵ; they vanish at the ends.

$$\epsilon\left(\frac{dL}{d\epsilon}\right)_{\epsilon=0} = \delta L$$

is the corresponding variation in the length. $\delta L = 0$ is the condition that the initial curve has a stationary length as compared with the other members of the family. If we use the symbol δQ in the same sense, we get

$$\delta L = \int_a^b \frac{\delta Q}{2\sqrt{Q}}\, ds = \tfrac{1}{2} \int_a^b \delta Q\, ds \tag{66}$$

since $Q = 1$ in the case of the initial curve. Now

$$\frac{dQ}{d\epsilon} = \frac{\partial g_{\alpha\beta}}{\partial x_i} \frac{dx_i}{d\epsilon} \frac{dx_\alpha}{ds} \frac{dx_\beta}{ds} + 2g_{ik} \frac{dx_k}{ds} \frac{d^2 x_i}{d\epsilon\, ds}$$

and hence (if we interchange "variation" and "differentiation," that is the differentiations with respect to ϵ and s) we get

$$\delta Q = \frac{\partial g_{\alpha\beta}}{\partial x_i} u^\alpha u^\beta \xi^i + 2g_{ik} u^k \frac{d\xi^i}{ds}.$$

If we substitute this in (66) and rewrite the second term by applying partial integration, and note that the ξ^i's vanish at the ends of the interval of integration, then

$$\delta L = \int_a^b \left(\tfrac{1}{2} \frac{\partial g_{\alpha\beta}}{\partial x_i} u^\alpha u^\beta - \frac{du_i}{ds} \right) \xi^i\, ds.$$

Hence the condition $\delta L = 0$ is fulfilled for any family of curves if, and only if, (63) holds. Indeed, if, for a value $s = s_0$ between a and b, one of these expressions, for example the first, namely, $i = 1$, differed from zero (were greater than zero), say, it would be possible to mark off a little interval around s_0 so small that, within it, the above expression would be always > 0. If we choose a non-negative function for ξ^1 such that it vanishes for points beyond this interval, all remaining ξ^i's, however, being $= 0$, we find the equation $\delta L = 0$ contradicted.

Moreover, it is evident from this proof that, of all the motions that lead from the same initial point to the same final point within the same interval of time $a \leq s \leq b$, a **translation** is distinguished by the property that $\int_a^b Q\,ds$ has a stationary value.

Although the author has aimed at lucidity of expression many a reader will have viewed with abhorrence the flood of formulæ and indices that encumber the fundamental ideas of infinitesimal geometry. It is certainly regrettable that we have to enter into the purely formal aspect in such detail and to give it so much space but, nevertheless, it cannot be avoided. Just as anyone who wishes to give expressions to his thoughts with ease must spend laborious hours learning language and writing, so here too the only way that we can lessen the burden of formulæ is to master the technique of tensor analysis to such a degree that we can turn to the real problems that concern us without feeling any encumbrance, our object being to get an insight into the nature of space, time, and matter so far as they participate in the structure of the external world. Whoever sets out in quest of this goal must possess a perfect mathematical equipment from the outset. Before we pass on after these wearisome preparations and enter into the sphere of physical knowledge along the route illumined by the genius of Einstein, we shall seek to obtain a clearer and deeper vision of metrical space. Our goal is to grasp the inner necessity and uniqueness of its metrical structure as expressed in Pythagoras' Law.

§ 18. Metrical Space from the Point of View of the Theory of Groups

Whereas the character of affine relationship presents no further difficulties—the postulate on page 184 to which we subjected the conception of parallel displacement, and which characterises it as a kind of **unaltered** transference, defines its character uniquely—we have not yet gained a view of metrical structure that takes us beyond experience. It was long accepted as a fact that a metrical character could be described by means of a quadratic differential form, but this fact was not clearly understood. Riemann many years ago pointed out that the metrical groundform might, with equal right essentially, be a homogeneous function of the fourth order in the differentials, or even a function built up in some other way, and that it need not even depend rationally on the differentials. But we dare not stop even at this point. The underlying general feature that determines the metrical structure at a point P is the **group of rotations**. The metrical constitution of the manifold at the point P is known if, among the linear transformations of the vector body (i.e. the totality of vectors), those are known that are **congruent** transformations of themselves. There are just as many different kinds of measure-determinations as there are essentially different groups of linear transformations (whereby essentially different groups are such as are distinguished not merely by the choice of co-ordinate system). In the case of **Pythagorean metrical space**, which we have alone investigated hitherto, the group of rotations consists of all linear transformations that convert the quadratic groundform into itself. But the group of rotations need not have an invariant at all in itself (that is, a function which is dependent on a single arbitrary vector and which remains unaltered after any rotations).

Let us reflect upon the natural requirements that may be imposed on the conception of rotation. At a single point, as long as the manifold has not yet a measure-determination, only the n-dimensional parallelepipeds can be compared with one another in respect to size. If $\mathbf{a}_i$ $(i = 1, 2, \ldots, n)$ are arbitrary vectors that are defined in terms of the initial unit vectors $\mathbf{e}_i$ according to the equations

$$\mathbf{a}_i = \alpha_i^k \mathbf{e}_k$$

then the determinant of the α_i^k's which, following Grassmann, we may conveniently denote by

$$\frac{[\mathbf{a}_1, \mathbf{a}_2, \ldots, \mathbf{a}_n]}{[\mathbf{e}_1, \mathbf{e}_2, \ldots, \mathbf{e}_n]}$$

is, according to definition, the volume of the parallelopiped mapped out by the n vectors $\mathbf{a}_i$. If we choose another system of unit vectors $\bar{\mathbf{e}}_i$ all the volumes become multiplied by a common constant factor, as we see from the "multiplication theorem of determinants," namely

$$\frac{[\mathbf{a}_1, \mathbf{a}_2, \ldots, \mathbf{a}_n]}{[\mathbf{e}_1, \mathbf{e}_2, \ldots, \mathbf{e}_n]} = \frac{[\mathbf{a}_1, \mathbf{a}_2, \ldots, \mathbf{a}_n]}{[\bar{\mathbf{e}}_1, \bar{\mathbf{e}}_2, \ldots, \bar{\mathbf{e}}_n]} \frac{[\bar{\mathbf{e}}_1, \bar{\mathbf{e}}_2, \ldots, \bar{\mathbf{e}}_n]}{[\mathbf{e}_1, \mathbf{e}_2, \ldots, \mathbf{e}_n]}.$$

The volumes are thus determined uniquely and independently of the co-ordinate system once the unit measure has been chosen. *Since a rotation is "not to alter" the vector body, it must obviously be a transformation that leaves the infinitesimal elements of volume unaffected.* Let the rotation that transforms the vector $\mathbf{x} = (\xi^i)$ into $\bar{\mathbf{x}} = (\bar{\xi}^i)$ be represented by the equations

$$\bar{\mathbf{e}}_i = \alpha_i^k \mathbf{e}_k \quad \text{or} \quad \xi^i = \alpha_k^i \bar{\xi}^k.$$

The determinant of the rotation matrix (α^i_k) then becomes equal to 1. This being the postulate that applies to a **single** rotation, we must demand of the rotations as a whole that they **form a group** in the sense of the definition given on page 13. Moreover, this group has to be a **continuous** one, that is the rotations are to be elements of a one-dimensional continuous manifold.

If a linear vector transformation be given by its matrix $A = (\alpha^i_k)$ in passing from one co-ordinate system $(\mathbf{e}_i)$ to another $(\bar{\mathbf{e}}_i)$ according to the equations

$$U : \bar{\mathbf{e}}_i = u^k_i \mathbf{e}_k, \tag{67}$$

then A becomes changed into UAU^{-1} (where U^{-1} denotes the inverse of U; UU^{-1} and $U^{-1}U$ are equal to identity E). Hence every group that is derived from a given matrix group $\mathbf{G}$ by applying the operation UGU^{-1} on every matrix G of $\mathbf{G}$ (U being the same for all G's) may be transformed into the given matrix group by an appropriate change of co-ordinate system. Such a group $U\mathbf{G}U^{-1}$ will be said to be of the same kind as $\mathbf{G}$ (or to differ from $\mathbf{G}$ only in orientation). If $\mathbf{G}$ is the group of rotation matrices at P and if $U\mathbf{G}U^{-1}$ is identical with $\mathbf{G}$ (this does not mean that G must again pass into G as a result of the operation UGU^{-1}, but all that is required is that G and UGU^{-1} belong to $\mathbf{G}$ simultaneously) then the expressions for the metrical structures of two co-ordinate systems (67), that are transformed into one another by U, are similar; U is a representation of the vector body on itself, such that it leaves all the metrical relations unaltered. This is the conception of **similar representation**. $\mathbf{G}$ is included in the group $\mathbf{G}^*$ of similar representations as a sub-group.

From the metrical structure at a single point we now pass on to "**metrical relationship**". The metrical relationship between the point P_0 and its immediate neighbourhood is given if a linear representation at $P_0 = x_i^0$ of the vector body on itself at an infinitely near point $P = (x_i^0 + dx_i)$ is a **congruent transference**. Together with A every representation (or transformation) AG_0, in which A is followed by a rotation G_0 at P_0, is likewise a congruent transference; thus, from one congruent transference A of the vector body from P_0 to P, we get all possible ones by making G_0 traverse the group of rotations belonging to P_0. If we consider the vector body belonging to the centre P_0 for two positions congruent to one another, they will resolve into two congruent positions at P if subjected to the same congruent transference A; for this reason, the group of rotations $\mathbf{G}$ at P is equal to $A\mathbf{G}_0 A^{-1}$. The metrical relationship thus tells us that the group of rotations at P differs from that at P_0 only in orientation. If we pass continuously from the point P_0 to any point of the manifold, we see that the groups of rotation are of a similar kind at all points of the manifold; thus there is homogeneity in this respect.

The only congruent transferences that we take into consideration are those in which the vector components ξ^i undergo changes $d\xi^i$ that are infinitesimal and of the same order as the displacement of the centre P_0,

$$d\xi^i = d\lambda_k^i \cdot \xi^k.$$

If L and M are two such transferences from P_0 to P, with coefficients $d\lambda_k^i$ and $d\mu_k^i$ respectively, then the rotation ML^{-1} is likewise infinitesimal: it is represented by the formula

$$d\xi^i = d\alpha_k^i \cdot \xi^k \quad \text{where} \quad d\alpha_k^i = d\mu_k^i - d\lambda_k^i. \tag{68}$$

The following will also be true. If an infinitesimal congruent transference consisting in the displacement (dx_i) of the centre P_0

is succeeded by one in which the centre is displaced by (δx_i), we get a congruent transference that is effected by the resultant displacement $dx_i + \delta x_i$ of the centre (plus an error which is infinitesimal compared with the magnitude of the displacements). Hence, if for the transition from $P_0 = (x_1^0, x_2^0, \ldots, x_n^0)$ to the point $(x_1^0 + \epsilon, x_2^0, \ldots, x_n^0)$, this being an infinitesimal change ϵ in the direction of the first co-ordinate axis,

$$d\xi^i = \epsilon \cdot \Lambda^i_k \xi^k$$

is a congruent transference, and if $\Lambda^i_{k2}, \ldots, \Lambda^i_{kn}$ have a corresponding meaning for the displacements of P_0 in the direction of the 2nd up to the nth co-ordinate in turn; then the equation

$$d\xi^i = \Lambda^i_{kr}\, dx_r \cdot \xi^k \tag{69}$$

gives a congruent transference for an arbitrary displacement having components dx_i.

Among the various kinds of metrical spaces we shall now designate by simple intrinsic relations the category to which, according to Pythagoras' and Riemann's ideas, real space belongs. The group of rotations that does not vary with position exhibits a property that belongs to space as a form of phenomena; it characterises the metrical nature of space. The metrical relationship,* from point to point, however, is *not* determined by the nature of space, nor by the mutual orientation of the groups of rotation at the various points of the manifold. The metrical relationship is dependent rather on the disposition of the material content, and is thus in itself free and capable of any "virtual" changes. We shall formulate the fact that it is subject to no limitation as our first axiom.

---

*Although, as will be shown later, it is everywhere of the same kind.

I. The Nature of Space Imposes no Restriction on the Metrical Relationship

It is **possible** to find a metrical relationship in space between the point P_0 and the points in its neighbourhood such that the formula (69) represents a system of congruent transferences to these neighbouring points **for arbitrarily given numbers** Λ^i_{kr}.

Corresponding to every co-ordinate system x_i at P_0 there is a possible conception of parallel displacement, namely, the displacement of the vectors from P_0 to the infinitely near points without the components undergoing a change in this co-ordinate system. Such a system of parallel displacements of the vector body from P_0 to all the infinitely near points is expressed, as we know, in terms of a definite co-ordinate system, selected once and for all by the formula

$$d\xi^i = -d\gamma^i \cdot \xi^k$$

in which the differential forms $d\gamma^i_k = \Gamma^i_{kr}\, dx_r$ satisfy the condition of symmetry

$$\Gamma^i_{kr} = \Gamma^i_{rk}. \tag{70}$$

And, indeed, a possible conception of parallel displacement corresponds to every system of symmetrical co-efficients Γ. For a given metrical relationship the further restriction that the "parallel displacements" shall simultaneously be congruent transferences must be imposed. The second postulate is the one enunciated above as the fundamental theorem of infinitesimal geometry; for a given metrical relationship there is always a **single** system of parallel displacements among the transferences of the vector body. We treated affine relationship in § 15 only provisionally as a rudimentary characteristic of space; the truth is, however, that parallel displacements, in virtue of their inherent properties, must be excluded from congruent transferences, and that the conception of parallel displacement is determined by the metrical relationship. This postulate may be enunciated thus:—

II. The Affine Relationship is Uniquely Determined by the Metrical Relationship

Before we can formulate it analytically we must deal with infinitesimal rotations. A continuous group **G** of r members is a continuous r-dimensional manifold of matrices. If $s_1, s_2, \ldots, s_r$ are co-ordinates in this manifold, then, corresponding to every value system of the co-ordinates there is a matrix $A(s_1, s_2, \ldots, s_r)$ of the group which depends on the value-system continuously. There is a definite value-system—we may assume for it that $s_1 = 0$—to which **identity**, E, corresponds. The matrices of the group that are infinitely near E differ from E by

$$\mathsf{A}_1\, ds_1 + \mathsf{A}_2\, ds_2 + \ldots + \mathsf{A}_r\, ds_r,$$

in which $\mathsf{A}_i = \left(\dfrac{\partial A}{\partial s_i}\right)_0$. We call a matrix A an infinitesimal operation of the group if the group contains a transformation (independent of ϵ) that coincides with E and $\epsilon\mathsf{A}$ to within an error that converges more rapidly towards zero than ϵ, for decreasing small values of ϵ. The infinitesimal operations of the group form the linear family

$$\mathbf{g}:\ \lambda_1\mathsf{A}_1 + \lambda_2\mathsf{A}_2 + \cdots + \lambda_r\mathsf{A}_r \quad (\lambda \text{ being arbitrary numbers}) \tag{71}$$

$\mathbf{g}$ is exactly r-dimensional and the A's are linearly independent of one another. For if A is an arbitrary matrix of the group, the group property expresses the transformations of the group which are infinitely near A in the formula $A(E+\epsilon\mathsf{A})$, in which ϵ is an infinitesimal factor and A traverses the group $\mathbf{g}$. If $\mathbf{g}$ were of less dimensions than r, the same would hold at each point of the manifold; for all values of s_i there would be linear relations between the derivatives $\frac{\partial A}{\partial s_i}$, and A would in reality depend on less than r parameters. The infinitesimal operations generate and determine the whole group. If we carry out the infinitesimal transformation $E+\frac{1}{n}\mathsf{A}$ (n being an infinitely great number) n times successively, we get a matrix (of the group) that is finite and different from E, namely,

$$A=\lim_{n\to\infty}\left(E+\frac{1}{n}\mathsf{A}\right)^n=E+\frac{\mathsf{A}}{1!}+\frac{\mathsf{A}^2}{2!}+\frac{\mathsf{A}^3}{3!}+\ldots;$$

and thus we get every matrix of the group (or at least every one that may be reached continuously in the group, by starting from identity) if we make A traverse the whole family $\mathbf{g}$. Not every arbitrarily given linear family(71) gives a group in this way, but only those in which the A's satisfy a certain condition of integrability. The latter is obtained by a method quite analogous to that by which, for example, the condition of integrability is obtained for parallel displacement in Euclidean space. If we pass from **Identity**, $E(s_i=0)$, by an infinitesimal change ds_i of the parameters, to the neighbouring matrix $A_d=E+dA$, and thence by a second infinitesimal change δs_i, from A_δ to $A_\delta A_d$ and then reverse these two operations whilst preserving the same order, we get $A_\delta^{-1}A_d^{-1}A_\delta A_d$, a matrix (of the group) differing by an infinitely small amount from E. Let d be the change in the direction of the first co-ordinate, and δ that in the direction of the second, then we are dealing with the matrix

$$A_{st} = A_t^{-1} A_s^{-1} A_t A_s$$

formed from

$$A_s = A(s, 0, 0, \ldots, 0) \quad \text{and} \quad A_t = A(0, t, 0, \ldots, 0).$$

Now, $A_{s0} = A_{0t} = E$, hence

$$\lim_{s \to 0, t \to 0} \frac{A_{st} - E}{s \cdot t} = \left(\frac{\partial^2 A_{st}}{\partial s \, \partial t} \right)_{\substack{s \to 0 \\ t \to 0}}.$$

Since A_{st} belongs to the group, this limit is an infinitesimal operation of the group. We find, however, that

$$\frac{\partial A_{st}}{\partial t} = -\mathsf{A}_2 + A_s^{-1} \mathsf{A}_2 A_s \quad \text{for} \quad t = 0;$$

leading to

$$\frac{\partial^2 A_{st}}{\partial s \, \partial t} = -\mathsf{A}_1 \mathsf{A}_2 + \mathsf{A}_2 \mathsf{A}_1 \quad \text{for} \quad t \to 0, s \to 0.$$

Accordingly $\mathsf{A}_1\mathsf{A}_2 - \mathsf{A}_2\mathsf{A}_1$, or, more generally, $\mathsf{A}_i\mathsf{A}_k - \mathsf{A}_k\mathsf{A}_i$ must be an infinitesimal operation of the group: or, what amounts to the same thing, if A and B are two infinitesimal operations of the group, then $\mathsf{AB} - \mathsf{BA}$ must also always be one. Sophus Lie, to whom we are indebted for the fundamental conceptions and facts of the theory of continuous transformation groups (*vide* note 12), has shown that this condition of integrability is not only necessary but also sufficient. Hence we may define an *r-dimensional linear family of matrices as an infinitesimal* *group having r members if, whenever any two matrices* A *and* B *belong to the family,* $\mathsf{AB} - \mathsf{BA}$ *also belongs to the family.* By introducing the infinitesimal operations of the group, the problem of continuous transformation groups becomes a linear question.

If all the transformations of the group leave the elements of volume unaltered, the "traces" of the infinitesimal operations = 0. For the development of the determinant of $E + \epsilon\mathsf{A}$ in powers of ϵ begins with the members $1 + \epsilon \cdot \text{trace}(\mathsf{A})$. U is a similar transformation, if, for every G of the group of rotations, UGU^{-1} or, what comes to the same thing, $UGU^{-1}G^{-1}$, belongs to the group of rotations $\mathbf{G}$. Accordingly, A_0^* is an infinitesimal operation of the group of similar transformations if, and only if, $\mathsf{A}_0^*\mathsf{A} - \mathsf{A}\mathsf{A}_0^*$ also belongs to $\mathbf{g}$, no matter which of the matrices A of the group of infinitesimal rotations is used.

The infinitesimal Euclidean rotations

$$d\xi^i = v_k^i \xi^k,$$

that is, the infinitesimal linear transformations that leave the unit quadratic form

$$Q_0 = (\xi^1)^2 + (\xi^2)^2 + \cdots + (\xi^n)^2$$

invariant, were determined on page 69. The condition which characterises them, namely,

$$\tfrac{1}{2}dQ_0 = \xi^i \, d\xi^i = 0, \quad \text{implies that} \quad v_i^k = -v_k^i.$$

Thus it is seen that we are dealing with the infinitesimal group δ of all skew-symmetrical matrices; it obviously has $\dfrac{n(n-1)}{2}$ members. It may be left to the reader to verify by direct calculation that it possesses the group property. If Q is any quadratic form that remains invariant during the infinitesimal Euclidean rotations, i.e. $dQ = 0$, then Q necessarily coincides with Q_0 except for a constant factor. Indeed, if

$$Q = a_{ik}\xi^i\xi^k \qquad (a_{ki} = a_{ik})$$

then for all skew-symmetrical number systems v_k^i the equation

$$a_{rk}v_i^k + a_{ri}v_k^r = 0 \tag{72}$$

must hold. If we assume $k = i$ and notice that the numbers $v_i^1, v_i^2, \ldots, v_i^n$ may be chosen arbitrarily for each particular i, excepting the case $v_i^i = 0$, we get $a_{ri} = 0$ for $r \neq i$. If we write a_{ii} for a_i, equation (72) becomes

$$v_i^k(a_i - a_k) = 0$$

from which we immediately deduce that all a_i's are equal. The corresponding group δ^* of similar transformations is derived from δ by "associating" the single matrix E; this here signifies $d\xi^i = \epsilon\xi^i$. For if the matrix $C = (c_i^k)$ belongs to δ^*, that is, if for every skew-symmetrical v_i^k, $c_r^i v_k^r - v_r^i c_k^r$ is also a skew-symmetrical number system, then the quantities $c_k^i + c_i^k = a_{ik}$ satisfy equation (72); whence it follows that $a_{ik} = 2a \cdot \delta_i^k$; that is, C is equal to aE *plus* a skew-symmetrical matrix.

More generally, let δ_Q denote the infinitesimal group of linear transformations that transform an arbitrary non-degenerate quadratic form Q into itself. δ_Q and $\delta_{Q'}$ are distinguished only by their orientation, if Q' is generated from Q by a linear transformation. Hence there are only a finite number of different kinds of infinitesimal groups δ_Q that differ from one another in the inertial index attached to the form Q. But even these differences are eliminated if, instead of confining ourselves to the realm of real quantities, we use that of complex members; in that case, every δ_Q is of the same type as δ.

These preliminary remarks enable us to formulate analytically the two postulates I and II. Let $\mathbf{g}$ be the group of infinitesimal rotations at P. We take Λ^i_{kr} to denote every system of n^3 numbers, A^i_{kr} to denote every system that is composed of matrices $(\mathsf{A}^i_{k1}), (\mathsf{A}^i_{k2}), \ldots, (\mathsf{A}^i_{kn})$ belonging to $\mathbf{g}$ and Γ^i_{kr} to denote an arbitrary system of numbers that satisfies the condition of symmetry (70). If the group of infinitesimal rotations has N members, these member systems form linear manifolds of n^3, nN and $n \cdot \dfrac{n(n+1)}{2}$ dimensions respectively. Since, according to I, if the metrical relationship runs through all possible values, any arbitrary number systems $\Lambda^i_{k1}, \Lambda^i_{k2}, \ldots, \Lambda^i_{kn}$ may occur as the coefficients of n infinitesimal congruent transferences in the n coordinate directions (cf. (69)), then, by II (cf. (68)) each Λ must be capable of resolution in one and only one way according to the formula

$$\Lambda^i_{kr} = \mathsf{A}^i_{kr} - \Gamma^i_{kr}.$$

This entails two results

1. $\quad n^3 = nN + n \cdot \dfrac{n(n+1)}{2} \quad$ or $\quad N = \dfrac{n(n-1)}{2};$

2. $\mathsf{A}^i_{kr} - \Gamma^i_{kr}$ is never equal to zero, unless all the A's and Γ's vanish; or, a non-vanishing system A can never fulfil the condition of symmetry, $\mathsf{A}^i_{kr} = \mathsf{A}^i_{rk}$. To enable us to formulate this condition invariantly let us define a symmetrical double matrix (an infinitesimal double rotation) belonging to $\mathbf{g}$ as a law expressed

by

$$\zeta^i = \mathsf{A}^i_{rs}\xi^r\eta^s \qquad (\mathsf{A}^i_{rs} = \mathsf{A}^i_{sr}),$$

which produces from two arbitrary vectors, ξ and η, a vector ζ as a bilinear symmetrical form, provided that for every fixed vector η, the transition $\xi \to \zeta$ (and hence also for every fixed vector ξ the transition $\eta \to \zeta$) is an operation of **g**. We may then summarise our results thus:—

The group of infinitesimal rotations has the following properties according to our axioms:

(*a*) *The trace of every matrix* $= 0$;

(*b*) *No symmetrical double matrix belongs to* **g** *except zero;*

(*c*) *The dimensional number of* **g** *is the highest that is still in agreement with postulate* (*b*), *namely,* $N = \dfrac{n(n-1)}{2}$.

These properties retain their meaning for complex quantities as well as for real ones. We shall just verify that they are true of the infinitesimal Euclidean group of rotations δ, that is, that n^3 numbers v^i_{kl} cannot simultaneously satisfy the conditions of symmetry

$$v^i_{lk} = v^i_{kl}, \qquad v^k_{il} = -v^i_{kl},$$

without all of them vanishing. This is evident from the calculation which was undertaken on page 184 to determine the affine relationship. For if we write down the three equations that we get from $v^i_{kl} + v^k_{il} = 0$ by interchanging the indices i, k, l cyclically, and then subtract the second from the sum of the first and the third, we get, as a result of the first condition of symmetry, $v^i_{kl} = 0$.

It seems highly probable to the author that δ is the only infinitesimal group that satisfies the postulates (*a*), (*b*), and (*c*); or, more exactly, in the case of complex quantities every such infinitesimal group may be made to coincide with δ by choosing the appropriate co-ordinate system. If this is true, then the group of infinitesimal rotations must be identical with a certain group δ_Q, in which Q is a non-degenerate quadratic form. Q itself is determined by **g** except for a constant of proportionality. It is real if **g** is real. For if we split Q (in which the variables are taken as real) into a real and an imaginary part $Q_1 + iQ_2$, then **g** leaves both these forms Q_1 and Q_2 invariant. Hence we must have

$$Q_1 = c_1 Q \qquad Q_2 = c_2 Q.$$

One of these two constants is certainly different from zero, since $c_1 + ic_2 = 1$, and hence Q must be a real form excepting for a constant factor. This would link up with the line of argument followed in the preceding paragraph and would complete the Analysis of Space; we should then be able to claim to have made intelligible the nature of space and the source of the validity of Pythagoras' Theorem, by having explored the ultimate grounds accessible to mathematical reasoning (*vide* note 13). If the supposed mathematical proposition is not true, definite characteristics and essentials of space will yet have escaped us. The author has proved that the proposition holds actually for the lowest dimensional numbers $n = 2$ and $n = 3$. It would lead too far to present these purely mathematical considerations here.

In conclusion, it will be advisable to call attention to two points. Firstly, axiom I is in no wise contradicted by the result of axiom II which states that not only the metrical structure, but also the metrical relationship is of the same kind at every point, namely, of the simplest type imaginable. For every point there is a geodetic co-ordinate system such that the shifting of all vectors at that point, which leaves its components unaltered, to a neighbouring point is always a congruent transference. Secondly, the possibility of grasping the unique significance of the metrical structure of Pythagorean space in the way here outlined depends solely on the circumstance that the quantitative metrical conditions admit of considerable virtual changes. This possibility stands or falls with the dynamical view of Riemann. It is this view, the truth of which can scarcely be doubted after the success that has attended Einstein's Theory of Gravitation (Chapter IV), that opens up the road leading to the discovery of the "Rationality of Space".

The investigations about space that have been conducted in Chapter II seemed to the author to offer a good example of the kind of analysis of the modes of existence (*Wesensanalyse*) which is the object of Husserl's phenomenological philosophy, an example that is typical of cases in which we are concerned with non-immanent modes. The historical development of the problem of space teaches how difficult it is for us human beings entangled in external reality to reach a definite conclusion. A prolonged phase of mathematical development, the great expansion of geometry dating from Euclid to Riemann, the discovery of the physical facts of nature and their underlying laws from the time of Galilei, together with the incessant impulses imparted by new empirical data, finally the genius of individual great minds—Newton, Gauss, Riemann, Einstein—all these factors were necessary to set us free from the external, accidental, non-essential characteristics which would otherwise have held us captive. Certainly, once the true point of view has been adopted reason becomes flooded with light, and it recognises and appreciates what is of itself intelligible to it. Nevertheless, although reason was, so to speak, always conscious of this point of view in the whole development of the problem, it had not the power to penetrate into it with one flash. This reproach must be directed at the impatience of those philosophers who believe it possible to describe adequately the mode of existence on the basis of a single act of typical presentation (*exemplarischer Vergegenwärtigung*): in principle they are right: yet from the point of view of human nature, how utterly they are wrong! The problem of space is at the same time a very instructive example of that question of phenomenology that seems to the author to be of greatest consequence, namely, in how far the delimitation of the essentialities perceptible in consciousness expresses the structure peculiar to the realm of presented objects, and in how far mere convention participates in this delimitation.

CHAPTER III

RELATIVITY OF SPACE AND TIME

§ 19. Galilei's Principle of Relativity

WE have already discussed in the introduction how it is possible to measure time by means of a clock and how, after an arbitrary initial point of time and a time-unit has been chosen, it is possible to characterise every point of time by a number t. But the **union of space and time** gives rise to difficult further problems that are treated in the theory of relativity. The solution of these problems, which is one of the greatest feats in the history of the human intellect, is associated above all with the names of **Copernicus** and **Einstein** (*vide* note 1).

By means of a clock we fix directly the time-conditions of only such events as occur just at the locality at which the clock happens to be situated. Inasmuch as I, as an unenlightened being, fix, without hesitation, the things that I see into the moment of their perception, I extend my time over the whole world. I believe that there is an objective meaning in saying of an event which is happening somewhere that it is happening "now" (at the moment at which I pronounce the word!); and that there is an objective meaning in asking which of two events that have happened at different places has occurred earlier or later than the other. **We shall for the present accept the point of view implied in these assumptions.** Every space-time event that is strictly localised, such as the flash of a spark that is instantaneously extinguished, occurs at a definite space-time-point or **world-point**, "here-now". As a result of the point of view enunciated above, to every world-point there corresponds a definite time-co-ordinate t.

We are next concerned with fixing the position of such a point-event in space. For example, we ascribe to two point-masses a distance separating them at a definite moment. We assume that the world-points corresponding to a definite moment t form a three-dimensional point-manifold for which Euclidean geometry holds. (In the present chapter we adopt the view of space set forth in Chapter I.) We choose a definite unit of length and a rectangular co-ordinate system at the moment t (such as the corner of a room). Every world-point whose time-co-ordinate is t then has three definite space-co-ordinates x_1, x_2, x_3.

Let us now fix our attention on another moment t'. We assume that there is a definite objective meaning in stating that measurements are carried out at the moment t' with the same unit length as that used at the moment t (by means of a "rigid" measuring staff that exists both at the time t and at the time t'). In addition to the unit of time we shall adopt a unit of length fixed once and for all (centimetre, second). We are then still free to choose the position of the Cartesian co-ordinate system independently of the choice of time t. Only when we believe that there is objective meaning in stating that two point-events happening at arbitrary moments take place at the **same** point of space, and in saying that a body is **at rest**, are we able to fix the position of the co-ordinate system for all times on the basis of the position chosen arbitrarily at a certain moment, without having to specify additional "individual objects"; that is, we accept the postulate that the co-ordinate system remains permanently at rest. After choosing an initial point in the time-scale and a definite co-ordinate system at this initial moment we then get four definite co-ordinates for every world-point. To be able to represent conditions graphically we suppress one space-co-ordinate, assuming

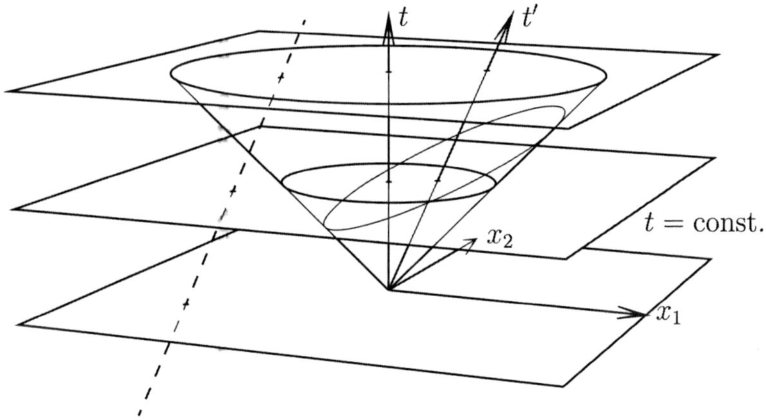

FIG. 7.

space to be only two-dimensional, a Euclidean plane.

We construct a graphical picture by representing in a space carrying the rectangular set of axes (x_1, x_2, t) the world-point by a "picture"-point with co-ordinates (x_1, x_2, t). We can then trace out graphically the "time-table" of all moving point-masses; the motion of each is represented by a "world-line," whose direction has always a positive component in the direction of the t-axis. The world-lines of point-masses that are at rest are parallels to the t-axis. The world-line of a point-mass which is in uniform translation is a straight line. On a section $t =$ constant we may read off the position of all the point-masses at the same time t. If we choose an initial point in the time-scale and also some other Cartesian co-ordinate system, and if (x_1, x_2, t), (x_1', x_2', t') are the co-ordinates of an arbitrary world-point in the first and second co-ordinate system respectively, the transformation formulæ

$$\left.\begin{aligned} x_1 &= \alpha_{11}x'_1 + \alpha_{12}x'_2 + \alpha_1 \\ x_2 &= \alpha_{21}x'_1 + \alpha_{22}x'_2 + \alpha_2 \\ t &= \qquad\qquad\quad t' + a \end{aligned}\right\} \qquad \text{(I)}$$

hold; in them, the α_i's and the a denote constants, the α_{ik}'s, in particular, are the co-efficients of an orthogonal transformation. The world-co-ordinates are thus fixed **except for an arbitrary transformation of this kind** in an objective manner without individual objects or events being specified. In this we have not yet taken into consideration the arbitrary choice of both units of measure. If the initial point remains unchanged both in space and in time, so that $\alpha_1 = \alpha_2 = a = 0$, then (x'_1, x'_2, t') are the co-ordinates with respect to a rectilinear system of axes whose t' axis coincides with the t-axis, whereas the axes x'_1, x'_2 are derived from x_1, x_2 by a rotation in their plane $t = 0$.

A moment's reflection suffices to show that one of the assumptions adopted is not true, namely, the one which states that the conception of rest has an objective content.* When I arrange to meet some one at the same place to-morrow as that at which we met to-day, this means in the same material surroundings, at the same building in the same street (which, according to Copernicus, may be in a totally different part of stellar space to-morrow). All this acquires meaning as a result of the fortunate circumstance that at birth we are introduced into an essentially stable world, in which changes occur in conjunction with a comparatively much more comprehensive set of permanent factors that preserve

*Even Aristotle was clear on this point, for he denotes "place" (τόπος) as the relation of a body to the bodies in its neighbourhood.

their constitution (which is partly perceived directly and partly deduced) unchanged or almost unchanged. The houses stand still; ships travel at so and so many knots: these things are always understood in ordinary life as referring to the firm ground on which we stand. **Only the motions of bodies (point-masses) relative to one another have an objective meaning**, that is, the distances and angles that are determined from simultaneous positions of the point-masses and their functional relation to the time-co-ordinate. The connection between the co-ordinates of the same world-point expressed in two different systems of this kind is given by formulæ

$$\left.\begin{aligned} x_1 &= \alpha_{11}(t')x_1' + \alpha_{12}(t')x_2' + \alpha_1(t') \\ x_2 &= \alpha_{21}(t')x_1' + \alpha_{22}(t')x_2' + \alpha_2(t') \\ t &= t' + a \end{aligned}\right\} \qquad \text{(II)}$$

in which the α_i's and α_{ik}'s may be any continuous functions of t', and the α_{ik}'s are the co-efficients of an orthogonal transformation for all values of t'. If we map out the surfaces $t' =$ const., as also $x_1' =$ const. and $x_2' =$ const. by our graphical method, then the surfaces of the first family are again planes that coincide with the planes $t =$ const.; on the other hand, the other two families of surfaces are curved surfaces. The transformation formulæ are no longer linear.

Under these circumstances we achieve an important aim, when investigating the motion of systems of point-masses, such as planets, by choosing the co-ordinate system so that the functions $x_1(t)$, $x_2(t)$ that express how the space-co-ordinates of the point-masses depend on the time become as simple as possible or at least satisfy laws of the greatest possible simplicity. This is the substance of the discovery of Copernicus that was afterwards elaborated to such an extraordinary degree by Kepler, namely, that there is in fact a co-ordinate system for which the laws of planetary motion assume a much simpler and more expressive form than if they are referred to a motionless earth. The work of Copernicus produced a revolution in the philosophic ideas about the world inasmuch as **he shattered the belief in the absolute importance of the earth**. His reflections as well as those of Kepler are purely **kinematical** in character. Newton crowned their work by discovering the true ground of the kinematical laws of Kepler to lie in the fundamental **dynamical** law of mechanics and in the law of attraction. Every one knows how brilliantly the mechanics of Newton has been confirmed both for celestial as well as for earthly phenomena. As we are convinced that it is valid universally and not only for planetary systems, and as its laws are by no means invariant with respect to the transformations (II), it enables us to fix the co-ordinate system in a manner independent of all individual specification and much more definitely than is possible on the kinematical view to which the principle of relativity (II) leads.

Galilei's Principle of Inertia (Newton's First Law of Motion) forms the foundation of mechanics. It states that a point-mass which is subject to no forces from without executes a uniform translation. Its world-line is consequently a straight line, and the space-co-ordinates x_1, x_2 of the point-mass are linear functions of the time t. If this principle holds for the two co-ordinate systems connected by (II), then x_1 and x_2 must become linear functions of t', when linear functions of t' are substituted for x'_1 and x'_2. It straightway follows from this that the α_{ik}'s must be constants, and that α_1 and α_2 must be linear functions of t; that is, the one Cartesian co-ordinate system (in space) must be moving uniformly in a straight line relatively to the other co-ordinate system. Conversely, it is easily shown that if $\mathbf{C}_1$, $\mathbf{C}_2$ are two **such** co-ordinate systems, then if the principle of inertia and Newtonian mechanics holds for $\mathbf{C}$ it will also hold for $\mathbf{C}'$. Thus, in mechanics, any two "allowable" co-ordinate systems are connected by formulæ

$$\left.\begin{aligned} x_1 &= \alpha_{11}x_1' + \alpha_{12}x_2' + \gamma_1 t' + \alpha_1 \\ x_2 &= \alpha_{21}x_1' + \alpha_{22}x_2' + \gamma_2 t' + \alpha_2 \\ t &= \qquad\qquad\qquad\quad\; t' + a \end{aligned}\right\} \qquad \text{(III)}$$

in which the α_{ik}'s are constant co-efficients of an orthogonal transformation, and a, α_i and γ_i are arbitrary constants. Every transformation of this kind represents a transition from one allowable co-ordinate system to another. (This is the **Principle of Relativity of Galilei and Newton**.) The essential feature of this transition is that, if we disregard the naturally arbitrary directions of the axis in space and the arbitrary initial point, there is invariance with respect to the transformations

$$x_1 = x_1' + \gamma_1 t', \qquad x_2 = x_2' + \gamma_2 t', \qquad t = t'. \qquad (1)$$

In our graphical representation (*vide* Fig. 7) x_1', x_2', t' would be the co-ordinates taken with respect to a rectilinear set of axes in which the x_1'-, x_2'-axes coincide with the x_1-, x_2-axes, whereas the new t'-axis has some new direction. The following considerations show that the laws of Newtonian mechanics are not altered in passing from one co-ordinate system $\mathbf{C}$ to another $\mathbf{C}'$. According to the law of attraction the gravitational force with which one point-mass acts on another at a certain moment is a vector, in space, which is independent of the co-ordinate system (as is also the vector that connects the simultaneous positions of both point-masses with one another). Every force, no matter what its physical origin, must be the same kind of magnitude; this is entailed in the assumptions of Newtonian mechanics, which demands a physics that satisfies this assumption in order to be able to give a content to its conception of force. We may prove, for example, in the theory of elasticity that the stresses (as a consequence of their relationship to deformation quantities) are of the required kind.

Mass is a scalar that is independent of the co-ordinate system. Finally, on account of the transformation formulæ that result from (1) for the motion of a point-mass,

$$\frac{dx_1}{dt} = \frac{dx_1'}{dt'} + \gamma_1, \quad \frac{dx_2}{dt} = \frac{dx_2'}{dt'} + \gamma_2; \qquad \frac{d^2x_1}{dt^2} = \frac{d^2x_1'}{dt'^2}, \quad \frac{d^2x_2}{dt^2} = \frac{d^2x_2'}{dt'^2}$$

not the velocity, but the acceleration is a vector (in space) independent of the co-ordinate system. Accordingly, the fundamental law: **mass** times **acceleration** = **force**, has the required invariant property.

According to Newtonian mechanics the centre of inertia of every isolated mass-system not subject to external forces moves in a straight line. If we regard the sun and his planets as such a system, there is no meaning in asking whether the centre of inertia of the solar system is at rest or is moving with uniform translation. The fact that astronomers, nevertheless, assert that the sun is moving towards a point in the constellation of Hercules, is based on the statistical observation that the stars in that region seem on the average to diverge from a certain centre—just as a cluster of trees appears to diverge as we approach them. If it is certain that the stars are on the average at rest, that is, that the centre of inertia of the stellar firmament is at rest, the statement about the sun's motion follows. It is thus merely an assertion about the relative motion of the centre of inertia and of that of the stellar firmament.

To grasp the true meaning of the principle of relativity, one must get accustomed to thinking not in "space," nor in "time," but "in the world," that is in **space-time**. Only the coincidence (or the immediate succession) of two events in space-time has a meaning that is directly evident, it is just the fact that in these cases space and time cannot be dissociated from one another absolutely that is asserted by the principle of relativity. Following the mechanistic view, according to which all physical happening can be traced back to mechanics, we shall assume that not only mechanics but the whole of the physical uniformity of Nature is subject to the principle of relativity laid down by Galilei and Newton, which states *that it is impossible to single out from the systems of reference that are equivalent for mechanics and of which each two are correlated by the formula of transformation* (III) *special systems without specifying* **individual objects**. These formulæ condition **the geometry of the four-dimensional world** in exactly the same way as the group of transformation substitutions connecting two Cartesian co-ordinate systems condition the Euclidean geometry of three-dimensional space. A relation between world-points has an objective meaning if, and only if, it is defined by such arithmetical relations between the co-ordinates of the points as are invariant with respect to the transformations (III). Space is said to be **homogeneous** at all points and homogeneous in all directions at every point. These assertions are, however, only parts of the **complete statement of homogeneity** that all Cartesian co-ordinate systems are equivalent. In the same way the principle of relativity determines exactly the sense in which the *world* (= space-time as the "form" of phenomena, not its "accidental" non-homogeneous material content) is homogeneous.

It is indeed remarkable that two mechanical events that are fully alike kinematically, may be different dynamically, as a comparison of the dynamical principle of relativity (III) with the much more general kinematical principle of relativity (II) teaches us. A rotating spherical mass of fluid existing all alone, or a rotating fly-wheel, cannot in itself be distinguished from a spherical fluid mass or a fly-wheel at rest; in spite of this the "rotating" sphere becomes flattened, whereas the one at rest does not change its shape, and stresses are called up in the rotating fly-wheel that cause it to burst asunder, if the rate of rotation be sufficiently great, whereas no such effect occurs in the case of a fly-wheel which is at rest. The cause of this varying behaviour can be found only in the "metrical structure of the world," that reveals itself in the centrifugal forces as an active agent. This sheds light on the idea quoted from Riemann above; if there corresponds to metrical structure (in this case that of the world and not the fundamental metrical tensor of space) something just as real, which acts on matter by means of forces, as the something which corresponds to Maxwell's stress tensor, then we must assume that, conversely, matter also reacts on this real something. We shall revert to this idea again later in Chapter IV.

For the present we shall call attention only to the linear character of the transformation formulæ (III); this signifies that **the world is a four-dimensional affine space**. To give a systematic account of its geometry we accordingly use **world-vectors** or displacements in addition to world-points. A displacement of the world is a transformation that assigns to every world-point P a world-point P', and is characterised by being expressible in an allowable co-ordinate system by means of equations of the form

$$x'_i = x_i + \alpha_i \qquad (i = 0, 1, 2, 3)$$

in which the x_i's denote the four space-time-co-ordinates of P (t being represented by x_0), and the x'_i's are those of P' in this co-ordinate system, whereas the α_i's are constants. This conception is independent of the allowable co-ordinate system selected. The displacement that transforms P into P' (or transfers P to P') is denoted by $\overrightarrow{PP'}$. The world-points and displacements satisfy all the axioms of the affine geometry whose dimensional number is $n = 4$. Galilei's Principle of Inertia (Newton's First Law of Motion) is an affine law; it states what motions realise the straight lines of our four-dimensional affine space ("world"), namely, those executed by point-masses moving under no forces.

From the **affine** point of view we pass on to the **metrical** one. From the graphical picture, which gave us an affine view of the world (one co-ordinate being suppressed), we can read off its essential metrical structure; this is quite different from that of Euclidean space. The world is "stratified"; the planes, $t =$ const., in it have an absolute meaning. After a unit of time has been chosen, each two world-points A and B have a definite time-difference, the time-component of the vector $\overrightarrow{AB} = \mathbf{x}$; as is generally the case with vector-components in an affine co-ordinate system, the time-component is a linear form $t(\mathbf{x})$ of the arbitrary vector $\mathbf{x}$. The vector $\mathbf{x}$ points into the past or the future according as $t(\mathbf{x})$ is negative or positive. Of two world-points A and B, A is earlier than, simultaneous with, or later than B, according as

$$t(\overrightarrow{AB}) > 0, \ = 0, \ \text{or} \ < 0.$$

Euclidean geometry, however, holds in each "stratum"; it is based on a definite quadratic form, which is in this case defined only for those world-vectors $\mathbf{x}$ that lie in one and the same stratum, that is, that satisfy the equation $t(\mathbf{x}) = 0$ (for there is sense only in speaking of the distance between **simultaneous** positions of two point-masses). Whereas, then, the **metrical structure** of Euclidean geometry is based on a definitely positive quadratic form, that **of Galilean geometry is based on**

1. *A linear form* $t(\mathbf{x})$ *of the arbitrary vector* $\mathbf{x}$ (the "duration" of the displacement $\mathbf{x}$).

2. *A definitely positive quadratic form* $(\mathbf{x}, \mathbf{x})$ (the square of the "length" of $\mathbf{x}$), *which is defined only for the three-dimensional linear manifold of all the vectors* $\mathbf{x}$ *that satisfy the equation* $t(\mathbf{x}) = 0$.

We cannot do without a definite space of reference, if we wish to form a picture of physical conditions. Such a space depends on the choice of an arbitrary displacement $\mathbf{e}$ in the world (within which the time-axis falls in the picture), and is then defined by the convention that all world-points that lie on a straight line of direction $\mathbf{e}$, meet at the **same point of space**. In geometrical language, we are merely dealing with the process of **parallel projection**. To arrive at an appropriate formulation we shall begin with some geometrical considerations that relate to an arbitrary n-dimensional affine space. To enable us to form a picture of the processes we shall confine ourselves to the case $n = 3$. Let us take a family of straight lines in space all drawn parallel to the vector $\mathbf{e}$ ($\neq \mathbf{0}$). If we look into space along these rays, all the space-points that lie behind one another in the direction of such a straight line would coincide; it is in no wise necessary to specify a plane on to which the points are projected. Hence our definition assumes the following form.

Let $\mathbf{e}$, a vector differing from $\mathbf{0}$, be given. If A and A' are two points such that $\overrightarrow{AA'}$ is a multiple of $\mathbf{e}$, we shall say that they pass into one and the same point $\mathbf{A}$ of the **minor space** defined by $\mathbf{e}$. We may represent $\mathbf{A}$ by the straight line parallel to $\mathbf{e}$, on which all these coincident points A, A', ... in the minor space lie. Since every displacement $\mathbf{x}$ of the space transforms a straight line parallel to $\mathbf{e}$ again into one parallel to $\mathbf{e}$, $\mathbf{x}$ brings about a definite displacement $\boldsymbol{x}$ of the minor space; but each two displacements $\mathbf{x}$ and $\mathbf{x}'$ become coincident in the minor space, if their difference is a multiple of $\mathbf{e}$. We shall denote the transition to the minor space, "the projection in the direction of $\mathbf{e}$," by printing the symbols for points and displacements in heavy oblique type. Projection converts

$$\lambda\mathbf{x},\ \mathbf{x}+\mathbf{y},\ \text{and}\ \overrightarrow{AB}\ \text{into}\ \lambda\boldsymbol{x},\ \boldsymbol{x}+\boldsymbol{y},\ \overrightarrow{\boldsymbol{AB}}$$

that is, the projection has a true affine character; this means that in the minor space affine geometry holds, of which the dimensions are less by one than those of the original "complete" space.

If the space is **metrical** in the Euclidean sense, that is, if it is based on a non-degenerate quadratic form which is its metrical groundform, $Q(\mathbf{x}) = (\mathbf{x}, \mathbf{x})$,—to simplify the picture of the process we shall keep the case for which Q is definitely positive in view, but the line of proof is applicable generally,—then we shall obviously ascribe to the two points of the minor space, which two straight lines parallel to $\mathbf{e}$ appear to be, when we look into the space in the direction of $\mathbf{e}$, a distance equal to the perpendicular distance between the two straight lines. Let us formulate this analytically. The assumption is that $(\mathbf{e}, \mathbf{e}) = e \neq \mathbf{0}$. Every displacement $\mathbf{x}$ may be split up uniquely into two summands

$$\mathbf{x} = \xi\mathbf{e} + \mathbf{x}^*, \tag{2}$$

of which the first is proportional to **e** and the second is perpendicular to it, viz.:—

$$(\mathbf{x}^*, \mathbf{e}) = 0, \qquad \xi = \frac{1}{e}(\mathbf{x}, \mathbf{e}). \tag{3}$$

We shall call ξ the **height** of the displacement $\mathbf{x}$ (it is the difference of height between A and B, if $\mathbf{x} = \overrightarrow{AB}$). We have

$$(\mathbf{x}, \mathbf{x}) = e\xi^2 + (\mathbf{x}^*, \mathbf{x}^*). \tag{4}$$

$\mathbf{x}$ is characterised fully, if its height ξ and the displacement x of the minor space produced by $\mathbf{x}$ are given; we write

$$\mathbf{x} = \xi \mid \mathsf{x}.$$

The "complete" space is "split up" into height and minor space, the "position-difference" $\mathbf{x}$ of two points in the complete space is split up into the difference of height ξ, and the difference of position x in the minor space. There is a meaning not only in saying that two points in space coincide, but also in saying that two points in the minor space coincide or have the same height, respectively. Every displacement x of the minor space is produced by one **and only one** displacement $\mathbf{x}^*$ of the complete space, this displacement being orthogonal to **e**. The relation between $\mathbf{x}^*$ and x is singly reversible and affine. The defining equation

$$(\mathsf{x}, \mathsf{x}) = (\mathbf{x}^*, \mathbf{x}^*)$$

endows the minor space with a metrical structure that is based on the quadratic groundform (x, x). This converts (4) into the fundamental equation of Pythagoras

$$(\mathbf{x}, \mathbf{x}) = e\xi^2 + (x, x) \tag{5}$$

which, for two displacements, may be generalised in the form

$$(\mathbf{x}, \mathbf{y}) = e\xi\eta + (x, y). \tag{5'}$$

Its symbolic form is clear.

These considerations, in so far as they concern affine space, may be applied directly. The complete space is the four-dimensional world: $\mathbf{e}$ is any vector pointing in the direction of the future: the minor space is what we generally call **space**. Each two world-points that lie on a world-line parallel to $\mathbf{e}$ project into the same space-point. This space-point may be represented graphically by the straight line parallel to $\mathbf{e}$ and may be indicated permanently by a point-mass at rest, that is, one whose world-line is just that straight line. The metrical structure, however, is, according to the Galilean principle of relativity, of a kind different from that we assumed just above. This necessitates the following modifications. Every world-displacement $\mathbf{x}$ has a definite duration $t(\mathbf{x}) = t$ (this takes the place of "height" in our geometrical argument) and produces a displacement x in the minor space; it splits up according to the formula

$$\mathbf{x} = t \mid x$$

corresponding to the resolution into space and time. In particular every space-displacement x may be produced by one and only one world-displacement $\mathbf{x}^*$, which satisfies the equation $t(\mathbf{x}^*) = 0$.

The quadratic form $(\mathbf{x}^*, \mathbf{x}^*)$ as defined for such vectors $\mathbf{x}^*$, impresses on space its Euclidean metrical structure

$$(x, x) = (\mathbf{x}^*, \mathbf{x}^*).$$

The space is dependent on the direction of projection. In actual cases the direction of projection may be fixed by any point-mass moving with uniform translation (or by the centre of mass of a closed isolated mass-system).

We have set forth these details with pedantic accuracy so as to be armed at least with a set of mathematical conceptions which have been sifted into a form that makes them immediately applicable to Einstein's principle of relativity for which our powers of intuition are much more inadequate than for that of Galilei.

To return to the realm of physics. The discovery **that light is propagated with a finite velocity** gave the death-blow to the natural view that things exist simultaneously with their perception. As we possess no means of transmitting time-signals more rapid than light itself (or wireless telegraphy) it is of course impossible to measure the velocity of light by measuring the time that elapses whilst a light-signal emitted from a station A travels to a station B. In 1675 Römer calculated this velocity from the apparent irregularity of the time of revolution of Jupiter's moons, which took place in a period which lasted exactly one year: he argued that it would be absurd to assume a mutual action between the earth and Jupiter's satellites such that the period of the earth's revolution caused a disturbance of so considerable an amount in the satellites. Fizeau confirmed the discovery by measurements carried out on the earth's surface. His method is based on the simple idea of making the transmitting station A and the receiving station B coincide by reflecting the ray, when it reaches B, back to A. According to these measurements we have to assume that the centre of the disturbances is propagated in concentric spheres with a constant velocity c. In our graphical picture (one space-co-ordinate again being suppressed) the propagation of a light-signal emitted at the world-point O is represented by the circular cone depicted, which has the equation

$$c^2t^2 - (x_1^2 + x_2^2) = 0. \tag{6}$$

Every plane given by $t =$ const. cuts the cone in a circle composed of those points which the light-signal has reached at the moment t. The equation (6) is satisfied by all and only by all those world-points reached by the light-signal (provided that $t > 0$). The question again arises on what space of reference this description of the event is based. The **aberration of the stars** shows that, relatively to this reference space, the earth moves in agreement with Newton's theory, that is, that it is identical with an allowable reference space as defined by Newtonian mechanics. The propagation in concentric spheres is, however, certainly not invariant with respect to the Galilei transformations (III); for a t'-axis that is drawn obliquely intersects the planes $t =$ const. at points that are excentric to the circles of propagation. Nevertheless, this cannot be regarded as an objection to Galilei's principle of relativity, if, accepting the ideas that have long held sway in physics, we assume that light is transmitted by a material medium, the **æther**, whose particles are movable with regard to one another. The conditions that obtain in the case of light are exactly similar to those that bring about concentric circles of waves on a surface of water on to which a stone has been dropped. The latter phenomenon certainly does not justify the conclusion that the equations of hydrodynamics are contrary to Galilei's principle of relativity. For the medium itself, the water or the æther respectively, whose particles are at rest with respect to one another, if we neglect the relatively small oscillations, furnishes us with the same system of reference as that to which the statement concerning the concentric transmission is referred.

To bring us into closer touch with this question we shall here insert an account of optics in the theoretical guise that it has preserved since the time of Maxwell under the name of the theory of moving electromagnetic fields.

§ 20. The Electrodynamics of Moving Fields Lorentz's Theorem of Relativity

In passing from stationary electromagnetic fields to moving electromagnetic fields (that is, to those that vary with the time) we have learned the following:—

1. The so-called electric current is actually composed of moving electricity: a charged coil of wire in rotation produces a magnetic field according to the law of Biot and Savart. If ρ is the density of charge, $\mathbf{v}$ the velocity, then clearly the density $\mathbf{s}$ of this convection current $= \rho\mathbf{v}$; yet, if the Biot-Savart Law is to remain valid in the old form, $\mathbf{s}$ must be measured in other units. Thus we must set $\mathbf{s} = \dfrac{\rho\mathbf{v}}{c}$, in which c is a universal constant having the dimensions of a velocity. The experiment carried out by Weber and Kohlrausch, repeated later by Rowland and Eichenwald, gave a value of c that was coincident with that obtained for the velocity of light, within the limits of errors of observation (*vide* note 2). We call $\dfrac{\rho}{c} = \rho'$ the electromagnetic measure of the charge-density and, so as to make the density of electric force $= \rho'\mathbf{E}'$ in electro-

magnetic units, too, we call $\mathbf{E}' = c\mathbf{E}$ the electromagnetic measure of the field-intensity.

2. A moving magnetic field induces a current in a homogeneous wire. It may be determined from the physical law $\mathbf{s} = \sigma\mathbf{E}$ and **Faraday's Law of Induction**; the latter asserts that the induced electromotive force is equal to the time-decrement of the magnetic flux through the conductor; hence we have

$$\int \mathbf{E}'\, d\mathbf{r} = -\frac{d}{dt}\int B_n\, do. \tag{7}$$

On the left there is the line-integral along a closed curve, on the right the surface-integral of the normal components of the magnetic induction $\mathbf{B}$, taken over a surface which fills the curve. The flux of induction through the conducting curve is uniquely determined because

$$\operatorname{div} \mathbf{B} = 0; \tag{8'}$$

that is, there is no real magnetism. By Stokes' Theorem we get from (7) the differential law

$$\operatorname{curl} \mathbf{E} + \frac{1}{c}\frac{\partial \mathbf{B}}{\partial t} = \mathbf{0}. \tag{8}$$

The equation $\operatorname{curl} \mathbf{E} = \mathbf{0}$, which holds for statistical cases, is hence increased by the term $\frac{1}{c}\frac{\partial \mathbf{B}}{\partial t}$ on the left, which is a derivative of the time. All our electro-technical sciences are based on it; thus the necessity for introducing it is justified excellently by actual experience.

3. On the other hand, in Maxwell's time, the term which was added to the fundamental equation of magnetism

$$\operatorname{curl} \mathbf{H} = \mathrm{s} \tag{9}$$

was purely hypothetical. In a moving field, such as in the discharge of a condenser, we cannot have $\operatorname{div}\mathbf{s} = 0$, but in place of it the "equation of continuity"

$$\frac{1}{c}\frac{\partial \rho}{\partial t} + \operatorname{div}\mathbf{s} = 0 \tag{10}$$

must hold. This gives expression to the fact that the current consists of moving electricity. Since $\rho = \operatorname{div}\mathbf{D}$, we find that not $\mathbf{s}$, but $\mathbf{s} + \frac{1}{c}\frac{\partial \mathbf{D}}{\partial t}$ must be irrotational, and this immediately suggests that instead of equation (9) we must write for moving fields

$$\operatorname{curl}\mathbf{H} - \frac{1}{c}\frac{\partial \mathbf{D}}{\partial t} = \mathbf{s}. \tag{11}$$

Besides this, we have just as before

$$\operatorname{div}\mathbf{D} = \rho. \tag{11'}$$

From (11) and (11′) we arrive conversely at the equation of continuity (10). It is owing to the additional member $\frac{1}{c}\frac{\partial \mathbf{D}}{\partial t}$ (Maxwell's **displacement current**), a differential co-efficient with respect to the time, that electromagnetic disturbances are propagated in the æther with the finite velocity c. It is the basis of the electromagnetic theory of light, which interprets optical phenomena with such wonderful success, and which is experimentally verified in the well-known experiments of Hertz and in wireless telegraphy, one of its technical applications. This also makes it clear that these laws are referred to the same reference-space as that for which the concentric propagation of light holds, namely, the "fixed" æther. The laws involving the specific characteristics of the matter under consideration have yet to be added to Maxwell's field-equations (8) and (8′), (11) and (11′).

We shall, however, here consider only the conditions in the æther; in it

$$\mathbf{D} = \mathbf{E} \quad \text{and} \quad \mathbf{H} = \mathbf{B},$$

and Maxwell's equations are

$$\operatorname{curl} \mathbf{E} + \frac{1}{c}\frac{\partial \mathbf{B}}{\partial t} = \mathbf{0}, \qquad \operatorname{div} \mathbf{B} = 0, \tag{12}$$

$$\operatorname{curl} \mathbf{B} - \frac{1}{c}\frac{\partial \mathbf{E}}{\partial t} = \mathbf{s}, \qquad \operatorname{div} \mathbf{E} = \rho. \tag{12'}$$

According to the atomic theory of electrons these are generally valid exact physical laws. This theory furthermore sets $\mathbf{s} = \frac{\rho \mathbf{v}}{c}$, in which $\mathbf{v}$ denotes the velocity of the matter with which the electric charge is associated.

The **force** which acts on the masses consists of components arising from the electrical and the magnetic field: its density is

$$\mathbf{p} = \rho \mathbf{E} + [\mathbf{s}, \mathbf{B}]. \tag{13}$$

Since $\mathbf{s}$ is parallel to $\mathbf{v}$, the work performed on the electrons per unit of time and of volume is

$$\mathbf{p} \cdot \mathbf{v} = \rho \mathbf{E} \cdot \mathbf{v} = c(\mathbf{s}, \mathbf{E}) = \mathbf{s} \cdot \mathbf{E}'.$$

It is used in increasing the kinetic energy of the electrons, which is partly transferred to the neutral molecules as a result of collisions. This augmented molecular motion in the interior of the conductor expresses itself physically as the heat arising during this phenomenon, as was pointed out by Joule. We find, in fact, experimentally that $\mathbf{s} \cdot \mathbf{E}'$ is the quantity of heat produced per unit of time and per unit of volume by the current. The energy used up in this way must be furnished by the instrument providing the current. If we multiply equation (12) by $-\mathbf{B}$, equation (12′) by $\mathbf{E}$ and add, we get

$$-c \cdot \operatorname{div}[\mathbf{E}, \mathbf{B}] - \frac{\partial}{\partial t}(\tfrac{1}{2}\mathbf{E}^2 + \tfrac{1}{2}\mathbf{B}^2) = c(\mathbf{s}, \mathbf{E}).$$

If we set

$$[\mathbf{E}, \mathbf{B}] = \mathrm{s}, \qquad \tfrac{1}{2}\mathbf{E}^2 + \tfrac{1}{2}\mathbf{B}^2 = W$$

and integrate over any volume V, this equation becomes

$$-\frac{d}{dt}\int_V W\,dV + c\int_\Omega S_n\,do = \int_V c(\mathbf{s}, \mathbf{E})\,dV.$$

The second member on the left is the integral, taken over the outer surface of V_1, of the component s_n of $\mathbf{s}$ along the inward normal. On the right-hand side we have the work performed on the volume V per unit of time. It is compensated by the decrease of energy $\int W\,dV$ contained in V and by the energy that flows into the portion of space V from without. Our equation is thus an expression of the **energy theorem**. **It confirms the assumption which we made initially about the density** W **of the field-energy**, and we furthermore see that $c\mathbf{s}$, familiarly known as Poynting's vector, represents the **energy stream or energy-flux**.

The field-equations (12), (12′) have been integrated by Lorentz in the following way, on the assumption that the distribution of charges and currents are known. The equation $\operatorname{div}\mathbf{B} = 0$ is satisfied by setting

$$-\mathbf{B} = \operatorname{curl}\mathbf{f} \tag{14}$$

in which $-\mathbf{f}$ is the vector potential. By substituting this in the first equation above we get that $\mathbf{E} - \frac{1}{c}\frac{\partial \mathbf{f}}{\partial t}$ is irrotational, so that we can set

$$\mathbf{E} - \frac{1}{c}\frac{\partial \mathbf{f}}{\partial t} = \operatorname{grad}\phi, \tag{15}$$

in which $-\phi$ is the scalar potential. We may make use of the arbitrary character yet possessed by $\mathbf{f}$ by making it fulfil the subsidiary condition

$$\frac{1}{c}\frac{\partial \phi}{\partial t} + \operatorname{div}\mathbf{f} = 0.$$

This is found to be expedient for our purpose (whereas for a stationary field we assumed $\operatorname{div}\mathbf{f} = 0$). If we introduce the potentials in the two latter equations, we find by an easy calculation

$$-\frac{1}{c^2}\frac{\partial^2 \phi}{\partial t^2} + \Delta\phi = \rho, \tag{16}$$

$$-\frac{1}{c^2}\frac{\partial^2 \mathbf{f}}{\partial t^2} + \Delta\mathbf{f} = \mathbf{s}. \tag{16'}$$

An equation of the form (16) denotes a wave disturbance travelling with the velocity c. In fact, just as Poisson's equation $\Delta\phi = \rho$ has the solution

$$-4\pi\phi = \int \frac{\rho}{r}\,dV$$

so (16) has the solution

$$-4\pi\phi = \int \frac{\rho\left(t - \frac{r}{c}\right)}{r}\,dV;$$

on the left-hand side of which ϕ is the value at a point O at time t; r is the distance of the source P, with respect to which we integrate, from the point of emergence O; and within the integral the value of ρ is that at the point P at time $t - \frac{r}{c}$. Similarly (16′) has the solution

$$-4\pi \mathbf{f} = \int \frac{\mathbf{s}\left(t - \frac{r}{c}\right)}{r}\, dV.$$

The field at a point does not depend on the distribution of charges and currents at the same moment, but the determining factor for every point is the moment that lies back just as many $\left(\frac{r}{c}\right)$'s as the disturbance propagating itself with the velocity c takes to travel from the source to the point of emergence.

Just as the expression for the potential (in Cartesian co-ordinates), namely,

$$\Delta\phi = \frac{\partial^2\phi}{\partial x_1^2} + \frac{\partial^2\phi}{\partial x_2^2} + \frac{\partial^2\phi}{\partial x_3^2}$$

is invariant with respect to linear transformations of the variables x_1, x_2, x_3, which are such that they convert the quadratic form

$$x_1^2 + x_2^2 + x_3^2$$

into itself, so the expression which takes the place of this expression for the potential when we pass from statical to moving fields, namely,

$$-\frac{1}{c^2}\frac{\partial^2\phi}{\partial t^2} + \frac{\partial^2\phi}{\partial x_1^2} + \frac{\partial^2\phi}{\partial x_2^2} + \frac{\partial^2\phi}{\partial x_3^2} \quad \textbf{(retarded potentials)}$$

is an invariant for those linear transformations of the four co-ordinates, t, x_1, x_2, x_3, the so-called Lorentz transformations, that transform the indefinite form

$$-c^2t^2 + x_1^2 + x_2^2 + x_3^2 \tag{17}$$

into itself. Lorentz and Einstein recognised that not only equation (16) but also the *whole system of electromagnetic laws for the æther has this property of invariance, namely, that these laws are the expression of invariant relations between tensors which exist in a four-dimensional affine space whose co-ordinates are* t, x_1, x_2, x_3 *and upon which a non-definite metrical structure is impressed by the form* (17). This is the **Lorentz-Einstein Theorem of Relativity**.

To prove the theorem we shall choose a new unit of time by putting $ct = x_0$. The co-efficients of the metrical groundform are then

$$g_{ik} = 0 \quad (i \neq k); \qquad g_{ii} = \epsilon_i,$$

in which $\epsilon_0 = -1$, $\epsilon_1 = \epsilon_2 = \epsilon_3 = +1$; so that in passing from components of a tensor that are co-variant with respect to an index i to the contra-variant components of that tensor we have only to multiply the ith component by the sign of ϵ_i. The question of continuity for electricity (10) assumes the desired invariant form

$$\sum_{i=0}^{3} \frac{\partial s^i}{\partial x_i} = 0$$

if we introduce $s^0 = \rho$, and s^1, s^2, s^3, which are equal to the components of $\mathbf{s}$, as the four contra-variant components of a vector in the above four-dimensional space, namely, of the "4-vector current". Parallel with this—as we see from (16) and (16′)—we must combine

$$\phi_0 = \phi \text{ and the components of } \mathbf{f}\text{, namely, } \phi^1,\ \phi^2,\ \phi^3,$$

to make up the contra-variant components of a four-dimensional vector, which we call the electromagnetic potential; of its co-variant components, the 0th, i.e. $\phi_0 = -\phi$, whereas the three others ϕ_1, ϕ_2, ϕ_3 are equal to the components of $\mathbf{f}$. The equations (14) and (15), by which the field-quantities $\mathbf{B}$ and $\mathbf{E}$ are derived from the potentials, may then be written in the invariant form

$$\frac{\partial \phi_i}{\partial x_k} - \frac{\partial \phi_k}{\partial x_i} = F_{ik} \tag{18}$$

in which we set

$$\mathbf{E} = (F_{10}, F_{20}, F_{30}), \qquad \mathbf{B} = (F_{23}, F_{31}, F_{12}).$$

This is then how we may combine electric and magnetic intensity of field to make up a single linear tensor of the second order F, the "field". From (18) we get the invariant equations

$$\frac{\partial F_{kl}}{\partial x_i} + \frac{\partial F_{li}}{\partial x_k} + \frac{\partial F_{ik}}{\partial x_l} = 0, \tag{19}$$

and this is Maxwell's first system of equations (12). We took a circuitous route in using Lorentz's solution and the potentials only so as to be led naturally to the proper combination of the three-dimensional quantities, which converts them into four-dimensional vectors and tensors. By passing over to contra-variant components we get

$$\mathbf{E} = (F^{01}, F^{02}, F^{03}), \qquad \mathbf{B} = (F^{23}, F^{31}, F^{12}).$$

Maxwell's second system, expressed invariantly in terms of four-dimensional tensors, is now

$$\sum_k \frac{\partial F^{ik}}{\partial x_k} = s^i. \tag{20}$$

If we now introduce the four-dimensional vector with the co-variant components

$$p_i = F_{ik} s^k \tag{21}$$

(and the contra-variant components $p^i = F^{ik} s_k$)—following our previous practice of omitting the signs of summation—then p^0 is the "work-density," that is, the work per unit of time and per unit of volume: $p^0 = (\mathbf{s}, \mathbf{E})$ [the unit of time is to be adapted to the new measure of time $x_0 = ct$], and p^1, p^2, p^3 are the components of the density of force.

This fully proves the Lorentz Theorem of Relativity. *We notice here that the laws that have been obtained are exactly the same as those which hold in the stationary magnetic field* (§ 9 (62)) *except that they have been transposed from three-dimensional to four-dimensional space.* There is no doubt that the real mathematical harmony underlying these laws finds as complete an expression as is possible in this formulation in terms of four-dimensional tensors.

Further, we learn from the above that, exactly as in the case of three-dimensions, we may derive the "4-force" $= p_i$ from a symmetrical four-dimensional "stress-tensor" S, thus

$$-p_i = \frac{\partial S_i^k}{\partial x_k} \quad \text{or} \quad -p^i = \frac{\partial S^{ik}}{\partial x_k}, \tag{22}$$

$$S_i^k = F_{ir} F^{kr} - \tfrac{1}{2} \delta_i^k |F|^2. \tag{22'}$$

The square of the numerical value of the field (which is not necessarily positive here) is

$$|F|^2 = \tfrac{1}{2} F_{ik} F^{ik}.$$

We shall verify formula (22) by direct calculation. We have

$$\frac{\partial S_i^k}{\partial x_k} = F_{ir} \frac{\partial F^{kr}}{\partial x_k} + F^{kr} \frac{\partial F_{ir}}{\partial x_k} - \tfrac{1}{2} F^{kr} \frac{\partial F_{kr}}{\partial x_i}.$$

The first term on the right gives us

$$-F_{ir}s^r = -p_i.$$

If we write the co-efficient of F^{kr} skew-symmetrically we get for the second term

$$\tfrac{1}{2}F^{kr}\left(\frac{\partial F_{ir}}{\partial x_k} - \frac{\partial F_{ik}}{\partial x_r}\right)$$

which, combined with the third, gives

$$-\tfrac{1}{2}F^{kr}\left(\frac{\partial F_{ik}}{\partial x_r} + \frac{\partial F_{kr}}{\partial x_i} + \frac{\partial F_{ri}}{\partial x_k}\right).$$

The expression consisting of three terms in the brackets $= 0$, by (19).

Now $|F|^2 = \mathbf{B}^2 - \mathbf{E}^2$. Let us examine what the individual components of S_{ik} signify, by separating the index 0 from the others 1, 2, 3, in conformity with the partition into space and time.

S^{00} = the energy-density $W = \frac{1}{2}(\mathbf{E}^2 + \mathbf{B}^2)$,

S^{0i} = the components of $\mathbf{S} = [\mathbf{E}, \mathbf{B}]$ $\quad i, k = (1, 2, 3)$,

S^{ik} = the components of the Maxwell stress-tensor, which is composed of the electrical and magnetic parts given in § 9. Accordingly the 0th equation of (22) expresses the law of energy. The 1st, 2nd, and 3rd have a fully analogous form. If, for a moment, we denote the components of the vector $\dfrac{1}{c}\mathbf{S}$ by G^1, G^2, G^3 and take $\mathbf{t}^{(i)}$ to stand for the vector with the components S^{i1}, S^{i2}, S^{i3} we get

$$-p_i = \frac{\partial G^i}{\partial t} + \operatorname{div}\mathbf{t}^{(i)}, \qquad (i = 1, 2, 3). \tag{23}$$

The force which acts on the electrons enclosed in a portion of space V produces an increase in time of momentum equal to itself numerically. This increase is balanced, according to (23), by a corresponding decrease of the **field-momentum** distributed in the field with a density $\frac{\mathbf{S}}{c}$, and the addition of field-momentum from without. The current of the ith component of momentum is given by $\mathbf{t}^{(i)}$, and thus the **momentum-flux** is nothing more than the Maxwell stress-tensor. *The Theorem of the Conservation of Energy is only one component, the time-component, of a law which is invariant for Lorentz transformations, the other components being the space-components which express the conservation of momentum.* The total energy as well as the total momentum remains unchanged: they merely stream from one part of the field to another, and become transformed from field-energy and field-momentum into kinetic-energy and kinetic-momentum of matter, and *vice versa*. That is the simple physical meaning of the formulæ (22). In accordance with it we shall in future refer to the tensor S of the four-dimensional world as the **energy-momentum-tensor** or, more briefly, as the **energy-tensor**. Its symmetry tells us that the **density of momentum** $= \frac{1}{c^2}$ *times* **the energy-flux**. The field-momentum is thus very weak, but, nevertheless, it has been possible to prove its existence by demonstrating the pressure of light on a reflecting surface.

A Lorentz transformation is linear. Hence (again suppressing one space co-ordinate in our graphical picture) we see that it is tantamount to introducing a new affine co-ordinate system. Let us consider how the fundamental vectors $\mathbf{e}'_0$, $\mathbf{e}'_1$, $\mathbf{e}'_2$ of the new co-ordinate system lie relatively to the original fundamental vectors $\mathbf{e}_0$, $\mathbf{e}_1$, $\mathbf{e}_2$, that is to the unit vectors in the direction of the x_0

(or t), x_1, x_2 axes. Since, for

$$\mathbf{x} = x_0\mathbf{e}_0 + x_1\mathbf{e}_1 + x_2\mathbf{e}_2 = x'_0\mathbf{e}'_0 + x'_1\mathbf{e}'_1 + x'_2\mathbf{e}'_2,$$

we must have

$$-x_0^2 + x_1^2 + x_2^2 = -x_0'^2 + x_1'^2 + x_2'^2 [= Q(\mathbf{x})]$$

we get $Q(\mathbf{e}'_0) = -1$. Accordingly, the vector $\mathbf{e}'_0$ starting from O (i.e. the t'-axis) lies within the cone of light-propagation; the parallel planes $t' = \text{const.}$ lie so that they cut ellipses from the cone, the middle points of which lie on the t'-axis (see Fig. 7); the x'_1-, x'_2-axis are in the direction of conjugate diameters of these elliptical sections, so that the equation of each is

$$x_1'^2 + x_2'^2 = \text{const.}$$

As long as we retain the picture of a material æther, capable of executing vibrations, we can see in Lorentz's Theorem of Relativity only a remarkable property of mathematical transformations; the relativity theorem of Galilei and Newton remains the truly valid one. We are, however, confronted with the task of interpreting not only optical phenomena but all electrodynamics and its laws as the result of a mechanics of the æther which satisfies Galilei's Theorem of Relativity. To achieve this we must bring the field-quantities into definite relationship with the density and velocity of the æther. Before the time of Maxwell's electromagnetic theory of light, attempts were made to do this for optical phenomena; these efforts were partly, but never wholly, crowned with success. This attempt was not carried on (*vide* note 3) in the case of the more comprehensive domain into which Maxwell relegated optical phenomena. On the contrary, **the idea of a field existing in empty space and not requiring a medium to sustain it** gradually began to win ground. Indeed, even Faraday had expressed in unmistakable language that not the field should derive its meaning through its association with matter, but, conversely, rather that particles of matter are nothing more than singularities of the field.

§ 21. Einstein's Principle of Relativity

Let us for the present retain our conception of the æther. It should be possible to determine the motion of a body, for example, the earth, relative to the fixed or motionless æther. We are not helped by aberration, for this only shows that this relative motion **changes** in the course of a year. Let A_1, O, A_2 be three fixed points on the earth that share in its motion. Suppose them to lie in a straight line along the direction of the earth's motion and to be equidistant, so that $A_1O = OA_2 = l$, and let v be the velocity of translation of the earth through the æther; let $\frac{v}{c} = q$, which we shall assume to be a very small quantity. A light-signal emitted at O will reach A_2 after a time $\frac{l}{c-v}$ has elapsed, and A_1 after a time $\frac{l}{c+v}$. Unfortunately, this difference cannot be demonstrated, as we have no signal that is more rapid than light and that we could use to communicate the time to another place.*

*It might occur to us to transmit time from one world-point to another by carrying a clock that is marking time from one place to the other. In practice, this process is not sufficiently accurate for our purpose. Theoretically, it is by no means certain that this transmission is independent of the traversed path. In fact, the theory of relativity proves that, on the contrary, they are

We have recourse to Fizeau's idea, and set up little mirrors at A_1 and A_2 which reflect the light-ray back to O. If the light-signal is emitted at the moment O, then the ray reflected from A_2 will reach A after a time

$$\frac{l}{c-v} + \frac{l}{c+v} = \frac{2lc}{c^2 - v^2}$$

whereas that reflected from A_1 reaches O after a time

$$\frac{l}{c+v} + \frac{l}{c-v} = \frac{2lc}{c^2 - v^2}.$$

There is now no longer a difference in the times. Let us, however, now assume a third point A which participates in the translational motion through the æther, such that $OA = l$, but that OA makes an angle θ with the direction of OA. In Fig. 8, O, O', O'' are the successive positions of the point O at the time 0 at which the signal is emitted, at the time t' at which it is reflected from the mirror A placed at A', and finally at the time $t' + t''$ at which it again reaches O, respectively. From the figure we get the proportion

$$OA' : O''A' = OO' : O''O'.$$

Consequently the two angles at A' are equal to one another. The reflecting mirror must be placed, just as when the system is at rest, perpendicularly to the rigid connecting line OA, in order that the light-ray may return to O. An elementary trigonometrical calculation gives for the **apparent rate of transmission in the direction** θ

$$\frac{2l}{t' + t''} = \frac{c^2 - v^2}{\sqrt{c^2 - v^2 \sin^2 \theta}}. \tag{24}$$

---

dependent on one another; cf. § 22.

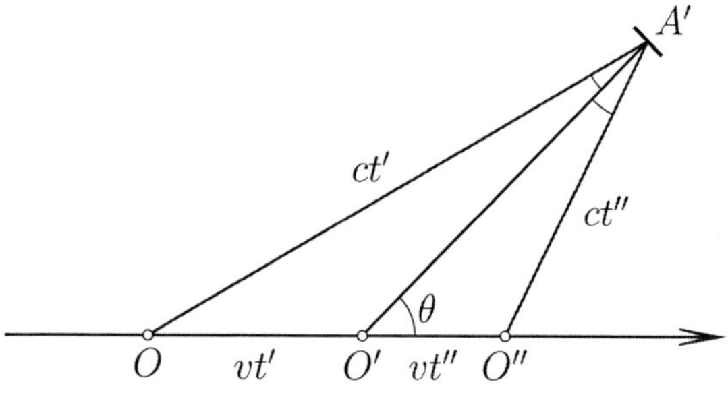

FIG. 8.

It is thus dependent on the angle θ, which gives the direction of transmission. Observations of the value of θ should enable us to determine the direction and magnitude of v.

These observations were attempted in the celebrated **Michelson-Morley experiment** (*vide* note 4). In this, two mirrors A, A' are rigidly fixed to O at distances l, l', the one along the line of motion the other perpendicular to it. The whole apparatus may be rotated about O. By means of a transparent glass plate, one-half of which is silvered and which bisects the right angle at O, a light-ray is split up into two halves, one of which travels to A, the other to A'. They are reflected at these two points; and at O, owing to the partly silvered mirror, they are again combined to a single composite ray. We take l and l' approximately equal; then, owing to the difference in path given by (24), namely,

$$\frac{2l}{1-q^2} - \frac{2l'}{\sqrt{1-q^2}},$$

interference occurs. If the whole apparatus is now turned slowly through 90° about O until A' comes into the direction of motion, this difference of path becomes

$$\frac{2l}{\sqrt{1-q^2}} - \frac{2l'}{1-q^2}.$$

Consequently, there is a shortening of the path by an amount

$$2(l+l')\left(\frac{1}{1-q^2} - \frac{1}{\sqrt{1-q^2}}\right) \sim (l+l')q^2.$$

This should express itself in a shift of the initial interference

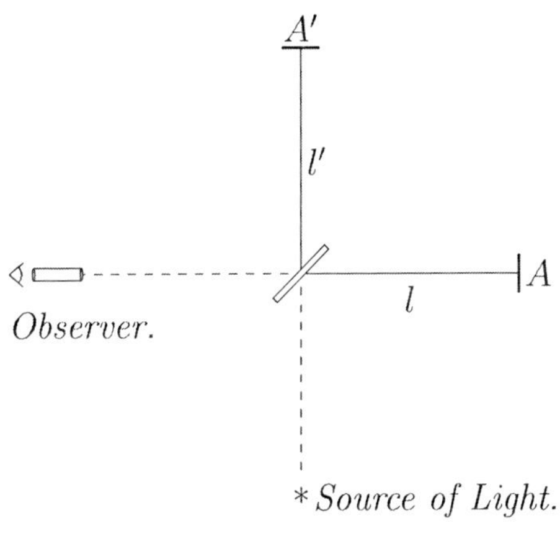

FIG. 9.

fringes. *Although conditions were such that, numerically, even only* 1 *per cent. of the displacement of the fringes expected by Michelson could not have escaped detection, no trace of it was to be found when the experiment was performed.*

Lorentz (and Fitzgerald, independently) sought to explain this strange result by the bold hypothesis that a rigid body in moving relatively to the æther undergoes a contraction in the direction of the line of motion in the ratio $1 : \sqrt{1-q^2}$. This would actually account for the null result of the Michelson-Morley experiment. For there, OA has in the first position the true length $l\sqrt{1-q^2}$, and OA' the length l', whereas in the second position OA has the true length l but OA' the length $l' \cdot \sqrt{1-q^2}$. The difference of path would, in **each** case, be $\dfrac{2(l-l')}{\sqrt{1-q^2}}$.

It was also found that, no matter into what direction a mirror rigidly fixed to O was turned, the same apparent velocity of transmission $\sqrt{c^2-v^2}$ was obtained for all directions; that is, that this velocity did not depend on the direction θ, in the manner given by (24). Nevertheless, theoretically, it still seemed possible to demonstrate the decrease of the velocity of transmission from c to $\sqrt{c^2-v^2}$. But if the æther shortens the measuring rods in the direction of motion in the ratio $1 : \sqrt{1-q^2}$, it need only retard clocks in the same ratio to hide this effect, too. *In fact, not only the Michelson-Morley experiment but a whole series of further experiments designed to demonstrate that the earth's motion has an influence on combined mechanical and electromagnetic phenomena, have led to a null result* (*vide* note 5). Æther mechanics has thus to account not only for Maxwell's laws but also for this remarkable interaction between matter and æther. It seems that the æther has betaken itself to the land of the shades in a final effort to elude the inquisitive search of the physicist!

The only reasonable answer that was given to the question as to why a translation in the æther cannot be distinguished from rest was that of Einstein, namely, that *there is no æther*! (The æther has since the very beginning remained a vague hypothesis and one, moreover, that has acted very poorly in the face of facts.) The position is then this: for mechanics we get Galilei's Theorem of Relativity, for electrodynamics, Lorentz's Theorem. If this is really the case, they neutralise one another and thereby define an absolute space of reference in which mechanical laws have the Newtonian form, electrodynamical laws that given by Maxwell. The difficulty of explaining the null result of the experiments whose purpose was to distinguish translation from rest, is overcome only by regarding one or other of these two principles of relativity as being valid for **all** physical phenomena. That of Galilei does not come into question for electrodynamics as this would mean that, in Maxwell's theory, those terms by which we distinguish moving fields from stationary ones would not occur: there would be no induction, no light, and no wireless telegraphy. On the other hand, even the contraction theory of Lorentz-Fitzgerald suggests that Newton's mechanics may be modified so that it satisfies the Lorentz-Einstein Theorem of Relativity, the deviations that occur being only of the order $\left(\frac{v}{c}\right)^2$; they are then easily within reach of observation for all velocities v of planets or on the earth. The solution of Einstein (*vide* note 6), which at one stroke overcomes all difficulties, is then this: *the world is a four-dimensional affine space whose metrical structure is determined by a non-definite quadratic form*

$$Q(\mathbf{x}) = (\mathbf{x}, \mathbf{x})$$

which has one negative and three positive dimensions. All physical quantities are scalars and tensors of this four-dimensional world, and all physical laws express invariant relations between them.

The simple concrete meaning of the form $Q(\mathbf{x})$ is that a light-signal which has been emitted at the world-point O arrives at all those and only those world-points A for which $\mathbf{x} = \overrightarrow{OA}$ belongs to the one of the two conical sheets defined by the equation $Q(\mathbf{x}) = 0$ (cf. § 4). Hence that sheet (of the two cones) which "opens into the future" namely, $Q(\mathbf{x}) \leq 0$ is distinguished objectively from that which opens into the past. By introducing an appropriate "normal" co-ordinate system consisting of the zero point O and the fundamental vectors $\mathbf{e}_i$, we may bring $Q(\mathbf{x})$ into the normal form

$$(\overrightarrow{OA}, \overrightarrow{OA}) = -x_0^2 + x_1^2 + x_2^2 + x_3^2,$$

in which the x_i's are the co-ordinates of A; in addition, the fundamental vector $\mathbf{e}_0$ is to belong to the cone opening into the future. **It is impossible to narrow down the selection from these normal co-ordinate systems any farther**: that is, none are specially favoured; they are all equivalent. If we make use of a particular one, then x_0 must be regarded as the time; x_1, x_2, x_3 as the Cartesian space co-ordinates; and all the ordinary expressions referring to space and time are to be used in this system of reference as usual. The adequate mathematical formulation of Einstein's discovery was first given by Minkowski (*vide* note 7): to him we are indebted for the idea of four-dimensional world-geometry, on which we based our argument from the outset.

How the null result of the Michelson-Morley experiment comes about is now clear. For if the interactions of the cohesive forces of matter as well as the transmission of light takes place according to Einstein's Principle of Relativity, measuring rods must behave so that no difference between rest and translation can be discovered by means of objective determinations. Seeing that Maxwell's equations satisfy Einstein's Principle of Relativity, as was recognised even by Lorentz, we must indeed regard *the Michelson-Morley experiment as a proof that the mechanics of rigid bodies must, strictly speaking, be in accordance not with that of Galilei's Principle of Relativity, but with that of Einstein.*

It is clear that this is mathematically much simpler and more intelligible than the former: world-geometry has been brought into closer touch with Euclidean space-geometry through Einstein and Minkowski. Moreover, as may easily be shown, Galilei's principle is found to be a limiting case of Einstein's world geometry by making c converge to ∞. The physical purport of this is that *we are to discard our belief in the objective meaning of simultaneity; it was the great achievement of Einstein in the field of the theory of knowledge that he banished this dogma from our minds*, and this is what leads us to rank his name with that of Copernicus. The graphical picture given at the end of the preceding paragraph discloses immediately that the planes $x'_0 =$ const. no longer coincide with the planes $x_0 =$ const. In consequence of the metrical structure of the world, which is based on $Q(\mathbf{x})$, each plane $x'_0 =$ const. has a measure-determination such that the ellipse in which it intersects the "light-cone," is a circle, and that Euclidean geometry holds for it. The point at which it is punctured by the x'_0-axis is the mid-point of the elliptical section. So the propagation of light takes place in the "accented" system of reference, too, in concentric circles.

We shall next endeavour to eradicate the difficulties that seem to our intuition, our inner knowledge of space and time, to be involved in the revolution caused by Einstein in the conception of time. According to the ordinary view the following is true. If I shoot bullets out with all possible velocities in all directions from a point O, they will all reach world-points that are later than O; I cannot shoot back into the past. Similarly, an event which happens at O has an influence only on what happens at later world-points, whereas "one can no longer undo" the past: the extreme limit is reached by gravitation, acting according to Newton's law of attraction, as a result of which, for example, by extending my arm, I at the identical moment produce an effect on the planets, modifying their orbits ever so slightly. If we again suppress a space-co-ordinate and use our graphical mode of representation, then the absolute meaning of the plane $t = 0$ which passes through O consists in the fact that it separates the "future" world-points, which can be influenced by actions at O, from the "past" world-points from which an effect may be conveyed to or conferred on O. According to Einstein's Principle of Relativity, we get in place of the plane of separation $t = 0$ the light cone

$$x_1^2 + x_2^2 - c^2t^2 = 0$$

(which degenerates to the above double plane when $c = \infty$). This makes the position clear in this way. The direction of all bodies projected from O must point into the forward-cone, opening into the future (so also the direction of the world-line of my own body, my "life-curve" if I happen to be at O). Events at O can influence only happenings that occur at world-points that lie within this forward-cone: the limits are marked out by the resulting propagation of light into empty space.* If I happen to be at O, then

*The propagation of gravitational force must, of course, likewise take place with the speed of light, according to Einstein's Theory of Relativity. The law for the gravitational potential must be modified in a manner analogous to that by which electrostatic potential was modified in passing from statical to moving fields.

O divides my life-curve into past and future; no change is thereby caused. As far as my relationship to the world is concerned, however, the forward-cone comprises all the world-points which are affected by my active or passive doings at O, whereas all events that are complete in the past, that can no longer be altered, lie externally to this cone. **The sheet of the forward-cone separates my active future from my active past.** On the other hand, the interior of the backward-cone includes all events in which I

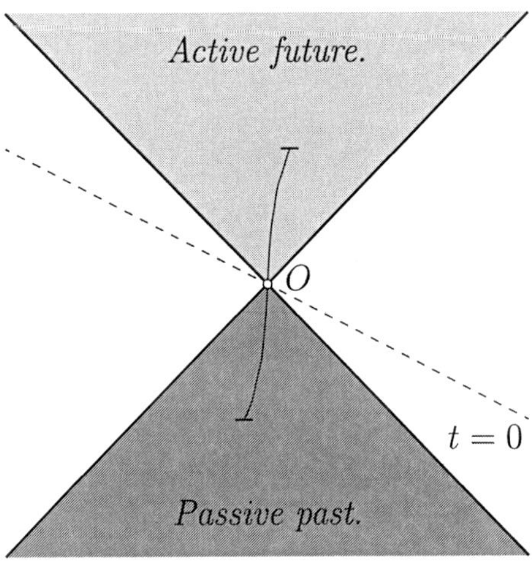

FIG. 10.

have participated (either actively or as an observer) or of which I have received knowledge of some kind or other, for only such events may have had an influence on me; outside this cone are all occurrences that I may yet experience or would yet experience if my life were everlasting and nothing were shrouded from my gaze. **The sheet of the backward-cone separates my passive past from my passive future.** The sheet itself contains everything on its surface that I see at this moment, or can see; it is thus properly the picture of my external surroundings. In the fact that we must in this way distinguish between **active** and **passive**, present, and future, there lies the fundamental importance of Römer's discovery of the finite velocity of light to which Einstein's Principle of Relativity first gave full expression. The plane $t = 0$ passing through O in an allowable co-ordinate system may be placed so that it cuts the light-cone $Q(\mathbf{x}) = 0$ only at O and thereby separates the cone of the active future from the cone of the passive past.

For a body moving with uniform translation it is always possible to choose an allowable co-ordinate system (= normal co-ordinate system) such that the body is at rest in it. The individual parts of the body are then separated by definite distances from one another, the straight lines connecting them make definite angles with one another, and so forth, all of which may be calculated by means of the formulæ of ordinary analytical geometry from the space-co-ordinates x_1, x_2, x_3 of the points under consideration in the allowable co-ordinate system chosen. I shall term them the **static measures** of the body (this defines, in particular, the **static length** of a measuring rod). If this body is a clock, in which a periodical event occurs, there will be associated with this period in the system of reference, in which the clock is at rest, a definite time, determined by the increase of the co-ordinate x_0 during a period; we shall call this the "proper time" of the clock. If we push the body at one and the same moment at different points, these points will begin to move, but as the effect can at most be propagated with the velocity of light, the motion will only gradually be communicated to the whole body. As long as the expanding spheres encircling each point of attack and travelling with the velocity of light do not overlap, the parts surrounding these points that are dragged along move independently of one another. It is evident from this that, according to the theory of relativity, there cannot be rigid bodies in the old sense; that is, no body exists which remains objectively always the same no matter to what influences it has been subjected. How is it that in spite of this we can use our measuring rods for carrying out measurements in space? We shall use an analogy. If a gas that is in equilibrium in a closed vessel is heated at various points by small flames and is then removed adiabatically, it will at first pass through a series of complicated stages, which will not satisfy the equilibrium laws of thermodynamics. Finally, however, it will attain a new state of equilibrium corresponding to the new quantity of energy it contains, which is now greater owing to the heating. We require of a rigid body that is to be used for purposes of measurement (in particular, a linear **measuring rod**) that, **after coming to rest in an allowable system of reference**, it shall always remain exactly the same as before, that is, that it shall have **the same static measures** (or **static length**); and we require of a **clock** that goes correctly **that it shall always have the same proper-time when it has come to rest** (as a whole) **in an allowable system of reference**. We may assume that the measuring rods and clocks which we shall use satisfy this condition to a sufficient degree of approximation. It is only when, in our analogy, the gas is warmed sufficiently slowly (strictly speaking, infinitely slowly) that it will pass through a series of thermodynamic states of equilibrium; only when we move the measuring rods and clocks steadily, without jerks, will they preserve their static lengths and proper-times. The limits of acceleration within which this assumption may be made without appreciable errors arising are certainly very wide. Definite and exact statements about this point can be made only when we have built up a **dynamics** based on physical and mechanical laws.

To get a clear picture of the Lorentz-Fitzgerald contraction from the point of view of Einstein's Theory of Relativity, we shall imagine the following to take place in a plane. In an allowable system of reference (co-ordinates t, x_1, x_2, one space-co-ordinate being suppressed), to which the following space-time expressions will be referred, there is at rest a plane sheet of paper (carrying rectangular co-ordinates x_1, x_2 marked on it), on which a closed curve **c** is drawn. We have, besides, a circular plate carrying a rigid clock-hand that rotates around its centre, so that its point traces out the edge of the plate if it is rotated slowly, thus proving that the edge is actually a circle. Let the plate now move along the sheet of paper with uniform translation. If, at the same time, the index rotates slowly, its point runs unceasingly along the edge of the plate: in this sense the disc is circular during translation too. Suppose the edge of the disc to coincide exactly with the curve **c** at a definite moment. If we measure **c** by means of measuring rods that are at rest, we find that **c** is not a circle but an ellipse. This phenomenon is shown graphically in Fig. 11. We have added the system of reference t', x_1', x_2' with respect to which the disc is at rest. Any plane $t' =$ const. intersects the light cone in this system of reference in a circle "that exists for a single moment". The cylinder above it erected in the direction of the t'-axis represents a circle that is at rest in the **accented** system, and hence marks off that part of the world which is passed over by our disc. The section of this cylinder and the plane $t = 0$ is not a circle but an ellipse. The right-angled cylinder constructed on it in the direction of the t-axis represents the constantly present curve traced on the paper.

If we now inquire what physical laws are necessary to distinguish normal co-ordinate systems from all other co-ordinate systems (in Riemann's sense), we learn that we require only Galilei's Principle of Relativity and the law of the propagation of light; by means of light-signals and point-masses moving under no forces—even if we have only small limits of velocity within which the latter may move—we are in a position to fix a co-ordinate system of this kind. To see this we shall next add a corollary to Galilei's Principle of Inertia. If a clock shares in the motion of the point-mass moving under no forces, then its time-data are a measure of the "proper-time" s of the motion. Galilei's principle states that the world-line of the point is a straight line; we elaborate this by stating further that the moments of the motion characterised by

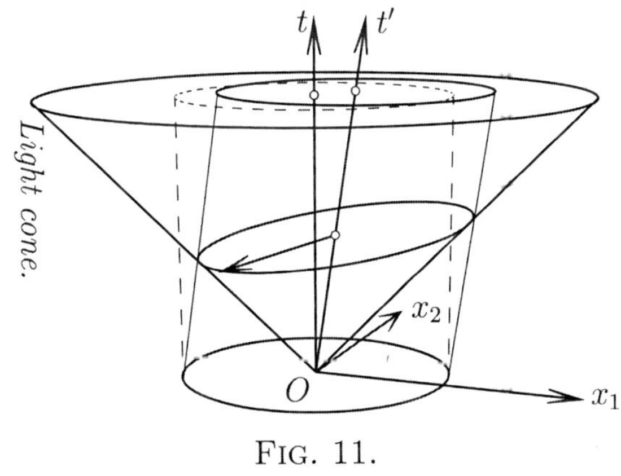

FIG. 11.

$s = 0, 1, 2, 3, \ldots$ (or by any arithmetical series of values of s) represent equidistant points along the straight line. By introducing the parameter of proper-time to distinguish the various stages of the motion we get not only a line in the four-dimensional world but also a "motion" in it (cf. the definition on page 155) and according to Galilei this motion is a translation.

The world-points constitute a four-dimensional manifold; this is perhaps the most certain fact of our empirical knowledge. We shall call a system of four co-ordinates x_i $(i = 0, 1, 2, 3)$, which are used to fix these points in a certain portion of the world, a **linear co-ordinate system**, if the motion of point-mass under no forces and expressed in terms of the parameter s of the proper-time be represented by formulæ in which the x_i's are linear functions of s. The fact that there are such co-ordinate systems is what the law of inertia really asserts. After this condition of linearity, all that is necessary to define the co-ordinate system fully is a linear transformation. That is, if x_i, x'_i are the co-ordinates respectively of one and the same world-point in two different linear co-ordinate systems, then the x'_i's a must be linear functions of the x's. By simultaneously interpreting the x_i's as Cartesian co-ordinates in a four-dimensional Euclidean space, the co-ordinate system furnishes us with a representation of the world (or of the portion of world in which the x_i's exist) on a Euclidean space of representation. We may, therefore, formulate our proposition thus. A representation of two Euclidean spaces by one another (or in other words a transformation from one Euclidean space to another), such that straight lines become straight lines and a series of equidistant points become a series of equidistant points is necessarily an affine transformation. Fig. 12 which represents Möbius' mesh-construction (*vide* note 8) may suffice to indicate the proof to the reader. It is obvious that this mesh-system may be arranged so that the three directions of the straight lines composing it may be derived from a given, arbitrarily thin, cone carrying these directions on it; the above geometrical theorem remains

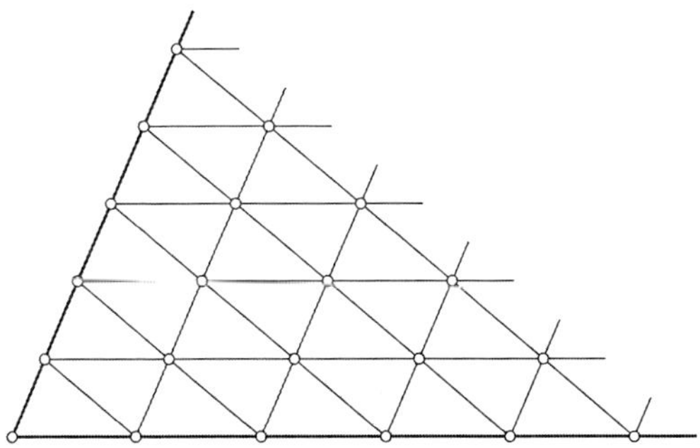

FIG. 12.

valid even if we only know that the straight lines whose directions belong to this cone become straight lines again as a result of the transformation.

Galilei's Principle of Inertia is sufficient in itself to prove conclusively that the world is affine in character: it will not, however, allow us deduce any further result. The metrical groundform $(\mathbf{x}, \mathbf{x})$ of the world is now accounted for by the process of light-propagation. A light-signal emitted from O arrives at the world-point A if, and only if, $\mathbf{x} = \overrightarrow{OA}$ belongs to one of the two conical sheets defined by $(\mathbf{x}, \mathbf{x}) = 0$. This determines the quadratic form except for a constant factor; to fix the latter we must choose an arbitrary unit-measure (cf. Appendix I).

§ 22. Relativistic Geometry, Kinematics, and Optics

We shall call a world-vector $\mathbf{x}$ **space-like** or **time-like**, according as $(\mathbf{x}, \mathbf{x})$ is positive or negative. Time-like vectors are divided into those that point into the **future** and those that point into the **past**. We shall call the invariant

$$\Delta s = \sqrt{-(\mathbf{x}, \mathbf{x})} \tag{25}$$

of a time-like vector $\mathbf{x}$ which points into the future its **proper-time**. If we set

$$\mathbf{x} = \Delta s \cdot \mathbf{e}$$

then $\mathbf{e}$, the direction of the time-like displacement, is a vector that points into the future, and that satisfies the condition of normality $(\mathbf{e}, \mathbf{e}) = -1$.

As in Galilean geometry, so in Einstein's world-geometry we must **resolve the world into space and time** by projection in the direction of a time-like vector $\mathbf{e}$ pointing into the future and normalised by the condition $(\mathbf{e}, \mathbf{e}) = -1$. The process of projection was discussed in detail in § 19. The fundamental formulæ (3), (5), (5′) that are set up must here be applied with $e = -1$.* World-points for which the vector connecting them is proportional to $\mathbf{e}$ coincide at a space-point which we may mark by means of a point-mass at rest, and which we may represent graphically by a world-line (straight) parallel to $\mathbf{e}$. The three-dimensional space $R_{\mathbf{e}}$ that is generated by the projection has a metrical character that is Euclidean since, for every vector $\mathbf{x}^*$ which is orthogonal to $\mathbf{e}$,

*Here the units of space and time are chosen so that the velocity of light *in vacuo* becomes equal to 1. To arrive at the ordinary units of the c.g.s. systems, the equation of normality $(\mathbf{e}, \mathbf{e}) = -1$ must be replaced by $(\mathbf{e}, \mathbf{e}) = -c^2$, and e must be taken equal to $-c^2$.

that is, every vector $\mathbf{x}^*$ that satisfies the condition $(\mathbf{x}^*, \mathbf{e}) = 0$, $(\mathbf{x}^*, \mathbf{x}^*)$ is a positive quantity (except in the case in which $\mathbf{x}^* = \mathbf{0}$; cf. § 4). Every displacement $\mathbf{x}$ of the world may be split up according to the formula

$$\mathbf{x} = \Delta t \mid \mathsf{x} :$$

Δt is its duration (called "height" in § 19): x is the displacement it produces in the space $R_{\mathbf{e}}$.

If e_1, e_2, e_3 form a co-ordinate system in $R_{\mathbf{e}}$, then the world-displacements $\mathbf{e}_1$, $\mathbf{e}_2$, $\mathbf{e}_3$ that are orthogonal to $\mathbf{e} = \mathbf{e}_0$, and that produce the three given space-displacements, form in conjunction with $\mathbf{e}_0$ a **co-ordinate system, which belongs to** $R_{\mathbf{e}}$, for the world-points. It is normal if the three vectors $\mathbf{e}_i$ in $R_{\mathbf{e}}$ form a Cartesian co-ordinate system. In every case the system of coefficients of the metrical groundform has, in it, the form

$$\begin{vmatrix} -1 & 0 & 0 & 0 \\ 0 & g_{11} & g_{12} & g_{13} \\ 0 & g_{21} & g_{22} & g_{23} \\ 0 & g_{31} & g_{32} & g_{33} \end{vmatrix}.$$

The proper time Δs of a time-like vector $\mathbf{x}$ pointing into the future (and for which $\mathbf{x} = \Delta s \cdot \mathbf{e}$) is equal to the duration of $\mathbf{x}$ in the space of reference $R_{\mathbf{e}}$, in which $\mathbf{x}$ calls forth no spatial displacement. In the sequel we shall have to contrast several ways of splitting up quantities into terms of the vectors $\mathbf{e}$, $\mathbf{e}'$, ...; $\mathbf{e}$ (with or without an index) is always to denote a time-like world-vector pointing into the future and satisfying the condition of normality $(\mathbf{e}, \mathbf{e}) = -1$.

Let K be a body at rest in $R_{\mathbf{e}}$, K' a body at rest in $R'_{\mathbf{e}}$. K' moves with uniform translation in $R_{\mathbf{e}}$. If, by splitting up $\mathbf{e}'$

into terms of $\mathbf{e}$, we get in $\mathsf{R}_\mathbf{e}$

$$e' = h \mid h\mathsf{v} \tag{26}$$

then K' undergoes the space-displacement $h\mathsf{v}$ during the time (i.e. with the duration) h in $\mathsf{R}_\mathbf{e}$. Accordingly, v is the velocity of K' in $\mathsf{R}_\mathbf{e}$ or **the relative velocity of K' with respect to K**. Its magnitude is determined by $v^2 = (\mathsf{v}, \mathsf{v})$. By (3) we have

$$h = -(\mathbf{e}', \mathbf{e}); \tag{27}$$

on the other hand, by (5)

$$1 = -(\mathbf{e}', \mathbf{e}') = h^2 - h^2(\mathsf{v}, \mathsf{v}) = h^2(1 - v^2),$$

thus we get

$$h = \frac{1}{\sqrt{1 - v^2}}. \tag{28}$$

If, between two moments of K''s motion, it undergoes the world-displacement $\Delta s \cdot \mathbf{e}'$, (26) shows that $h \cdot \Delta s = \Delta t$ is the duration of this displacement in $\mathsf{R}_\mathbf{e}$. The proper time Δs and the duration Δt of the displacement in $\mathsf{R}_\mathbf{e}$ are related by

$$\Delta s = \Delta t \sqrt{1 - v^2}. \tag{29}$$

Since (27) is symmetrical in $\mathbf{e}$ and $\mathbf{e}'$, (28) teaches us that the **magnitude of the relative velocity of K' with respect to K is equal to that of K with respect to K'**. The vectorial relative velocities **cannot** be compared with one another since the one exists in the space $\mathsf{R}_\mathbf{e}$, the other in the space $\mathsf{R}'_\mathbf{e}$.

Let us consider a partition into three quantities $\mathbf{e}$, $\mathbf{e}_1$, $\mathbf{e}_2$. Let K_1, K_2 be two bodies at rest in $\mathsf{R}_{\mathbf{e}_1}$, $\mathsf{R}_{\mathbf{e}_2}$ respectively. Suppose we have in $\mathsf{R}_{\mathbf{e}}$

$$\mathbf{e}_1 = h_1 \mid h_1 \mathsf{v}_1 \qquad h_1 = \frac{1}{\sqrt{1-v_1^2}},$$

$$\mathbf{e}_2 = h_2 \mid h_2 \mathsf{v}_2 \qquad h_2 = \frac{1}{\sqrt{1-v_2^2}}.$$

Then

$$-(\mathbf{e}_1, \mathbf{e}_2) = h_1 h_2 \{1 - (v_1 v_2)\}.$$

Hence, if K_1 and K_2 have velocities v_1, v_2 respectively in $\mathsf{R}_{\mathbf{e}}$, with numerical values v_1, v_2, then if these velocities v_1, v_2 make an angle θ with each other, and if $v_{12} = v_{21}$ is the magnitude of the velocity of K_2 relatively to K_1 (or *vice versa*), we find that the formula

$$\frac{1 - v_1 v_2 \cos\theta}{\sqrt{1-v_1^2}\sqrt{1-v_2^2}} = \frac{1}{\sqrt{1-v_{12}^2}} \tag{30}$$

holds: **it shows how the relative velocity of two bodies is determined from their given velocities**. If, using hyperbolic functions, we set $v = \tanh v$ for each of the values v of the velocity (v being < 1), we get

$$\cosh u_1 \cosh u_2 - \sinh u_1 \sinh u_2 \cos\theta = \cosh u_{12}.$$

This formula becomes the cosine theorem of spherical geometry if we replace the hyperbolic functions by their corresponding trigonometrical functions; thus u_{12} is the side opposite the angle θ in a triangle on the Bolyai-Lobatschefsky plane, the two remaining

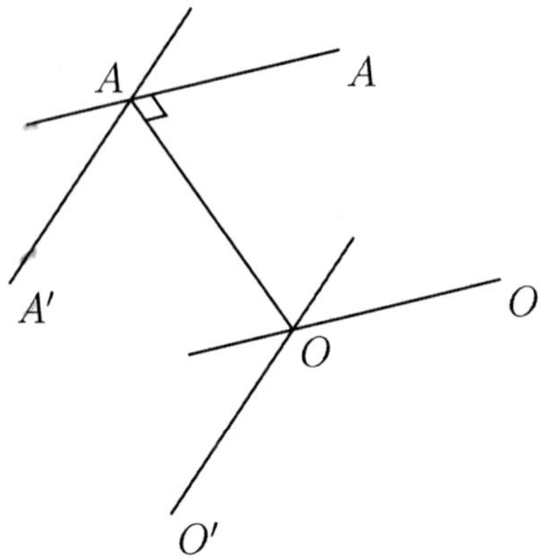

FIG. 13.

sides being u_1 and u_2.

Analogous to the relationship (29) between time and proper-time, there is one between length and statical-length. We shall use R_e as our space of reference. Let the individual point-masses of the body at a **definite** moment be at the world-points O, A, ... The space-points O, A, ... at R_e at which they are situated form a figure in R_e, on which we can confer duration, by making the body leave behind it a copy of itself at the moment under consideration in the space R_e; an example of this was presented in the illustration given at the close of the preceding paragraph. If, on the other hand, the world-points O, A, ... are at the space-points O', A', ... in the space R_e in which K' is at rest, then O', A', ... constitute the statical shape of the body K' (cf. Fig. 13, in which orthogonal world-distances are drawn perpendicularly). There is a transformation that connects the part of R_e, which receives the imprint or copy, and the statical shape of the body in R'_e. This transformation transforms the points A, A' into one another. It is obviously affine (in fact, it is nothing more than an orthogonal projection). Since the world-points O, A are **simultaneous** for the partition into $\mathbf{e}$, we have

$$\overrightarrow{OA} = \mathbf{x} = \mathbf{0} \mid x \text{ in } R_{\mathbf{e}}, \text{ and } x = \overrightarrow{OA}.$$

By formula (5)

$$\begin{aligned}\overrightarrow{OA}^2 &= (x, x) = (\mathbf{x}, \mathbf{x}),\\ \overrightarrow{O'A'}^2 &= (\mathbf{x}, \mathbf{x}) + (\mathbf{x}, \mathbf{e}')^2.\end{aligned}$$

If, however, we determine $(\mathbf{x}, \mathbf{e}')$ in $R_{\mathbf{e}}$ by $(5')$ we get

$$(\mathbf{x}, \mathbf{e}') = h(x, v),$$

and hence

$$\overrightarrow{O'A'}^2 = (x, x) + \frac{(x, v)^2}{1 - v^2}.$$

If we use a Cartesian co-ordinate system x_1, x_2, x_3 in $R_{\mathbf{e}}$ with O as origin, and having its x_1-axis in the direction of the velocity v, then if x_1, x_2, x_3 are the co-ordinates of A, we have

$$\begin{aligned}\overrightarrow{OA}^2 &= x_1^2 + x_2^2 + x_3^2,\\ \overrightarrow{O'A'}^2 &= \frac{x_1^2}{1 - v^2} + x_2^2 + x_3^2 = x_1'^2 + x_2'^2 + x_3'^2,\end{aligned}$$

in the last term of which we have set

$$x_1' = \frac{x_1}{\sqrt{1 - v^2}}, \qquad x_2' = x_2, \qquad x_3' = x_3. \tag{31}$$

By assigning to every point in R_e with co-ordinates (x_1, x_2, x_3) the point with co-ordinates (x'_1, x'_2, x'_3) as given by (31), we effect a dilatation of the imprinted copy in the ratio $1 : \sqrt{1 - v^2}$ along the direction of the body's motion. Our formulæ assert that the copy thereby assumes a shape congruent to that of the body when at rest; this is the **Lorentz-Fitzgerald contraction**. In particular, the volume V that the body K' occupies at a definite moment in the space R_e is connected to its statical volume V_0 by the relation

$$V = V_0\sqrt{1 - v^2}.$$

Whenever we measure angles by optical means we determine the angles formed by the light-rays for the system of reference in which the (rigid) measuring instrument is at rest. *Again, when our eyes take the place of these instruments it is these angles that determine the visual form of objects that lie within the field of vision.* To establish the relationship between geometry and the observation of geometrical magnitudes, we must therefore take optical considerations into account. The solution of Maxwell's equations for light-rays in the æther as well as in a homogeneous medium, which is at rest in an allowable reference system, is of a form such that the component of the "phase" quantities (in complex notation) are all

$$= \text{const. } e^{2\pi i\Theta(P)}$$

in which $\Theta = \Theta(P)$ is, with the omission of an additive constant, the phase determined by the conditions set down; it is a function of the world-point which here occurs as the argument. If the world co-ordinates are transformed linearly in any way, the components in the new co-ordinate system will again have the same form with the same phase-function Θ. The phase is accordingly an invariant.

For a plane wave it is a **linear** and (if we exclude absorbing media) real function of the world-co-ordinates of P; hence the phase-difference at two arbitrary points $\Theta(B) - \Theta(A)$ is a linear form of the arbitrary displacement $\mathbf{x} = \overrightarrow{AB}$, that is, a co-variant world-vector. If we represent this by the corresponding displacement $\mathbf{l}$ (we shall allude to it briefly as the light-ray $\mathbf{l}$) then

$$\Theta(B) - \Theta(A) = (\mathbf{l}, \mathbf{x}).$$

If we split it up by means of the time-like vector $\mathbf{e}$ into space and time and set

$$\mathbf{l} = \nu \mid \frac{\nu}{q} \mathsf{a} \tag{32}$$

so that the space-vector a in $R_{\mathbf{e}}$ is of unit length

$$\mathbf{x} = \Delta t \mid \mathsf{x},$$

then the phase-difference is

$$\nu \left\{ \frac{(\mathsf{a}, \mathsf{x})}{q} - \Delta t \right\}.$$

From this we see that ν signifies the frequency, q the velocity of transmission, and a the direction of the light-ray in the space $R_{\mathbf{e}}$. Maxwell's equations tell us that in the æther the velocity of transmission $q = 1$, or that

$$(\mathbf{l}, \mathbf{l}) = 0.$$

If we split the world up into space and time in two ways, firstly by means of $\mathbf{e}$, secondly by means of $\mathbf{e}'$, and distinguish the magnitudes derived from the second process by accents we immediately find as a result of the invariance of $(\mathbf{l}, \mathbf{l})$ the law

$$\nu^2 \left(\frac{1}{q^2} - 1 \right) = \nu'^2 \left(\frac{1}{q'^2} - 1 \right). \tag{33}$$

If we fix our attention on two light-rays $\mathbf{l}_1$, $\mathbf{l}_2$ with frequencies ν_1, ν_2 and velocities of transmission q_1, q_2 then

$$(\mathbf{l}_1, \mathbf{l}_2) = \nu_1\nu_2 \left\{ \frac{\mathsf{a}_1 \mathsf{a}_2}{q_1 q_2} - 1 \right\}.$$

If they make an angle ω to with one another, then

$$\nu_1\nu_2 \left\{ \frac{\cos\omega}{q_1 q_2} - 1 \right\} = \nu_1'\nu_2' \left\{ \frac{\cos\omega'}{q_1' q_2'} - 1 \right\}. \tag{34}$$

For the æther, these equations become

$$q = q' \ (= 1), \qquad \nu_1\nu_2 \sin^2\frac{\omega}{2} = \nu_1'\nu_2' \sin^2\frac{\omega'}{2}. \tag{35}$$

Finally, to get the relationship between the frequencies ν and ν' we assume a body that is at rest in $R_\mathbf{e}'$; let it have the velocity v in the space $R_\mathbf{e}$, then, as before, we must set

$$\mathbf{e}' = h \mid h\mathsf{v} \text{ in } R_\mathbf{e}. \tag{26}$$

From (26) and (32) it follows that

$$\nu' = -(\mathbf{l}, \mathbf{e}') = \nu h \left\{ 1 - \frac{(\mathsf{a}, \mathsf{v})}{q} \right\}.$$

Accordingly, if the direction of the light-ray in $R_\mathbf{e}$ makes an angle θ with the velocity of the body, then

$$\frac{\nu'}{\nu} = \frac{1 - \dfrac{v\cos\theta}{q}}{\sqrt{1 - v^2}}. \tag{36}$$

(36) is Doppler's Principle. For example, since a sodium-molecule which is at rest in an allowable system remains objectively the same, this relationship (36) will exist between the frequency ν' of a sodium-molecule which is at rest and ν the frequency of a sodium-molecule moving with a velocity ν, both frequencies being observed in a spectroscope which is at rest; θ is the angle between the direction of motion of the molecule and the light-ray which enters the spectroscope. If we substitute (36) in (33) we get an equation between q and q' which enables us to calculate the velocity of propagation q in a moving medium from the velocity of propagation q' in the same medium at rest; for example, in water, v now represents the rate of flow of the water; θ represents the angle that the direction of flow of the water makes with the light-rays. If we suppose these two directions to coincide, and then neglect powers of v higher than the first (since v is in practice very small compared with the velocity of light), we get

$$q = q' + v(1 - q'^2);$$

that is, **not** the whole of the velocity v of the medium is added to the velocity of propagation, but only the fraction $1 - \frac{1}{n^2}$ (in which $n = \frac{1}{q'}$ is the index of refraction of the medium). Fresnel's "convection-co-efficient" $1 - \frac{1}{n^2}$ was determined experimentally by Fizeau long before the advent of the theory of relativity by making two light-rays from the same source interfere, after one had travelled through water which was at rest whilst the other had travelled through water which was in motion (*vide* note 9). The fact that the theory of relativity accounts for this remarkable result shows that it is valid for the optics and electrodynamics of moving media (and also that in such cases the relativity principle, which is derived from that of Lorentz and Einstein by putting q for c, does not hold; one might be tempted to believe this erroneously from the equation of wave-motion that holds in such cases). We shall find the special form of (34) for the *æther*, in which $q = q' = 1$ (cf. (35)), to be

$$\sin^2 \frac{\omega}{2} = \frac{(1 - v\cos\theta_1)(1 - v\cos\theta_2)}{1 - v^2} \sin^2 \frac{\omega'}{2}.$$

If the reference-space R_{e} happens to be the one on which the theory of planets is commonly founded (and in which the centre of mass of the solar system is at rest), and if the body in question is the earth (on which an observing instrument is situated), v its velocity in R_{e}, ω the angle in R_{e} that two rays which reach the solar system from two infinitely distant stars make with one another, θ_1, θ_2 the angles which these rays make with the direction of motion of the earth in R_{e}, then the angle ω', at which the stars are observed from the earth, is determined by the preceding equation. We cannot, of course, measure ω, but we note the changes in ω' (the **aberration**) by taking account of the changes in θ_1 and θ_2 in the course of a year.

The formulæ which give the relationship between time, proper-time, volume and statical volume are also valid in the case of **non-uniform motion**. If $d\mathbf{x}$ is the infinitesimal displacement that a moving point-mass experiences during an infinitesimal length of time in the world, then

$$d\mathbf{x} = ds \cdot \mathbf{u}, \qquad (\mathbf{u}, \mathbf{u}) = -1, \qquad ds > 0$$

give the proper-time ds and the world-direction $\mathbf{u}$ of this displacement. The integral

$$\int ds = \int \sqrt{-(d\mathbf{x}, d\mathbf{x})}$$

taken over a portion of the world-line is the proper-time that elapses during this part of the motion: it is independent of the manner in which the world has been split up into space and time and, provided the motion is not too rapid, will be indicated by a clock that is rigidly attached to the point-mass. If we use any linear co-ordinates x_i whatsoever in the world, and the proper time s as our parameters to represent our world-line analytically (just as we use length of arc in three-dimensional geometry), then

$$\frac{dx_i}{ds} = u^i$$

are the (contra-variant) components of $\mathbf{u}$, and we get $\sum_i u_i u^i = -1$. If we split up the world into space and time by means of $\mathbf{e}$, we find

$$\mathbf{u} = \frac{1}{\sqrt{1-v^2}} \,\Big|\, \frac{\mathsf{v}}{\sqrt{1-v^2}} \text{ in } \mathsf{R}_\mathbf{e}$$

in which v is the velocity of the mass-point; and we find that the time dt that elapses during the displacement $d\mathbf{x}$ in $\mathsf{R}_\mathbf{e}$ and the proper-time ds are connected by

$$ds = dt\sqrt{1-v^2}. \tag{37}$$

If two world-points A, B are so placed with respect to one another that $\overrightarrow{AB}$ is a time-like vector pointing into the future, then A and B may be connected by world-lines, whose directions all likewise satisfy this condition: in other words, point-masses that leave A may reach B. The proper-time necessary for them to do this is dependent on the world-line; it is longest for a point-mass that passes from A to B by uniform translation. For if we split up the world into space and time in such a way that A and B occupy the same point in space, this motion degenerates simply to rest, and we derive the proposition (37) which states that the proper-time lags behind the time t. The life-processes of mankind may well be compared to a clock. Suppose we have two twin-brothers who take leave from one another at a world-point A, and suppose one remains at home (that is, permanently at rest in an allowable reference-space), whilst the other sets out on voyages, during which he moves with velocities (relative to "home") that approximate to that of light. When the wanderer returns home in later years he will appear appreciably younger than the one who stayed at home.

An element of mass dm (of a continuously extended body) that moves with a velocity whose numerical value is v occupies at a particular moment a volume dV which is connected with its statical volume dV_0 by the formula

$$dV = dV_0\sqrt{1 - v^2}.$$

Accordingly, we have the relation between the density $\dfrac{dm}{dV} = \mu$ and the statical density $\dfrac{dm}{dV_0} = \mu_0$:

$$\mu_0 = \mu\sqrt{1 - v^2}.$$

μ_0 is an invariant, and $\mu_0\mathbf{u}$ with components $\mu_0 u^i$ is thus a contravariant vector, the "flux of matter," which is determined by the motion of the mass independently of the co-ordinate system. It satisfies the equation of continuity

$$\sum_i \frac{\partial(\mu_0 u^i)}{\partial x_i} = 0.$$

The same remarks apply to electricity. If it is associated with matter so that de is the electric charge of the element of mass dm, then the statical density $\rho_0 = \frac{de}{dV_0}$ is connected to the density $\rho = \frac{de}{dV}$ by

$$\rho_0 = \rho\sqrt{1 - v^2},$$

then

$$s^i = \rho_0 u^i$$

are the contra-variant components of the electric current (4-vector); this corresponds exactly to the results of § 20. In Maxwell's phenomenological theory of electricity, the concealed motions of the electrons are not taken into account as motions of matter, consequently electricity is not supposed attached to matter in his theory. The only way to explain how it is that a piece of matter carries a certain charge is to say this charge is that which is simultaneously in the portion of space that is occupied by the matter at the moment under consideration. From this we see that the charge is not, as in the theory of electrons, an invariant determined by the portion of matter, but is dependent on the way the world has been split up into space and time.

§ 23. The Electrodynamics of Moving Bodies

By splitting up the world into space and time we split up all tensors. We shall first of all investigate purely mathematically how this comes about, and shall then apply the results to derive the fundamental equations of electrodynamics for moving bodies. Let us take an n-dimensional metrical space, which we shall call "world," based on the metrical groundform $(\mathbf{x}, \mathbf{x})$. Let $\mathbf{e}$ be a vector in it, for which $(\mathbf{e}, \mathbf{e}) = e \neq 0$. We split up the world in the usual way into space $R_{\mathbf{e}}$ and time in terms of $\mathbf{e}$. Let e_1, $e_2, \ldots, e_{n-1}$ be any co-ordinate system in the space $R_{\mathbf{e}}$, and let $\mathbf{e}_1$, $\mathbf{e}_2, \ldots, \mathbf{e}_{n-1}$ be the displacements of the world that are orthogonal to $\mathbf{e} = \mathbf{e}_0$ and that are produced in $R_{\mathbf{e}}$ by $e_1, e_2, \ldots, e_{n-1}$. In the co-ordinate system $\mathbf{e}_i$ $(i = 0, 1, 2, \ldots, n-1)$ "belonging to $R_{\mathbf{e}}$" and representing the world, the scheme of the co-variant components of the metrical ground-tensor has the form

$$\begin{vmatrix} e & 0 & 0 \\ 0 & g_{11} & g_{12} \\ 0 & g_{21} & g_{22} \end{vmatrix} \qquad (n = 3).$$

As an example, we shall consider a tensor of the second order and suppose it to have components T_{ik} in this co-ordinate system. Now, we assert that it splits up, in a manner dependent only on $\mathbf{e}$, according to the following scheme:

$$\begin{array}{|c|cc|} \hline T_{00} & T_{01} & T_{02} \\ \hline T_{10} & T_{11} & T_{12} \\ T_{20} & T_{21} & T_{22} \\ \hline \end{array}$$

that is, into a scalar, two vectors and a tensor of the second order existing in $R_{\mathbf{e}}$, which are here characterised by their components in the co-ordinate system e_i $(i = 1, 2, \ldots, n-1)$.

For if the arbitrary world-displacement $\mathbf{x}$ splits up in terms of $\mathbf{e}$ thus

$$\mathbf{x} = \xi \mid \mathsf{x}$$

and if, when we divide $\mathbf{x}$ into two factors, one of which is proportional to $\mathbf{e}$ and the other orthogonal to $\mathbf{e}$, we have

$$\mathbf{x} = \xi\mathbf{e} + \mathbf{x}^*$$

then, if $\mathbf{x}$ has components ξ^i, we get

$$\mathbf{x} = \sum_{i=0}^{n-1} \xi^i \mathbf{e}_i, \qquad \xi = \xi^0, \qquad \mathbf{x}^* = \sum_{i=1}^{n-1} \xi^i \mathbf{e}_i, \qquad \mathsf{x} = \sum_{i=1}^{n-1} \xi^i e_i.$$

Thus, without using a co-ordinate system we may represent the splitting up of a tensor in the following manner. If $\mathbf{x}$, $\mathbf{y}$ are any two arbitrary displacements of the world, and if we set

$$\mathbf{x} = \xi\mathbf{e} + \mathbf{x}^*, \qquad \mathbf{y} = \eta\mathbf{e} + \mathbf{y}^*, \tag{38}$$

so that $\mathbf{x}^*$ and $\mathbf{y}^*$ are orthogonal to $\mathbf{e}$, then the bilinear form belonging to the tensor of the second order is

$$T(\mathbf{x}, \mathbf{y}) = \xi\eta T(\mathbf{e}, \mathbf{e}) + \eta T(\mathbf{x}^*, \mathbf{e}) + \xi T(\mathbf{e}, \mathbf{y}^*) + T(\mathbf{x}^*, \mathbf{y}^*).$$

Hence, if we interpret $\mathbf{x}^*$, $\mathbf{y}^*$ as the displacements of the world orthogonal to $\mathbf{e}$, which produce the two arbitrary displacements x, y of the space, we get

1. a scalar $T(\mathbf{e}, \mathbf{e}) = J = \mathsf{J}$,
2. two linear forms (vectors) in the space $\mathsf{R}_{\mathbf{e}}$, defined by

$$\mathsf{L}(\mathbf{x}) = T(\mathbf{x}^*, \mathbf{e}), \qquad \mathsf{L}'(\mathsf{x}) = T(\mathbf{e}, \mathbf{x}^*),$$

3. a bilinear form (tensor) in the space $R_{\mathbf{e}}$, defined by

$$T(x, y) = T(\mathbf{x}^*, \mathbf{y}^*).$$

If $\mathbf{x}$, $\mathbf{y}$ are arbitrary world-displacements that produce x, y, respectively in $R_{\mathbf{e}}$ we must replace $\mathbf{x}^*$, $\mathbf{y}^*$ in this definition by $\mathbf{x}-\xi\mathbf{e}$, $\mathbf{y}-\eta\mathbf{e}$ in accordance with (38); in these,

$$\xi = \frac{1}{e}(\mathbf{x}, \mathbf{e}), \qquad \eta = \frac{1}{e}(\mathbf{y}, \mathbf{e}).$$

If we now set

$$T(\mathbf{x}, \mathbf{e}) = L(\mathbf{x}), \qquad T(\mathbf{e}, \mathbf{x}) = L'(\mathbf{x}),$$

we get

$$\left.\begin{aligned} L(x) &= L(\mathbf{x}) - \frac{J}{e}(\mathbf{x}, \mathbf{e}), \qquad L'(x) = L'(\mathbf{x}) - \frac{J}{e}(\mathbf{x}, \mathbf{e}), \\ T(x, y) &= T(\mathbf{x}, \mathbf{y}) - \frac{1}{e}(\mathbf{y}, \mathbf{e})L(\mathbf{x}) - \frac{1}{e}(\mathbf{x}, \mathbf{e})L'(\mathbf{y}) + \frac{J}{e^2}(\mathbf{x}, \mathbf{e})(\mathbf{y}, \mathbf{e}). \end{aligned}\right\} \tag{39}$$

The linear and bilinear forms (vectors and tensors) of $R_{\mathbf{e}}$ on the left may be represented by the world-vectors and world-tensors on the right which are derived uniquely from them. In the above representation by means of components, this amounts to the following: that, for example,

$$T = \begin{vmatrix} T_{11} & T_{12} \\ T_{21} & T_{22} \end{vmatrix} \quad \text{is represented by} \quad \begin{vmatrix} 0 & 0 & 0 \\ 0 & T_{11} & T_{12} \\ 0 & T_{21} & T_{22} \end{vmatrix}.$$

It is immediately clear that in all calculations the tensors of space may be replaced by the representative world-tensors. We shall, however, use this device only in the case when, if one space-tensor is λ times another, the same is true of the representative world-tensors.

If we base our calculations of components on an **arbitrary** co-ordinate system, in which

$$\mathbf{e} = (e^0, e^1, \ldots, e^{n-1})$$

then the invariant is

$$J = T_{ik}e^i e^k \quad \text{and} \quad e = e^i e_i.$$

But the two vectors and the tensor in $R_{\mathbf{e}}$ have as their representatives in the world, according to (39), the two vectors and the tensor with components:

$$\begin{aligned}
\mathsf{L} &: L_i - \frac{J}{e}e_i \qquad L_i = T_{ik}e^k, \\
\mathsf{L}' &: L'_i - \frac{J}{e}e_i \qquad L'_i = T_{ki}e^k; \\
\mathsf{T} &: T_{ik} - \frac{e_k L_i + e_i L'_k}{e} + \frac{J}{e^2}e_i e_k.
\end{aligned}$$

In the case of a skew-symmetrical tensor, J becomes $= 0$ and $\mathsf{L}' = -\mathsf{L}$; our formulæ degenerate into

$$\begin{aligned}
\mathsf{L} &: L_i = T_{ik}e^k, \\
\mathsf{T} &: T_{ik} + \frac{e_i L_k - e_k L_i}{e}.
\end{aligned}$$

A linear world-tensor of the second order splits up in space into a vector and a linear space-tensor of the second order.

Maxwell's field-equations for bodies at rest have been set out in § 20. H. Hertz was the first to attempt to extend them so that they might apply generally for moving bodies. Faraday's Law of Induction states that the time-decrement of the flux of induction enclosed in a conductor is equal to the induced electromotive force, that is

$$-\frac{1}{c}\frac{d}{dt}\int B_n\,do = \int \mathbf{E}\,d\mathbf{r}. \tag{40}$$

The surface-integral on the left, if the conductor be in motion, must be taken over a surface stretched out inside the conductor and moving with it. Since Faraday's Law of Induction has been proved for just those cases in which the time-change of the flux of induction within the conductor is brought about by the motion of the conductor, Hertz did not doubt that this law was equally valid for the case, too, when the conductor was in motion. The equation $\operatorname{div}\mathbf{B} = 0$ remains unaffected. From vector analysis we know that, taking this equation into consideration, the law of induction (40) may be expressed in the differential form:

$$\operatorname{curl}\mathbf{E} = -\frac{1}{c}\frac{\partial \mathbf{B}}{\partial t} + \frac{1}{c}\operatorname{curl}[\mathbf{v}, \mathbf{B}] \tag{41}$$

in which $\frac{\partial \mathbf{B}}{\partial t}$ denotes the differential co-efficient of $\mathbf{B}$ with respect to the time for a fixed point in space, and $\mathbf{v}$ denotes the velocity of the matter.

Remarkable inferences may be drawn from (41). As in Wilson's experiment (*vide* note 10), we suppose a homogeneous dielectric between the two plates of a condenser, and assume that this dielectric moves with a constant velocity of magnitude $\mathbf{v}$ between these plates, which we shall take to be connected by means of a conducting wire. Suppose, further, that there is a homogeneous magnetic field H parallel to the plates and perpendicular to $\mathbf{v}$. We shall imagine the dielectric separated from the plates of the condenser by a narrow empty space, whose thickness we shall assume $\rightarrow 0$ in the limit. It then follows from (41) that, in the space between the plates, $\mathbf{E} - \frac{1}{c}[\mathbf{v}, \mathbf{B}]$ is derivable from a potential; since the latter must be zero at the plates which are connected by a conducting wire it is easily seen that we must have $\mathbf{E} = \frac{1}{c}[\mathbf{v}, \mathbf{B}]$. Hence a homogeneous electric field of intensity $E = \frac{\mu}{c} vH$ (in which μ denotes permeability) arises which acts perpendicularly to the plates. Consequently, a statical charge of surface-density $\frac{\epsilon\mu}{c} vH$ ($\epsilon =$ dielectric constant) must be called up on the plates. If the dielectric is a gas, this effect should manifest itself, no matter to what degree the gas has been rarefied, since $\epsilon\mu$ converges, **not** towards 0, but towards 1, at infinite rarefaction. This can have only one meaning if we are to retain our belief in the æther, namely, that the effect must occur if the æther between the plates is moving relatively to the plates and to the æther outside them. To explain induction we should, however, be compelled to assume that the æther is dragged along by the connecting wire.* General observations, Fizeau's experiment dealing with the propagation of light in flowing water, and Wilson's experiment itself, prove that

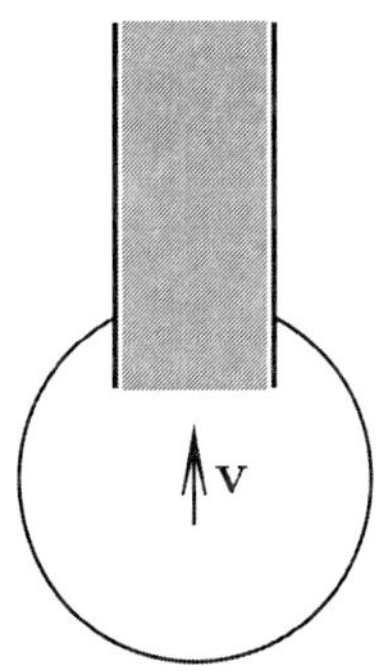

FIG. 14.

*In (41) $\mathbf{v}$ signified the velocity of the æther, **not** relative to the matter but relative to what?

this assumption is incorrect. Just as in Fizeau's experiment the convection-co-efficient $1 - \dfrac{1}{n^2}$ appears, so in the present experiment we observe only a change of magnitude

$$\frac{\epsilon\mu - 1}{c} vH$$

which vanishes when $\epsilon\mu = 1$. This seems to be an inexplicable contradiction to the phenomenon of induction in the moving conductor.

The theory of relativity offers a full explanation of this. If, as in § 20, we again set $ct = x_0$, and if we again build up a field F out of **E** and **B**, and a skew-symmetrical tensor H of the second order out of **D** and **H**, we have the field-equations

$$\left.\begin{aligned} \frac{\partial F_{kl}}{\partial x_i} + \frac{\partial F_{li}}{\partial x_k} + \frac{\partial F_{ik}}{\partial x_l} &= 0, \\ \sum_k \frac{\partial H^{ik}}{\partial x_k} &= s^i. \end{aligned}\right\} \qquad (42)$$

These hold if we regard the F_{ik}'s as co-variant, the H^{ik}'s as contra-variant components, in each case, of a tensor of the second order, but the s^i's as the contra-variant components of a vector in the four-dimensional world, since the latter are invariant in any arbitrary linear co-ordinate system. The laws of matter

$$\mathbf{D} = \epsilon\mathbf{E}, \qquad \mathbf{B} = \mu\mathbf{H}, \qquad \mathbf{s} = \sigma\mathbf{E}$$

signify, however, that if we split up the world into space and time in such a way that matter is at rest, and if F splits up into $\mathbf{E} \mid \mathbf{B}$, H into $\mathbf{D} \mid \mathbf{H}$, and s into $\rho \mid \mathbf{s}$, then the above relations hold. If we now use any arbitrary co-ordinate system, and if the world-direction of the matter has the components u^i in it then, after our explanations above, these facts assume the form

$$(a) \qquad H_i^* = \epsilon F_i^* \tag{43}$$

in which

$$F_i^* = F_{ik}u^k \quad \text{and} \quad H_i^* = H_{ik}u^k;$$

$$(b) \qquad F_{ik} - (u_i F_k^* - u_k F_i^*) = \mu\{H_{ik} - (u_i H_k^* - u_k H_i^*)\}; \tag{44}$$

$$\text{and } (c) \qquad s_i + u_i(s_k u^k) = \sigma F_i^*. \tag{45}$$

This is the invariant form of these laws. For purposes of calculation it is convenient to replace (44) by the equations

$$F_{kl}u_i + F_{li}u_k + F_{ik}u_l = \mu\{H_{kl}u_i + H_{li}u_k + H_{ik}u_l\} \tag{46}$$

which are derived directly from them. Our manner of deriving them makes it clear that they hold only for matter which is in uniform translation. We may, however, consider them as being valid also for a single body in uniform translation, if it is separated by empty space from bodies moving with velocities differing from its own.* Finally, they may also be considered to hold for matter moving in any manner whatsoever, provided that its velocity does

*This is the essential point in most applications. By applying Maxwell's statical laws to a region composed, in each case, of a body K and the empty space surrounding it and referred to the system of reference in which K is at rest, we find no discrepancies occurring in empty space when we derive results from different bodies moving relatively to one another, **because the principle of relativity holds for empty space**.

not fluctuate too rapidly. After having obtained the invariant form in this way, we may now split up the world in terms of any arbitrary **e**. Suppose the measuring instruments that are used to determine the ponderomotive effects of field to be at rest in $R_{\mathbf{e}}$. We shall use a co-ordinate system belonging to $R_{\mathbf{e}}$ and thus set

$$\begin{array}{lll}
(F_{10}, & F_{20}, & F_{30}) = (E_1, \quad E_2, \quad E_3) = \mathbf{E}, \\
(F_{23}, & F_{31}, & F_{12}) = (B_{23}, \quad B_{31}, \quad B_{12}) = \mathbf{B}, \\
\hline
(H_{10}, & H_{20}, & H_{30}) = (D_1, \quad D_2, \quad D_3) = \mathbf{D}, \\
(H_{23}, & H_{31}, & H_{12}) = (H_{23}, \quad H_{31}, \quad H_{12}) = \mathbf{H},
\end{array}$$

$$s^0 = \rho; \qquad (s^1, s^2, s^3) = (s^1, s^2, s^3) = \mathbf{s},$$

$$u^0 = \frac{1}{\sqrt{1-v^2}} \qquad (u^1, u^2, u^3) = \frac{(v^1, v^2, v^3)}{\sqrt{1-v^2}} = \frac{\mathbf{v}}{\sqrt{1-v^2}},$$

we hereby again arrive at **Maxwell's field-equations, which are thus valid in a totally unchanged form, not only for static, but also for moving matter**. Does this not, however, conflict violently with the observations of induction, which appear to require the addition of a term as in (41)? No; for these observations do not really determine the intensity of field **E**, but only the current which flows in the conductor; for moving bodies, however, the connection between the two is given by a different equation, namely, by (45).

If we write down those equations of (43), (45), which correspond to the components with indices $i = 1, 2, 3$, and those of (46), which correspond to

$$(i, k, l) = (2, 3, 0), \quad (3, 1, 0), \quad (1, 2, 0)$$

(the others are superfluous), the following results, as is easily seen, come about. If we set

$$\mathbf{E} + [\mathbf{v}, \mathbf{B}] = \mathbf{E}^*, \qquad \mathbf{D} + [\mathbf{v}, \mathbf{H}] = \mathbf{D}^*,$$
$$\mathbf{B} - [\mathbf{v}, \mathbf{E}] = \mathbf{B}^*, \qquad \mathbf{H} - [\mathbf{v}, \mathbf{D}] = \mathbf{H}^*,$$

then

$$\mathbf{D}^* = \epsilon \mathbf{E}^*, \qquad \mathbf{B}^* = \mu \mathbf{H}^*.$$

If, in addition, we resolve $\mathbf{s}$ into the "convection-current" $\mathbf{c}$ and the "conduction-current" $\mathbf{s}^*$, that is,

$$\mathbf{s} = \mathbf{c} + \mathbf{s}^*,$$

$$\mathbf{c} = \rho^* \mathbf{v}, \qquad \rho^* = \frac{\rho - (\mathbf{v}, \mathbf{s})}{1 - v^2} = \rho - (\mathbf{v}, \mathbf{s}^*),$$

then

$$\mathbf{s}^* = \frac{\sigma \mathbf{E}^*}{\sqrt{1 - v^2}}.$$

Everything now becomes clear: the current is composed partly of a convection-current which is due to the motion of charged matter, and partly of a conduction-current, which is determined by the conductivity σ of the substance. The conduction-current is calculated from Ohm's Law, if the electromotive force is defined by the line-integral, not of $\mathbf{E}$, but of $\mathbf{E}^*$. An equation exactly analogous to (41) holds for $\mathbf{E}^*$, namely:

$$\operatorname{curl} \mathbf{E}^* = -\frac{\partial \mathbf{B}}{\partial t} + \operatorname{curl}[\mathbf{v}, \mathbf{B}] \quad \text{(we now always take } c = 1\text{)}$$

or expressed in integrals, as in (40),

$$-\frac{d}{dt} \int B_n \, do = \int \mathbf{E}^* \, d\mathbf{r}.$$

This explains fully Faraday's phenomenon of induction in moving conductors. For Wilson's experiment, according to the present theory, curl $\mathbf{E} = \mathbf{0}$, that is, $\mathbf{E}$ will be zero between the plates. This gives us the constant values of the individual vectors (of which the electrical ones are perpendicular to the plates, whilst the magnetic ones are directed parallel to the plates and perpendicular to the velocity): these values are:

$$E^* = vB^* = v\mu H^* = \mu v(H + vD),$$
$$D = D^* - vH = \epsilon E^* - vH.$$

If we substitute the expression for E^* in the first equation, we get

$$D = v\{(\epsilon\mu - 1)H + \epsilon\mu v D\},$$
$$D = \frac{\epsilon\mu - 1}{1 - \epsilon\mu v^2} vH.$$

This is the value of the superficial density of charge that is called up on the condenser plates: it agrees with our observations since, on account of v being very small, the denominator in our formula differs very little from unity.

The boundary conditions at the boundary between the matter and the æther are obtained from the consideration that the field-magnitudes F and H must not suffer any sudden (discontinuous) changes in moving along with the matter; but, in general, they will undergo a sudden change, at some fixed space-point imagined in the æther for the sake of clearness, at the instant at which the matter passes over this point. If s is the proper-time of an elementary particle of matter then

$$\frac{dF_{ik}}{ds} = \frac{\partial F_{ik}}{\partial x_l} u^l$$

must remain finite everywhere. If we set

$$\frac{\partial F_{ik}}{\partial x_l} = -\left(\frac{\partial F_{kl}}{\partial x_i} + \frac{\partial F_{li}}{\partial x_k}\right)$$

we see that this expression

$$= \frac{\partial F_i^*}{\partial x_k} - \frac{\partial F_k^*}{\partial x_i}.$$

Consequently, $\mathbf{E}^*$ cannot have a surface-curl (and $\mathbf{B}$ cannot have a surface-divergence).

The fundamental equations for moving bodies were deduced by Lorentz from the theory of electrons in a form equivalent to the above before the discovery of the principle of relativity. This is not surprising, seeing that Maxwell's fundamental laws for the æther satisfy the principle of relativity, and that the theory of electrons derives those governing the behaviour of matter by building up mean values from these laws. Fizeau's and Wilson's experiments and another analogous one, that of Röntgen and Eichwald (*vide* note 11), prove that the electromagnetic behaviour of matter is in accordance with the principle of relativity; the problems of the electrodynamics of moving bodies first led Einstein to enunciate it. We are indebted to Minkowski for recognising clearly that the fundamental equations for moving bodies are determined uniquely by the principle of relativity if Maxwell's theory for matter at rest is taken for granted. He it was, also, who formulated it in its final form (*vide* note 12).

Our next aim will be to subjugate **mechanics**, which does not obey the principle in its classical form, to the principle of relativity of Einstein, and to inquire whether the modifications that the latter demands can be made to harmonise with the facts of experiment.

§ 24. Mechanics according to the Principle of Relativity

On the theory of electrons we found the mechanical effect of the electromagnetic field to depend on a vector **p** whose contravariant components are

$$p^i = F^{ik} s_k = \rho_0 F^{ik} u_k.$$

It therefore satisfies the equation

$$p^i u_i = (\mathbf{p}, \mathbf{u}) = 0 \tag{47}$$

in which **u** is the world-direction of the matter. If we split up **p** and **u** in any way into space and time thus

$$\left.\begin{aligned} \mathbf{u} &= h \mid h\mathsf{v}, \\ \mathbf{p} &= \lambda \mid \mathsf{p}, \end{aligned}\right\} \tag{48}$$

we get p as the force-density and, as we see from (47) or from

$$h\{\lambda - (\mathsf{p}, \mathsf{v})\} = 0$$

that λ is the work-density.

We arrive at the fundamental law of the mechanics which agrees with Einstein's Principle of Relativity by the same method as that by which we obtain the fundamental equations of electromagnetics. We assume that Newton's Law remains valid in the system of reference in which the matter is at rest. We fix our attention on the point-mass m, which is situated at a definite world-point O and split up our quantities in terms of its world-direction **u** into space and time. m is momentarily at rest in $R_{\mathbf{u}}$. Let μ_0 be the density in $R_{\mathbf{u}}$ of the matter at the point O. Suppose that, after an infinitesimal element of time ds has elapsed,

m has the world-direction $\mathbf{u} + d\mathbf{u}$. It follows from $(\mathbf{u}, \mathbf{u}) = -1$ that $(\mathbf{u} \cdot d\mathbf{u}) = 0$. Hence, splitting up with respect to $\mathbf{u}$, we get

$$\mathbf{u} = 1 \mid \mathsf{O}, \qquad d\mathbf{u} = 0 \mid d\mathsf{v}, \qquad \mathbf{p} = 0 \mid \mathsf{p}.$$

It follows from

$$\mathbf{u} + d\mathbf{u} = 1 \mid d\mathsf{v}$$

that $d\mathsf{v}$ is the relative velocity acquired by m (in $\mathsf{R}_\mathbf{u}$) during the time ds. Thus there can be no doubt that the fundamental law of mechanics is

$$\mu_0 \frac{d\mathsf{v}}{ds} = \mathsf{p}.$$

From this we derive at once the invariant form

$$\mu_0 \frac{d\mathbf{u}}{ds} = \mathbf{p}, \tag{49}$$

which is quite independent of the manner of splitting up. In it, μ_0 is the **statical density**, that is, the density of the mass when at rest; ds is the **proper-time** that elapses during the infinitesimal displacement of the particle of matter, during which its world-direction increases by $d\mathbf{u}$.

Resolution into terms of $\mathbf{u}$ is a partition which would alter during the motion of the particle of matter. If we now split up our quantities, however, into space and time by means of some fixed time-like vector $\mathbf{e}$ that points into the future and satisfies the condition of normality $(\mathbf{e}, \mathbf{e}) = -1$, then, by (48), (49) resolves into

$$\left.\begin{aligned} \mu_0 \frac{d}{ds}\left(\frac{1}{\sqrt{1-v^2}}\right) &= \lambda, \\ \mu_0 \frac{d}{ds}\left(\frac{\mathsf{v}}{\sqrt{1-v^2}}\right) &= \mathsf{p}. \end{aligned}\right\} \tag{50}$$

If, in this partition or resolution, t denotes the time, dV the volume, and dV_0 the static volume of the particle of matter at a definite moment, its mass, however, being $m = \mu_0\, dV_0$, and if

$$\mathsf{p}\, dv = \mathsf{P}, \qquad \lambda\, dV = \mathsf{L}$$

denotes the force acting on the particle and its work, respectively, then if we multiply our equations by dV and take into account that

$$\mu_0\, dV \cdot \frac{d}{ds} = m\sqrt{1-v^2} \cdot \frac{d}{ds} = m \cdot \frac{d}{dt}$$

and that the mass m remains constant during the motion, we get finally

$$\frac{d}{dt}\left(\frac{m}{\sqrt{1-v^2}}\right) = \mathsf{L}, \tag{51}$$

$$\frac{d}{dt}\left(\frac{m\mathsf{v}}{\sqrt{1-v^2}}\right) = \mathsf{P}. \tag{52}$$

These are the equations for the mechanics of the point-mass. The equation of momentum (52) differs from that of Newton only in that the (kinetic) momentum of the point-mass is not $m\mathsf{v}$ but $= \dfrac{m\mathsf{v}}{\sqrt{1-v^2}}$. The equation of energy (51) seems strange at first: if we expand it into powers of v, we get

$$\frac{m}{\sqrt{1-v^2}} = m + \frac{mv^2}{2} + \dots,$$

so that if we neglect higher powers of v and also the constant m we find that the expression for the kinetic energy degenerates into the one given by classical mechanics.

This shows that the deviations from the mechanics of Newton are, as we suspected, of only the second order of magnitude in the velocity of the point-masses as compared with the velocity of light. Consequently, in the case of the small velocities with which we usually deal in mechanics, no difference can be demonstrated experimentally. It will become perceptible only for velocities that approximate to that of light; in such cases the inertial resistance of matter against the accelerating force will increase to such an extent that the possibility of actually reaching the velocity of light is excluded. **Cathode rays** and the β-radiations emitted by radioactive substances have made us familiar with free negative electrons whose velocity is comparable to that of light. Experiments by Kaufmann, Bucherer, Ratnowsky, Hupka, and others, have shown in actual fact that the longitudinal acceleration caused in the electrons by an electric field or the transverse acceleration caused by a magnetic field is just that which is demanded by the theory of relativity. A further confirmation based on the motion of the electrons circulating in the atom has been found recently in the *fine structure* of the spectral lines emitted by the atom (*vide* note 13). Only when we have added to the fundamental equations of the electron theory, which, in § 20, was brought into an invariant form agreeing with the principle of relativity, the equation $s^i = \rho_0 u^i$, namely, the assertion that electricity is associated with matter, and also the fundamental equations of mechanics, do we get a complete cycle of connected laws, in which a statement of the actual unfolding of natural phenomena is contained, independent of all conventions of notation. Now that this final stage has been carried out, we may at last claim to have proved the validity of the principle of relativity for a certain region, that of electromagnetic phenomena.

In the electromagnetic field the ponderomotive vector p_i is derived from a tensor S_{ik}, dependent only on the local values of the phase-quantities, by the formulæ:

$$p^i = -\frac{\partial S_i^k}{\partial x_k}.$$

In accordance with the universal meaning ascribed to the conception *energy* in physics, we must assume that this holds not only for the electromagnetic field but for every region of physical phenomena, and that it is expedient to regard this tensor instead of the ponderomotive force as the primary quantity. Our purpose is to discover for every region of phenomena in what manner the energy-momentum-tensor (whose components S_{ik} must always satisfy the condition of symmetry) depends on the characteristic field- or phase-quantities. The left-hand side of the mechanical equations

$$\mu_0 \frac{du^i}{ds} = p_i$$

may be reduced directly to terms of a "kinetic" energy-momentum-tensor thus:

$$U_{ik} = \mu_0 u_i u_k.$$

For

$$\frac{\partial U_i^k}{\partial x_k} = u_i \frac{\partial(\mu_0 u^k)}{\partial x_k} + \mu_0 u^k \frac{\partial u_i}{\partial x_k}.$$

The first term on the right $= 0$, on account of the equation of continuity for matter; the second $= \mu_0 \dfrac{du^i}{ds}$ because

$$u^k \frac{\partial u_i}{\partial x_k} = \frac{\partial u_i}{\partial x_k}\frac{\partial x_k}{\partial s} = \frac{du_i}{ds}.$$

Accordingly, the equations of mechanics assert that the complete energy-momentum-tensor $T_{ik} = U_{ik} + S_{ik}$ composed of the kinetic tensor U and the potential tensor S satisfies the theorems of conservation

$$\frac{\partial T_i^k}{\partial x_k} = 0.$$

The Principle of the Conservation of Energy is here expressed in its clearest form. But, according to the theory of relativity, it is indissolubly connected with the principle of the conservation of momentum and **the conception *momentum* (or *impulse*) must claim just as universal a significance as that of energy**. If we express the kinetic tensor at a world-point in terms of a normal co-ordinate system such that, relatively to it, the matter itself is momentarily at rest, its components assume a particularly simple form, namely, $U_{00} = \mu_0$ (or $= c^2\mu_0$, if we use the c.g.s. system, in which c is not $= 1$), and all the remaining components vanish. This suggests the idea that mass is to be regarded as concentrated potential energy that moves on through space.

§ 25. Mass and Energy

To interpret the idea expressed in the preceding sentence we shall take up the thread by returning to the consideration of the motion of the electron. So far, we have imagined that we have to write for the force $\mathbf{P}$ in its equation of motion (52) the following:

$$\mathbf{P} = e\big(\mathbf{E} + [\mathbf{v}, \mathbf{H}]\big) \quad (e = \text{charge of the electron})$$

that is, that $\mathbf{P}$ is composed of the impressed electric and magnetic fields $\mathbf{E}$ and $\mathbf{H}$. Actually, however, the electron is subject not only to the influence of these external fields during its motion but also to the accompanying field which it itself generates. A difficulty arises, however, in the circumstance that we do not know the constitution of the electron, and that we do not know the nature and laws of the cohesive pressure that keeps the electron together against the enormous centrifugal forces of the negative charge compressed in it. In any case the electron at rest and its electric field (which we consider as part of it) is a physical system, which is in a state of statical equilibrium—and that is the essential point. Let us choose a normal co-ordinate system in which the electron is at rest. Suppose its energy-tensor to have components t_{ik}. The fact that the electron is at rest is expressed by the vanishing of the energy-flux of whose components are t_{0i} $(i = 1, 2, 3)$. The 0th condition of equilibrium

$$\frac{\partial t_i^k}{\partial x_k} = 0 \tag{53}$$

then tells us that the energy-density t_{00} is independent of the time x_0. On account of symmetry the components t_{i0} $(i = 1, 2, 3)$ of the momentum-density each also vanish. If $\mathbf{t}^{(1)}$ is the vector whose components are t_{11}, t_{12}, t_{13}, the condition for equilibrium (53), $(i = 1)$, gives

$$\operatorname{div} \mathbf{t}^{(1)} = 0.$$

Hence we have, for example,

$$\operatorname{div}(x_2 \mathbf{t}^{(1)}) = x_2 \operatorname{div} \mathbf{t}^{(1)} + t_{12} = t_{12}$$

and since the integral of a divergence is zero (we may assume that the t's vanish at infinity at least as far as to the fourth order) we get

$$\int t_{12}\, dx_1\, dx_2\, dx_3 = 0.$$

In the same way we find that, although the t_{ik}'s (for $i, k = 1, 2, 3$) do not vanish, their volume integrals $\int t_{ik}\, dV_0$ do so. We may regard these circumstances as existing for every system in statical equilibrium. The result obtained may be expressed by invariant formulæ for the case of any arbitrary co-ordinate system thus:

$$\int t_{ik}\, dV_0 = E_0 u_i u_k \quad (i, k = 0, 1, 2, 3). \tag{54}$$

E_0 is the energy-content (measured in the space of reference for which the electron is at rest), u_i are the co-variant components of the world-direction of the electron, and dV_0 the statical volume of an element of space (calculated on the supposition that the whole of space participates in the motion of the electron). (54) is rigorously true for uniform translation. We may also apply the formula in the case of non-uniform motion if $\mathbf{u}$ does not change too suddenly in space or in time. The components

$$\bar{p}^i = -\frac{\partial t^{ik}}{\partial x_k}$$

of the ponderomotive effect, exerted on the electron by itself, are however, then no longer $= 0$.

If we assume the electron to be entirely without mass, and if p^i is the "4-force" acting from without, then equilibrium demands that

$$\bar{p}^i + p^i = 0. \tag{55}$$

We split up $\mathbf{u}$ and $\mathbf{p}$ into space and time in terms of a fixed $\mathbf{e}$, getting

$$\mathbf{u} = h \mid h\mathsf{v}, \qquad \mathbf{p} = (p^i) = \lambda \mid \mathsf{p}$$

and we integrate (55) with respect to the volume $dV = dV_0\sqrt{1-v^2}$. Since, if we use a normal co-ordinate system corresponding to R_{e}, we have

$$\int \vec{p}^i \, dV = \int \vec{p}^i \, dx_1 \, dx_2 \, dx_3 = -\frac{d}{dx_0} \int t^{i0} \, dx_1 \, dx_2 \, dx_3$$
$$= -\frac{d}{dx_0}(E_0 u^0 u^i \sqrt{1-v^2}) = -\frac{d}{dt}(E_0 u^i)$$

(in which $x_0 = t$, the time), we get

$$\frac{d}{dt}\left(\frac{E_0}{\sqrt{1-v^2}}\right) = \mathsf{L} \, \left(= \int \lambda \, dV\right),$$
$$\frac{d}{dt}\left(\frac{E_0 \mathsf{v}}{\sqrt{1-v^2}}\right) = \mathsf{P} \, \left(= \int \mathsf{p} \, dV\right).$$

These equations hold if the force P acting from without is not too great compared with $\frac{E_0}{a}$, a being the radius of the electron, and if its density in the neighbourhood of the electron is practically constant. They agree exactly with the fundamental equations of mechanics if the mass m is replaced by E. In other words, **inertia is a property of energy**. In mechanics we ascribe to every material body an invariable mass m which, in consequence of the manner in which it occurs in the fundamental law of mechanics, represents the inertia of matter, that is, its resistance to the accelerating forces. Mechanics accepts this inertial mass as given and as requiring no further explanation. We now recognise that the potential energy contained in material bodies is the cause of this inertia, and that the value of the mass corresponding to the energy E_0 expressed in the c.g.s. system, in which the velocity of light is **not** unity, is

$$m = \frac{E_0}{c^2}. \tag{56}$$

We have thus attained a new, purely dynamical view of matter.* Just as the theory of relativity has taught us to reject the belief that we can recognise one and the same point in space at different times, **so now we see that there is no longer a meaning in speaking of the same position of matter at different times**. The electron, which was formerly regarded as a body of foreign substance in the non-material electromagnetic field, now no longer seems to us a very small region marked off distinctly from the field, but to be such that, for it, the field-quantities and the electrical densities assume enormously high values. An "energy-knot" of this type propagates itself in empty space in a manner no different from that in which a water-wave advances over the surface of the sea; there is no "one and the same substance" of which the electron is composed at all times. There is only a potential; and no kinetic energy-momentum-tensor becomes added to it. The resolution into these two, which occurs in mechanics, is only the separation of the thinly distributed energy in the field from that concentrated in the energy-knots, electrons and atoms; the boundary between the two is quite indeterminate. The theory of fields has to explain why the field is granular in structure and why these energy-knots preserve themselves permanently from energy and momentum in their passage to and fro (although they do not remain fully unchanged, they retain their identity to an extraordinary degree of accuracy); therein lies the **problem of**

*Even Kant in his *Metaphysischen Anfangsgründen der Naturwissenschaft*, teaches the doctrine that matter fills space not by its mere existence but in virtue of the repulsive forces of all its parts.

matter. The theory of Maxwell and Lorentz is incapable of solving it for the primary reason that the force of cohesion holding the electron together is wanting in it. **What is commonly called matter is by its very nature atomic**; for we do not usually call diffusely distributed energy matter. **Atoms and electrons are not**, of course, **ultimate invariable elements**, which natural forces attack from without, pushing them hither and thither, but they are themselves distributed continuously and subject to minute changes of a fluid character in their smallest parts. It is not the field that requires matter as its carrier in order to be able to exist itself, but **matter** is, on the contrary, **an offspring of the field**. The formulæ that express the components of the energy-tensor T_{ik} in terms of phase-quantities of the field tell us *the laws according to which* the field is associated with energy and momentum, that is, with matter. Since there is no sharp line of demarcation between diffuse field-energy and that of electrons and atoms, we must broaden our conception of matter, if it is still to retain an *exact* meaning. In future we shall assign the term matter to that real thing, which is represented by the energy-momentum-tensor. In this sense, the optical field, for example, is also associated with matter. Just as in this way matter is merged into the field, so mechanics is expanded into physics. For the law of conservation of matter, the fundamental law of mechanics

$$\frac{\partial T_i^k}{\partial x_k} = 0, \tag{57}$$

in which the T_{ik}'s are expressed in terms of the field-quantities, represents a differential relationship between these quantities, and must therefore follow from the field-equations. In the wide sense, in which we now use the word, matter is that of which we take cognisance directly through our senses. If I seize hold of a piece of ice, I experience the energy-flux flowing between the ice and my body as warmth, and the momentum-flux as pressure. The energy-flux of light on the surface of the epithelium of my eye determines the optical sensations that I experience. Hidden behind the matter thus revealed directly to our organs of sense there is, however, the **field**. To discover the laws governing the latter itself and also the laws by which it determines matter we have a first brilliant beginning in Maxwell's Theory, but this is not our final destination in the quest of knowledge.*

To account for the inertia of matter we must, according to formula (56), ascribe a very considerable amount of energy-content to it: one kilogram of water is to contain $9 \cdot 10^{23}$ ergs. A small portion of this energy is energy of cohesion, that keeps the molecules or atoms associated together in the body. Another portion is the chemical energy that binds the atoms together in the molecule and the sudden liberation of which we observe in an explosion (in solid bodies this chemical energy cannot be distinguished from the energy of cohesion). Changes in the chemical constitution of bodies or in the grouping of atoms or electrons involve the energies due to the electric forces that bind together the negatively charged electrons and the positive nucleus; all ionisation phenomena are included in this category. The energy of the composite atomic nucleus, of which a part is set free during radioactive disintegration, far exceeds the amounts mentioned above. The greater part of this, again, consists of the intrinsic energy of the elements of the atomic nucleus and of the electrons. We know of it only through inertial effects as we have hitherto—owing to a merciful

*Later we shall once again modify our views of matter; the idea of the existence of substance has, however, been finally quashed.

Providence—not discovered a means of bringing it to "explosion". **Inertial mass varies with the contained energy.** If a body is heated, its inertial mass increases; if it is cooled, it decreases; this effect is, of course, too small to be observed directly.

The foregoing treatment of systems in statical equilibrium, in which we have in general followed Laue,* was applied to the electron with special assumptions concerning its constitution, even before Einstein's discovery of the principle of relativity. The electron was assumed to be a sphere with a uniform charge either on its surface or distributed evenly throughout its volume, and held together by a cohesive pressure composed of forces equal in all directions and directed towards the centre. The resultant "electromagnetic mass" $\frac{E_0}{c^2}$ agrees numerically with the results of observation, if one ascribes a radius of the order of magnitude 10^{-13} cms. to the electron. There is no cause for surprise at the fact that even before the advent of the theory of relativity this interpretation of electronic inertia was possible; for, in treating electrodynamics after the manner of Maxwell, one was already unconsciously treading in the steps of the principle of relativity as far as this branch of phenomena is concerned. We are indebted to Einstein and Planck, above all, for the enunciation of the inertia of energy (*vide* note 15). Planck, in his development of dynamics, started from a "test body" which, contrary to the electron, was fully known although it was not in the ordinary sense material, namely, cavity-radiation in thermodynamical equilibrium, as produced according to Kirchoff's Law, in every cavity enclosed by walls at the same uniform temperature.

In the phenomenological theories in which the atomic structure

---

* *Vide* note 14.

of matter is disregarded we imagine the energy that is stored up in the electrons, atoms, etc., to be distributed uniformly over the bodies. We need take it into consideration only by introducing the statical density of mass μ as the density of energy in the energy-momentum-tensor—referred to a co-ordinate system in which the matter is at rest. Thus, if in hydrodynamics we limit ourselves to adiabatic phenomena, we must set

$$|T_i^k| = \left|\begin{array}{c|ccc} -\mu_0 & 0 & 0 & 0 \\ \hline 0 & p & 0 & 0 \\ 0 & 0 & p & 0 \\ 0 & 0 & 0 & p \end{array}\right|$$

in which p is the homogeneous pressure; the energy-flux is zero in adiabatic phenomena. To enable us to write down the components of this tensor in any arbitrary co-ordinate system, we must set $\mu_0 = \mu^* - p$, in addition. We then get the invariant equations

$$\begin{aligned} T_i^k &= \mu^* u_i u^k + p\delta_i^k, \\ \text{or}\quad T_{ik} &= \mu^* u_i u_k + p \cdot g_{ik}. \end{aligned} \tag{58}$$

The statical density of mass is

$$T_{ik} u^i u^k = \mu^* - p = \mu_0$$

and hence we must put μ_0, and **not** μ^*, equal to a constant in the case of incompressible fluids. If no forces act on the fluid, the hydrodynamical equations become

$$\frac{\partial T_i^k}{\partial x_k} = 0.$$

Just as is here done for hydrodynamics so we may find a form for the theory of elasticity based on the principle of relativity (*vide* note 16). There still remains the task of making the law of gravitation, which, in Newton's form, is entirely bound to the principle of relativity of Newton and Galilei, conform to that of Einstein. This, however, involves special problems of its own to which we shall return in the last chapter.

§ 26. Mie's Theory

The theory of Maxwell and Lorentz cannot hold for the interior of the electron; therefore, from the point of view of the ordinary theory of electrons we must treat the electron as something given *a priori*, as a foreign body in the field. A more general theory of electrodynamics has been proposed by **Mie**, by which it seems possible to derive the matter from the field (*vide* note 17). We shall sketch its outlines briefly here—as an example of a physical theory fully conforming with the new ideas of matter, and one that will be of good service later. It will give us an opportunity of formulating the problem of matter a little more clearly.

We shall retain the view that the following phase-quantities are of account: (1) the four-dimensional current-vector s, the "electricity"; (2) the linear tensor of the second order F, the "field". Their properties are expressed in the equations

$$(1) \qquad \frac{\partial s^i}{\partial x_i} = 0,$$

$$(2) \qquad \frac{\partial F_{kl}}{\partial x_i} + \frac{\partial F_{li}}{\partial x_k} + \frac{\partial F_{ik}}{\partial x_l} = 0.$$

Equations (2) hold if F is derivable from a vector ϕ_i according to the formulæ

$$(3) \quad F_{ik} = \frac{\partial \phi_i}{\partial x_k} - \frac{\partial \phi_k}{\partial x_i}.$$

Conversely, it follows from (2) that a vector ϕ must exist such that equations (3) hold. In the same way (1) is fulfilled if s is derivable from a skew-symmetrical tensor H of the second order according to

$$(4) \qquad s^i = \frac{\partial H^{ik}}{\partial x_k}.$$

Conversely, it follows from (1) that a tensor H satisfying these conditions must exist. Lorentz assumed generally, not only for the æther, but also for the domain of electrons, that $H = F$. Following Mie, we shall make the more general assumption that H is not a mere number of calculation but has a real significance, and that its components are, therefore, universal functions of the primary phase-quantities s and F. To be logical we must then make the same assumption about ϕ. The resultant scheme of quantities

$$\begin{array}{c|c} \phi & F \\ \hline s & H \end{array}$$

contains the quantities of intensity in the first row; they are connected with one another by the differential equations (3). In the second row we have the quantities of magnitude, for which the differential quantities (4) hold. If we perform the resolution into space and time and use the same terms as in § 20 we arrive at the well-known equations

$$(1) \quad \frac{d\rho}{dt} + \operatorname{div} s \;\; = 0,$$

$$(2) \quad \frac{d\mathsf{B}}{dt} + \operatorname{curl} \mathsf{E} = \mathbf{0} \qquad (\operatorname{div} \mathsf{B} = 0),$$

$$(3) \quad \frac{df}{dt} + \operatorname{grad} \phi = \mathsf{E} \quad (-\operatorname{curl} f = \mathsf{B}),$$
$$(4) \quad \frac{d\mathsf{D}}{dt} - \operatorname{curl} \mathsf{H} = -s \quad (\operatorname{div} \mathsf{D} = \rho).$$

If we know the universal functions, which express ϕ and H in terms of s and F, then, excluding the equations in brackets, and counting each component separately, we have ten "principal equations" before us, in which the derivatives of the ten phase-quantities with respect to the time are expressed in relation to themselves and their spatial derivatives; that is, we have physical laws in the form that is demanded by the **principle of causality**. The principle of relativity that here appears as an antithesis, in a certain sense, to the principle of causality, demands that the principal equations be accompanied by the bracketed "subsidiary equations," in which no time derivatives occur. The conflict is avoided by noticing that the subsidiary equations are superfluous. For it follows from the principal equations (2) and (3) that

$$\frac{\partial}{\partial t}(\mathsf{B} + \operatorname{curl} f) = \mathbf{0},$$

and from (1) and (4) that

$$\frac{\partial \rho}{\partial t} = \frac{\partial}{\partial t}(\operatorname{div} \mathsf{D}).$$

It is instructive to compare Mie's Theory with Lorentz's fundamental equations of the theory of electrons. In the latter, (1), (2), and (4) occur, whilst the law by which H is determined from the primary phase-quantities is simply expressed by $\mathsf{D} = \mathsf{E}$, $\mathsf{H} = \mathsf{B}$. On the other hand, in Mie's theory, ϕ and f are defined in (3) as the result of a *process of calculation*, and there is no law that determines how these potentials depend on the phase-quantities of the field and on the electricity. In place of this we find the formula giving the density of the mechanical force and the law of mechanics, which governs the motion of electrons under the influence of this force. Since, however, according to the new view which we have put forward, the mechanical law must follow from the field-equations, an addendum becomes necessary; for this purpose, Mie makes the assumption that ϕ and f acquire a physical meaning in the sense indicated. We may, however, enunciate Mie's equation (3) in a form fully analogous to that of the fundamental law of mechanics. We contrast the ponderomotive force occurring in it with the "electrical force" E in this case. In the statical case (3) states that

$$\mathsf{E} - \operatorname{grad} \phi = \mathbf{0}, \tag{59}$$

that is, the electric force E is counterbalanced in the æther by an "**electrical pressure**" ϕ. In general, however, a resulting electrical force arises which, by (3), now belongs to the magnitude f as the "**electrical momentum**". It inspires us with wonder to see how, in Mie's Theory, the fundamental equation of electrostatics (59) which stands at the commencement of electrical theory, suddenly acquires a much more vivid meaning by the appearance of potential as an electrical pressure; this is the required cohesive pressure that keeps the electron together.

The foregoing presents only an empty scheme that has to be filled in by the yet unknown universal functions that connect the quantities of magnitude with those of intensity. Up to a certain degree they may be determined purely speculatively by means of the postulate that the theorem of conservation (57) must hold for the energy-momentum-tensor T_{ik} (that is, that the principle of energy must be valid). For this is certainly a necessary condition, if we are to arrive at some relationship with experiment at all. The energy-law must be of the form

$$\frac{\partial W}{\partial t} + \operatorname{div} s = 0$$

in which W is the density of energy, and s the energy-flux. We get at Maxwell's Theory by multiplying (2) by H and (4) by E, and then adding, which gives

$$H\frac{\partial B}{\partial t} + E\frac{\partial D}{\partial t} + \operatorname{div}[E, H] = -(E, s). \tag{60}$$

In this relation (60) we have also on the right, the work, which is used in increasing the kinetic energy of the electrons or, according to our present view, in increasing the potential energy of the field of electrons. Hence this term must also be composed of a term differentiated with respect to the time, and of a divergence. If we now treat equations (1) and (3) in the same way as we just above treated (2) and (4), that is, multiply (1) by ϕ and (3) scalarly by s, we get

$$\phi\frac{\partial \rho}{\partial t} + s\frac{\partial f}{\partial t} + \operatorname{div}(\phi s) = (E, s). \tag{61}$$

(60) and (61) together give the energy theorem; accordingly the energy-flux must be

$$S = [E, H] + \phi s,$$

and

$$\phi\,\delta\rho + s\,\delta f + H\,\delta B + E\,\delta D = \delta W$$

is the total differential of the energy-density. It is easy to see why a term proportional to s, namely ϕs, has to be added to the term (E, H) which holds in the æther. For when the electron that generates the convection-current s moves, its energy-content flows also. In the æther the term (E, H) is overpowered by S, but in the electron the other ϕs easily gains the upper hand. The quantities ρ, f, B, D occur in the formula for the total differential of the energy-density as independent differentiated phase-quantities. For the sake of clearness we shall introduce ϕ and E as independent variables in place of ρ and D. By this means all the quantities of intensity are made to act as independent variables. We must build up

$$L = W - ED - \rho\psi, \tag{62}$$

and then we get

$$\delta L = (H\,\delta B - D\,\delta E) + (s\,\delta f - \rho\,\delta\phi).$$

If L is known as a function of the quantities of intensity, then these equations express the quantities of magnitude as functions of the quantities of intensity. **In place of the ten unknown universal functions we have now only one**, L; this is accomplished by the **principle of energy**.

Let us again return to four-dimensional notation, we then have

$$\delta L = \tfrac{1}{2}H^{ik}\,\delta F_{ik} + s^i\,\delta\phi_i. \tag{63}$$

From this it follows that δL, and hence L, the "**Hamiltonian Function**" is an invariant. The simplest invariants that may be formed from a vector having components ϕ_i and a linear tensor of the second order having components F_{ik} are the squares of the following expressions:

the vector ϕ^i, $\qquad \phi_i\phi^i$,

the tensor F_{ik}, $\qquad 2L^0 = \tfrac{1}{2}F_{ik}F^{ik}$, the linear tensor of the fourth order with components $\sum \pm F_{ik}F_{lm}$ (the summation extends over the 24 permutations of the indices i, k, l, m; the upper sign applies to the even permutations, the lower ones to the odd); and finally of the vector $F_{ik}\phi^k$.

Just as in three-dimensional geometry the most important theorem of congruence is that a vector-pair **a**, **b** is fully characterised in respect to congruence by means of the invariants $\mathbf{a}^2$, **ab**, $\mathbf{b}^2$, so it may be shown in four-dimensional geometry that the invariants quoted determine fully in respect to congruence the figure composed of a vector ϕ and a linear tensor of the second order F. Every invariant, in particular the Hamiltonian Function L, must therefore be expressible algebraically in terms of the above four quantities. Mie's Theory thus resolves the problem of matter into a determination of this expression. Maxwell's Theory of the æther which, of course, precludes the possibility of electrons, is contained in it as the special case $L = L^0$. If we also express W and the components of S in terms of four-dimensional quantities, we see that they are the negative (0th) row in the scheme

$$T_i^k = F_{ir}H^{kr} + \phi_i s^k - L \cdot \delta_i^k. \tag{64}$$

The T_i^k's are thus the mixed components of the energy-momentum-tensor, which, according to our calculations, fulfil the theorem of conservation (57) for $i = 0$ and hence also for $i = 1, 2, 3$. In the next chapter we shall add the proof that its co-variant components satisfy the condition of symmetry $T_{ki} = T_{ik}$.

The laws for the field may be summarised in a very simple principle of variation, Hamilton's Principle. For this we regard only the potential with components ϕ_i as an independent phase-quantity, and *define* the field by the equation

$$F_{ik} = \frac{\partial \phi_i}{\partial x_k} - \frac{\partial \phi_k}{\partial x_i}.$$

Hamilton's invariant function L which depends on the potential and the field enters into these laws. We *define* the current-vector s and the skew-symmetrical tensor H by means of (63). If in an arbitrary linear co-ordinate system

$$d\omega = \sqrt{g}\, dx_0\, dx_1\, dx_2\, dx_3$$

is the four-dimensional "volume-element" of the world ($-g$ is the determinant of the metrical groundform) then the integral $\int L\, d\omega$ taken over any region of the world is an invariant. It is called the **Action** contained in the region in question. Hamilton's Principle states that the change in the total *Action* for each infinitesimal variation of the state of the field, which vanishes outside a finite region, is zero, that is,

$$\delta \int L\, d\omega = \int \delta L\, d\omega = 0. \tag{65}$$

This integral is to be taken over the whole world or, what comes to the same thing, over a finite region beyond which the variation of the phase vanishes. This variation is represented by the infinitesimal increments $\delta\phi_i$ of the potential-components and the accompanying infinitesimal change of the field

$$\delta F_{ik} = \frac{\partial(\delta\phi_i)}{\partial x_k} - \frac{\partial(\delta\phi_k)}{\partial x_i}$$

in which $\delta\phi_i$ are space-time functions that only differ from zero within a finite region. If we insert for δL the expression (63), we get

$$\delta L = s^i \, \delta\phi_i + H^{ik} \, \frac{\partial(\delta\phi_i)}{\partial x_k}.$$

By the principle of partial integration (*vide* page 165) we get

$$\int H^{ik} \, \frac{\partial(\delta\phi_i)}{\partial x_k} \, d\omega = - \int \frac{\partial H^{ik}}{\partial x_k} \, \delta\phi_i \, d\omega,$$

and, accordingly,

$$\delta \int L \, d\omega = \int \left\{ s^i - \frac{\partial H^{ik}}{\partial x_k} \right\} \delta\phi_i \, d\omega. \tag{66}$$

Whereas (3) is given by definition, we see that Hamilton's Principle furnishes the field-equations (4). In point of fact, if, for instance,

$$s - \frac{\partial H^{ik}}{\partial x_k} \neq 0$$

but is > 0 at a certain point, then we could mark off a small region encircling this point, such that, for it, this difference is positive throughout. If we then choose a non-negative function for $\delta\phi_1$ that vanishes outside the region marked off, and if $\delta\phi_2 = \delta\phi_3 = \delta\phi_4 = 0$, we arrive at a contradiction to equation (65)—(1) and (2) follow from (3) and (4).

We find, then, **that Mie's Electrodynamics exists in a compressed form in Hamilton's Principle** (65)—analogously to the manner in which the development of mechanics attains its zenith in the principle of action. Whereas in mechanics, however, a definite function L of action corresponds to every given mechanical system and has to be deduced from the constitution of the system, we are here concerned with a single system, the world. This is where the real problem of matter takes its beginning: we have to determine the "function of action," the world-function L, belonging to the world. For the present it leaves us in perplexity. If we choose an arbitrary L, we get a "possible" world governed by this function of action, which will be perfectly intelligible to us—more so than the actual world—provided that our mathematical analysis does not fail us. We are, of course, then concerned in discovering the only existing world, the **real** world for us. Judging from what we know of physical laws, we may expect the L which belongs to it to be distinguished by having simple mathematical properties. Physics, this time as a physics of fields, is again pursuing the object of reducing the totality of natural phenomena to **a single physical law**: it was believed that this goal was almost within reach once before when Newton's *Principia*, founded on the physics of mechanical point-masses was celebrating its triumphs. But the treasures of knowledge are not like ripe fruits that may be plucked from a tree.

For the present we do not yet know whether the phase-quantities on which Mie's Theory is founded will suffice to describe matter or whether matter is purely "electrical" in nature. Above all, the ominous clouds of those phenomena that we are with varying success seeking to explain by means of the quantum of action, are throwing their shadows over the sphere of physical knowledge, threatening no one knows what new revolution.

Let us try the following hypothesis for L:

$$L = \frac{1}{2}|F|^2 + w(\sqrt{-\phi_i\phi^i}) \tag{67}$$

(w is the symbol for a function of one variable); it suggests itself as being the simplest of those that go beyond Maxwell's Theory. We have no grounds for assuming that the world-function has actually this form. We shall confine ourselves to a consideration of statical solutions, for which we have

$$\begin{aligned} &\mathsf{B} = \mathsf{H} = \mathbf{0}, && \mathsf{s} = \mathsf{f} = \mathbf{0}, \\ &\mathsf{E} = \operatorname{grad}\phi, && \operatorname{div}\mathsf{D} = \rho, \\ &\mathsf{D} = \mathsf{E}, && \rho = -w'(\phi) \end{aligned}$$

(the accent denoting the derivative). In comparison with the ordinary electrostatics of the æther we have here the new circumstance that the density ρ is a universal function of the potential, the electrical pressure ϕ. We get for Poisson's equation

$$\Delta\phi + w'(\phi) = 0. \tag{68}$$

If $w(\phi)$ is not an even function of ϕ, this equation no longer holds after the transition from ϕ to $-\phi$; this would account for **the difference between the natures of positive and negative electricity**. Yet it certainly leads to a remarkable difficulty in the case of non-statical fields. If charges having opposite signs are to occur in the latter, the root in (67) must have different signs at different points of the field. Hence there must be points in the field, for which $\phi_i\phi^i$ vanishes. In the neighbourhood of such a point $\phi_i\phi^i$ must be able to assume positive and negative values (this does not follow in the statical case, as the minimum of the function ϕ_0^2 for ϕ_0 is zero). The solutions of our field-equations must, therefore, become imaginary at regular distances apart. It would be difficult to interpret a degeneration of the field into separate portions in this way, each portion containing only charges of one sign, and separated from one another by regions in which the field becomes imaginary.

A solution (vanishing at infinity) of equation (68) represents a possible state of electrical equilibrium, or a possible corpuscle capable of existing individually in the world that we now proceed to construct. The equilibrium can be stable, only if the solution is radially symmetrical. In this case, if r denotes the radius vector, the equation becomes

$$\frac{1}{r^2}\frac{d}{dr}\left(r^2\frac{d\phi}{dr}\right)+w'(\phi)=0. \tag{69}$$

If (69) is to have a regular solution

$$-\phi=\frac{e_0}{r}+\frac{e_1}{r^2}+\dots \tag{70}$$

at $r=\infty$, we find by substituting this power series for the first term of the equation that the series for $w'(\phi)$ begins with the power r^{-4} or one with a still higher negative index, and hence that $w(x)$ must be a zero of at least the fifth order for $x=0$. On this assumption the equations must have a single infinity of regular solutions at $r=0$ and also a single infinity of regular solutions at $r=\infty$. We may (in the "general" case) expect these two **one-dimensional** families of solutions (included in the two-dimensional complete family of all the solutions) to have a finite or, at any rate, a discrete number of solutions. These would represent the various possible corpuscles. (Electrons and elements of the atomic nucleus?) *One* electron or *one* atomic nucleus does not, of course, exist alone in the world; but the distances between them are so great in comparison with their own size that they do not bring about an appreciable modification of the structure of the field within the i interior of an individual electron or atomic nucleus. If ϕ is a solution of (69) that represents such a corpuscle in (70) then its total charge

$$= 4\pi \int_0^\infty w'(\phi) r^2\, dr = -4\pi \cdot r^2 \left.\frac{d\phi}{dr}\right|_{r=\infty} = 4\pi c_0,$$

but its mass is calculated as the integral of the energy-density W that is given by (62):

$$\begin{aligned} \text{Mass} &= 4\pi \int_0^\infty \left\{\tfrac{1}{2}(\operatorname{grad}\phi)^2 + w(\phi) - \phi w'(\phi)\right\} r^2\, dr \\ &= 4\pi \int_0^\infty \left\{w(\phi) - \tfrac{1}{2}\phi w'(\phi)\right\} r^2\, dr. \end{aligned}$$

These physical laws, then, enable us to calculate the mass and charge of the electrons, and the atomic weights and atomic charges of the individual existing elements whereas, hitherto, we have always accepted these ultimate constituents of matter as things given with their numerical properties. All this, of course, is merely a suggested *plan of action* as long as the world-function L is not known. The special hypothesis (67) from which we just now started was assumed only to show what a deep and thorough knowledge of matter and its constituents as based on laws would be exposed to our gaze if we could but discover the action-function. For the rest, the discussion of such arbitrarily chosen hypotheses cannot lead to any proper progress; new physical knowledge and principles will be required to show us the right way to determine the Hamiltonian Function.

To make clear, *ex contrario*, the nature of pure physics of fields, which was made feasible by Mie for the realm of electrodynamics as far as its general character furnishes hypotheses, the principle of action (65) holding in it will be contrasted with that by which the theory of Maxwell and Lorentz is governed; the latter theory recognises, besides the electromagnetic field, a substance moving in it. This substance is a three-dimensional continuum; hence its parts may be referred in a continuous manner to the system of values of three co-ordinates α, β, γ. Let us imagine the substance divided up into infinitesimal elements. Every element of substance has then a definite invariable positive mass dm and an invariable electrical charge de. As an expression of its history there corresponds to it then a world-line with a definite direction of traverse or, in better words, an infinitely thin "world-filament". If we again divide this up into small portions, and if

$$ds = \sqrt{-g_{ik}\,dx_i\,dx_k}$$

is the proper-time length of such a portion, then we may introduce the space-time function μ_0 of the statical mass-density by means of the invariant equation

$$dm\,ds = \mu_0\,d\omega. \tag{71}$$

We shall call the integral

$$\int_{\mathfrak{X}} \mu_0\,d\omega = \int dm\,ds = \int dm \int \sqrt{-g_{ik}\,dx_i\,dx_k}$$

taken over a region $\mathfrak{X}$ of the world the **substance-action of mass**. In the last integral the inside integration refers to that part of the world-line of any arbitrary element of substance of mass dm, which belongs to the region $\mathfrak{X}$, the outer integral signifies summation taken for all elements of the substance. In purely mathematical language this transition from substance-proper-time integrals to space-time integrals occurs as follows. We first introduce the substance-density ν of the mass thus:

$$dm = \nu \, d\alpha \, d\beta \, d\gamma$$

(ν behaves as a scalar-density for arbitrary transformations of the substance co-ordinates α, β, γ). On each world-line of a substance-point α, β, γ we reckon the proper-time s from a definite initial point (which must, of course, vary **continuously** from substance-point to substance-point). The co-ordinates x_i of the world-point at which the substance-point α, β, γ happens to be at the moment s of its motion (after the proper-time s has elapsed), are then continuous functions of α, β, γ, s, whose functional determinant

$$\frac{\partial(x_0, x_1, x_2, x_3)}{\partial(\alpha, \beta, \gamma, s)}$$

we shall suppose to have the absolute value Δ. The equation (71) then states that

$$\mu_0 \sqrt{g} = \frac{\nu}{\Delta}.$$

In an analogous manner we may account for the statical density ρ_0 of the electrical charge. We shall set down

$$\int \left(de \int \phi_i \, dx_i \right)$$

as **substance-action of electricity**; in it the outer integration is again taken over all the substance-elements, but the inner one in each case over that part of the world-line of a substance-element carrying the charge de whose path lies in the interior of the world-region $\mathfrak{X}$. We may therefore also write

$$\int de \, ds \cdot \phi u = \int \rho_0 u^i \phi_i \, d\omega = \int s^i \phi_i \, d\omega$$

if $u^i = \dfrac{dx_i}{ds}$ are the components of the world-direction, and $s^i = \rho_0 u^i$ are the components of the 4-current (a pure convection current). Finally, in addition to the substance-action there is also a **field-action of electricity**, for which Maxwell's Theory makes the simple convention

$$\tfrac{1}{4}\int F_{ik}F^{ik}\,d\omega \qquad \left(F_{ik} = \frac{\partial \phi_i}{\partial x_k} - \frac{\partial \phi_k}{\partial x_i}\right).$$

Hamilton's Principle, which gives a condensed statement of the Maxwell-Lorentz Laws, may then be expressed thus:

The total action, that is, the sum of the field-action and substance-action of electricity plus the substance-action of the mass for any arbitrary variation (vanishing for points beyond a finite region) of the field-phase (of the ϕ_i's) and for a similarly conditioned space-time displacement of the world-lines described by the individual substance-points undergoes no change.

This principle clearly gives us the equations

$$\frac{\partial F^{ik}}{\partial x_k} = s^i = \rho_0 u^i,$$

if we vary the ϕ_i's. If, however, we keep the ϕ_i's constant, and perform variations on the world-lines of the substance-points, we get, by interchanging differentiation and variation (as in § 17 in determining the shortest lines), and then integrating partially:

$$\begin{aligned}\int \phi_i\,dx_i &= \int (\delta\phi_i\,dx_i + \phi_i\,d\delta\phi_i) = \int (\delta\phi_i\,dx_i - \delta x_i\,d\phi_i) \\ &= \int \left(\frac{\partial \phi_i}{\partial x_k} - \frac{\partial \phi_k}{\partial x_i}\right)\delta x_k \cdot dx_i.\end{aligned}$$

In this the δx_i's are the components of the infinitesimal displacement, which the individual points of the world-line undergo. Accordingly, we get

$$\delta \int \left(de \int \phi_i \, dx_i \right) = \int de \, ds \cdot F_{ik} u^i \, \delta x_k = \int \rho_0 F_{ik} u^i \, \delta x_k \cdot d\omega.$$

If we likewise perform variation on the substance-action of the mass (this has already been done in § 17 for a more general case, in which the g_{ik}'s were variable), we arrive at the mechanical equations which are added to the field-equations in Maxwell's Theory; namely

$$\mu_0 \frac{du_i}{ds} = p_i \qquad p_i = \rho_0 F_{ik} u^k = F_{ik} s^k.$$

This completes the cycle of laws which were mentioned on page 297. This theory does not, of course, explain the existence of the electron, since the cohesive forces are lacking in it.

A striking feature of the principle of action just formulated is that a field-action does not associate itself with the substance-action of the mass, as happens in the case of electricity. This gap will be filled in the next chapter, in which it will be shown that the **gravitational field** is what corresponds to mass in the same way as the electromagnetic field corresponds to the electrical charge.

The great advance in our knowledge described in this chapter consists in recognising that the scene of action of reality is not a three-dimensional Euclidean space but rather a **four-dimensional world, in which space and time are linked together indissolubly**. However deep the chasm may be that separates the intuitive nature of space from that of time in our experience, nothing of this qualitative difference enters into the objective world which physics endeavours to crystallise out of direct experience. It is a four-dimensional continuum, which is neither "time" nor "space". Only the consciousness that passes on in one portion of this world experiences the detached piece which comes to meet it and passes behind it, as **history**, that is, as a process that is going forward in time and takes place in space.

This four-dimensional space is **metrical** like Euclidean space, but the quadratic form which determines its metrical structure is not definitely positive, but has **one** negative dimension. This circumstance is certainly of no mathematical importance, but has a deep significance for reality and the relationship of its action. It was necessary to grasp the idea of the metrical four-dimensional world, which is so simple from the mathematical point of view, not only in isolated abstraction but also to pursue the weightiest inferences that can be drawn from it towards setting up the view of physical phenomena, so that we might arrive at a proper understanding of its content and the range of its influence: that was what we aimed to do in a short account. It is remarkable that the three-dimensional geometry of the statical world that was put into a complete axiomatic system by Euclid has such a translucent character, whereas we have been able to assume command over the four-dimensional geometry only after a prolonged struggle and by referring to an extensive set of physical phenomena and empirical data. Only now the theory of relativity has succeeded in enabling our knowledge of physical nature to get a full grasp of the fact of motion, of change in the world.

CHAPTER IV

THE GENERAL THEORY OF RELATIVITY

§ 27. The Relativity of Motion, Metrical Fields, Gravitation*

HOWEVER successfully the Principle of Relativity of Einstein worked out in the preceding chapter marshals the physical laws which are derived from experience and which define the relationship of action in the world, we cannot express ourselves as satisfied from the point of view of the theory of knowledge. Let us again revert to the beginning of the foregoing chapter. There we were introduced to a "kinematical" principle of relativity; x_1, x_2, x_3, t were the space-time co-ordinates of a world-point referred to a definite permanent Cartesian co-ordinate system in space; x'_1, x'_2, x'_3, t' were the co-ordinates of the same point relative to a second such system, that may be moving arbitrarily with respect to the first; they are connected by the transformation formulæ (II), page 225. It was made quite clear that two series of physical states or phases cannot be distinguished from one another in an objective manner, if the phase-quantities of the one are represented by the same mathematical functions of x'_1, x'_2, x'_3, t' as those that describe the first series in terms of the arguments x_1, x_2, x_3, t. Hence the physical laws must have exactly the same form in the one system of independent space-time arguments as in the other. It must certainly be admitted that the facts of dynamics are apparently in direct contradiction to Einstein's postulate, and it is just these facts that, since the time of Newton, have forced us to attribute an absolute meaning, not to translation, but to rotation. Yet our

* *Vide* note 1.

minds have never succeeded in accepting unreservedly this torso thrust on them by reality (in spite of all the attempts that have been made by philosophers to justify it, as, for example, Kant's *Metaphysische Anfangsgründe der Naturwissenschaften*), and the problem of centrifugal force has always been felt to be an unsolved enigma (*vide* note 2).

Where do the centrifugal and other inertial forces take their origin? Newton's answer was: in absolute space. The answer given by the special theory of relativity does not differ essentially from that of Newton. It recognises as the source of these forces the metrical structure of the world and considers this structure as a formal property of the world. But that which expresses itself as force must itself be real. We can, however, recognise the metrical structure as something real, if it is itself capable of undergoing changes and reacts in response to matter. Hence our only way out of the dilemma—and this way, too, was opened up by Einstein—is to apply Riemann's ideas, as set forth in Chapter II, to the four-dimensional Einstein-Minkowski world which was treated in Chapter III instead of to three-dimensional Euclidean space. In doing this we shall not for the present make use of the most general conception of the metrical manifold, but shall retain Riemann's view. According to this, we must assume the world-points to form a four-dimensional manifold, on which a measure-determination is impressed by a non-degenerate quadratic differential form Q having one positive and three negative dimensions.* In any co-

*We have made a change in the notation, as compared with that of the preceding chapter, by placing reversed signs before the metrical groundform. The former convention was more convenient for representing the splitting up of the world into space and time, the present one is found more expedient in the general theory.

ordinate system x_i $(i = 0, 1, 2, 3)$, in Riemann's sense, let

$$Q = \sum_{i,k} g_{ik}\, dx_i\, dx_k. \qquad (1)$$

Physical laws will then be expressed by tensor relations that are invariant for arbitrary continuous transformations of the arguments x_i. In them the co-efficients g_{ik} of the quadratic differential form (1) will occur in conjunction with the other physical phase-quantities. Hence we shall satisfy the postulate of relativity enunciated above, without violating the facts of experience, **if we regard the g_{ik}'s** in exactly the same way as we regarded the components ϕ_i of the electromagnetic potential (which are formed by the co-efficients of an invariant **linear** differential form $\sum \phi_i\, dx_i$), **as physical phase-quantities, to which there corresponds something real, namely, the "metrical field"**. Under these circumstances invariance exists not only with respect to the transformations mentioned (II), which have a fully arbitrary (non-linear) character only for the time-co-ordinate, but for any transformations whatsoever. This special distinction conferred on the time-co-ordinate by (II), is, indeed, incompatible with the knowledge gained from Einstein's Principle of Relativity. By allowing any arbitrary transformations in place of (II), that is, also such as are non-linear with respect to the space-co-ordinates, we affirm that Cartesian co-ordinate systems are in no wise more favoured than any "curvilinear" co-ordinate system. **This seals the doom of the idea that a geometry may exist independently of physics** in the traditional sense, and it is just because we had not emancipated ourselves from the dogma that such a geometry existed that we arrived by logical considerations at the relativity principle (II), and not at once at the principle of invariance for arbitrary transformations of the four world-co-ordinates. Actually, however, spatial measurement is based on a physical event: the reaction of light-rays and rigid measuring rods on our whole physical world. We have already encountered this view in § 21, but we may, above all, take up the thread from our discussion in § 12, for we have, indeed, here arrived at Riemann's "dynamical" view as a necessary consequence of the relativity of all motion. The behaviour of light-rays and measuring rods, besides being determined by their own natures, is also conditioned by the "metrical field," just as the behaviour of an electric charge depends not only on it, itself, but also on the electric field. Again, just as the electric field, for its part, depends on the charges and is instrumental in producing a mechanical interaction between the charges, so we must assume here that **the metrical field** (or, in mathematical language, the tensor with components g_{ik}) **is related to the material content filling the world**. We again call attention to the principle of action set forth at the conclusion of the preceding paragraph; in both of the parts which refer to substance, the metrical field takes up the same position towards mass as the electrical field does towards the electric charge. The assumption, which was made in the preceding chapter, concerning the metrical structure of the world (corresponding to that of Euclidean geometry in three-dimensional space), namely, that there are specially favoured co-ordinate systems, "linear" ones, in which the metrical groundform has constant co-efficients, can no longer be maintained in the face of this view.

A simple illustration will suffice to show how geometrical conditions are involved when motion takes place. Let us set a plane disc spinning uniformly. I affirm that if we consider Euclidean geometry valid for the reference-space relative to which we speak of uniform rotation, then it is no longer valid for the rotating disc itself, if the latter be measured by means of measuring rods moving with it. For let us consider a circle on the disc described with its centre at the centre of rotation. Its radius remains the same no matter whether the measuring rods with which I measure it are at rest or not, since its direction of motion is perpendicular to the measuring rod when in the position required for measuring the radius, that is, along its length. On the other hand, I get a value greater for the circumference of the circle than that obtained when the disc is at rest when I apply the measuring rods, owing to the Lorentz-Fitzgerald contraction which the latter undergoes. The Euclidean theorem which states that the circumference of the circle $= 2\pi$ times the radius thus no longer holds on the disc when it rotates.

The falling over of glasses in a dining-car that is passing round a sharp curve and the bursting of a fly-wheel in rapid rotation are not, according to the view just expressed, effects of "an absolute rotation" as Newton would state but whose existence we deny; they are effects of the "metrical field" or rather of the affine relationship associated with it. Galilei's principle of inertia shows that there is a sort of "forcible guidance" which compels a body that is projected with a definite velocity to move in a definite way which can be altered only by external forces. This "guiding field," which is physically real, was called "affine relationship" above. When a body is diverted by external forces the guidance by forces such as centrifugal reaction asserts itself. In so far as the state of the guiding field does not persist, and the present one has emerged from the past ones under the influence of the masses existing in the world, namely, the fixed stars, the phenomena cited above are partly an effect of the fixed stars, **relative to which** the rotation takes place.*

Following Einstein by starting from the special theory of relativity described in the preceding chapter, we may arrive at the general theory of relativity in two successive stages.

I. In conformity with the principle of continuity we take the same step in the four-dimensional world that, in Chapter II, brought us from Euclidean geometry to Riemann's geometry. This causes a quadratic differential form (1) to appear. There is no difficulty in adapting the physical laws to this generalisation. It is expedient to represent the magnitude quantities by tensor-densities instead of by tensors as in Chapter III; we can do this by multiplying throughout by $\sqrt{g}$ (in which g is the negative determinant of the g_{ik}'s). Thus, in particular, the mass- and charge-densities μ and ρ, instead of being given by formula (71) of § 26, will be given by

$$dm\, ds = \mu\, dx, \qquad de\, ds = \rho\, dx \qquad (dx = dx_0\, dx_1\, dx_2\, dx_3).$$

The proper time ds along the world-line is determined from

$$ds^2 = g_{ik}\, dx_i\, dx_k.$$

*We say "partly" because the distribution of matter in the world does not define the "guiding field" uniquely, for both are **at one moment** independent of one another and accidental (analogously to charge and electric field). Physical laws tell us merely how, when such an initial state is given, all other states (past and future) necessarily arise from them. At least, this is how we must judge, if we are to maintain the standpoint of pure physics of fields. The statement that the world in the form we perceive it taken as a whole is stationary (i.e. at rest) can be interpreted, if it is to have a meaning at all, as signifying that it is in statistical equilibrium. Cf. § 34.

Maxwell's equations will be

$$F_{ik} = \frac{\partial \phi_i}{\partial x_k} - \frac{\partial \phi_k}{\partial x_i}, \qquad \frac{\partial \mathbf{F}^{ik}}{\partial x_k} = \mathbf{s}^i,$$

in which the ϕ_i's are the co-efficients of an invariant linear differential form $\phi_i \, dx_i$, and $\mathbf{F}^{ik}$ denotes $\sqrt{g} \cdot F^{ik}$ according to our convention above. In Lorentz's Theory we set

$$\mathbf{s}^i = \rho u^i \qquad \left(u^i = \frac{dx_i}{ds} \right).$$

The mechanical force per unit of volume (a co-variant vector-density in the four-dimensional world) is given by:*

$$\mathbf{p}_i = -F_{ik}\mathbf{s}^k, \tag{2}$$

and the mechanical equations are in general

$$\mu \left(\frac{du_i}{ds} - \left\{ \begin{matrix} i\beta \\ \alpha \end{matrix} \right\} u_\alpha u^\beta \right) = \mathbf{p}_i \tag{3}$$

with the condition that $\mathbf{p}_i u^i$ always $= 0$. We may put them into the same form as we found for them earlier by introducing, in addition to the $\mathbf{p}_i$'s, the quantities

$$\left\{ \begin{matrix} i\beta \\ \alpha \end{matrix} \right\} \cdot \mu u_\alpha u^\beta = \tfrac{1}{2} \frac{\partial g_{\alpha\beta}}{\partial x_i} \cdot \mu u^\alpha u^\beta \tag{4}$$

(cf. § 17, equation (64)) as the density components $\bar{\mathbf{p}}_i$ of a "pseudo-force" (force of reaction of the guiding field). The equations then become

$$\mu \frac{du_i}{ds} = \mathbf{p}_i + \bar{\mathbf{p}}_i.$$

*The sign is reversed on account of the reversal of sign in the metrical groundform.

The simplest examples of such "pseudo-forces" are centrifugal forces and Coriolis forces. If we compare formula (4) for the "pseudo-force" arising from the metrical field with that for the mechanical force of the electromagnetic field, we find them fully analogous. For just as the vector-density with the contra-variant components $\mathbf{s}^i$ characterises electricity so, as we shall presently see, moving matter is described by the tensor-density which has the components $\mathbf{T}_i^k = \mu u_i u^k$. The quantities

$$\Gamma^{\alpha}_{i\beta} = \begin{Bmatrix} i\beta \\ \alpha \end{Bmatrix}$$

correspond as components of the metrical field to the components F_{ik} of the electric field. Just as the field-components F are derived by differentiation from the electromagnetic potential ϕ_i, so also the Γ's from the g_{ik}'s; these thus constitute the potential of the metrical field. The force-density is the product of the electric field and electricity on the one hand, and of the metrical field and matter on the other, thus

$$\mathbf{p}_i = -F_{ik}\mathbf{s}^k, \qquad \bar{\mathbf{p}}_i = \Gamma^{\alpha}_{i\beta}\mathbf{T}^{\beta}_{\alpha}.$$

If we abandon the idea of a substance existing independently of physical states, we get instead the general energy-momentum-density $\mathbf{T}_i^k$ which is determined by the state of the field. According to the special theory of relativity it satisfies the Law of Conservation

$$\frac{\partial \mathbf{T}_i^k}{\partial x_k} = 0.$$

This equation is now to be replaced, in accordance with formula (37) § 14, by the general invariant

$$\frac{\partial \mathbf{T}_i^k}{\partial x_k} - \Gamma^{\alpha}_{i\beta}\mathbf{T}^{\beta}_{\alpha} = 0. \tag{5}$$

If the left-hand side consisted only of the first member, **T** would now again satisfy the laws of conservation. But we have, in this case, a second term. The "real" total force

$$\mathbf{p}_i = -\frac{\partial \mathbf{T}_i^k}{\partial x_k}$$

does not vanish but must be counterbalanced by the "pseudo-force" which has its origin in the metrical field, namely

$$\bar{\mathbf{p}}_i = \Gamma_{i\beta}^{\alpha} \mathbf{T}_{\alpha}^{\beta} = \tfrac{1}{2} \frac{\partial g_{\alpha\beta}}{\partial x_i} \mathbf{T}^{\alpha\beta}. \tag{6}$$

These formulæ were found to be expedient in the special theory of relativity when we used curvilinear co-ordinate systems, or such as move curvilinearly or with acceleration. To make clear the simple meaning of these considerations we shall use this method to determine the **centrifugal force** that asserts itself in a rotating system of reference. If we use a normal co-ordinate system for the world, namely, t, x_1, x_2, x_3, but introduce r, z, θ, in place of the Cartesian space co-ordinates, we get

$$ds^2 = dt^2 - (dz^2 + dr^2 + r^2\, d\theta^2).$$

Using ω to denote a constant angular velocity, we make the substitution

$$\theta = \theta' + \omega t', \qquad t = t'$$

and, after the substitution, drop the accents. We then get

$$ds^2 = dt^2(1 - r^2\omega^2) - 2r^2\omega\, d\theta\, dt - (dz^2 + dr^2 + r^2\, d\theta^2).$$

If we now put

$$t = x_0, \qquad \theta = x_1, \qquad z = x_2, \qquad r = x_3,$$

we get for a point-mass which is at rest in the system of reference now used

$$u^1 = u^2 = u^3 = 0; \quad \text{and hence } (u^0)^2(1 - r^2\omega^2) = 1.$$

The components of the centrifugal force satisfy formula (4)

$$\bar{\mathbf{p}}_i = \tfrac{1}{2}\frac{\partial g_{00}}{\partial x_i} \cdot \mu(u^0)^2$$

and since the derivatives with respect to x_0, x_1, x_2 of g_{00}, which is equal to $1 - r^2\omega^2$, vanish and since

$$\frac{\partial g_{00}}{\partial x_3} = \frac{\partial g_{00}}{\partial r} = -2r\omega^2$$

then, if we return to the usual units, in which the velocity of light is **not** unity, and if we use contra-variant components instead of co-variant ones, and instead of the indices $0, 1, 2, 3$ the more indicative ones t, θ, z, r, we obtain

$$\bar{\mathbf{p}}^t = \bar{\mathbf{p}}^\theta = \bar{\mathbf{p}}^z = 0, \qquad \bar{\mathbf{p}}^r = \frac{\mu r\omega^2}{1 - \left(\frac{r\omega}{c}\right)^2}. \tag{7}$$

Two closely related circumstances characterise the "pseudo-forces" of the metrical field. *Firstly*, the acceleration which they impart to a point-mass situated at a definite space-time point (or, more exactly, one passing through this point with a definite velocity) is independent of its mass, i.e. the force itself is proportional to the inertial mass of the point-mass at which it acts. *Secondly*, if we use an appropriate co-ordinate system, namely, a geodetic one, at a definite space-time point, these forces vanish (cf. § 14). If the special theory of relativity is to be maintained, this vanishing can be effected simultaneously for all space-time points by the introduction of a linear co-ordinate system, but in the general case it is possible to make the whole 40 components $\Gamma^{\alpha}_{i\beta}$ of the affine relationship vanish at least for each individual point by choosing an appropriate co-ordinate system at this point.*

Now the two related circumstances just mentioned are true, as we know, of the **force of gravitation**. The fact that a given gravitational field imparts the same acceleration to every mass that is brought into the field constitutes the real essence of the problem of gravitation. In the electrostatic field a slightly charged particle is acted on by the force $e \cdot \mathbf{E}$, the electric charge e depending only on the particle, and $\mathbf{E}$, the electric intensity of field, depending only on the field. If no other forces are acting, this force imparts to the particle whose inertial mass is m an acceleration which is given by the fundamental equation of mechanics $m\mathbf{b} = e\mathbf{E}$. There is something fully analogous to this in the gravitational field. The force that acts on the particle is equal to $g\mathbf{G}$, in which g, the "gravitational charge," depends only on the particle, whereas $\mathbf{G}$ depends only on the field: the acceleration is determined here again by the equation $m\mathbf{b} = g\mathbf{G}$. The curious fact now manifests itself that the "gravitational charge" or **the "gravitational mass"** g **is equal to the "inertial mass"** m. Eötvös has comparatively recently tested the accuracy of this law by actual experiments of the greatest refinement (*vide* note 3). The centrifugal force imparted to a body at the earth's surface by the earth's rotation is proportional to its inertial mass but its weight is proportional

*Hence we see that it is in the nature of the metrical field that it cannot be described by a field-tensor Γ which is invariant with respect to arbitrary transformations.

to its gravitational mass. The resultant of these two, the *apparent* weight, would have different directions for different bodies if gravitational and inertial mass were not proportional throughout. The absence of this difference of direction was demonstrated by Eötvös by means of the exceedingly sensitive instrument known as the torsion-balance: it enables the inertial mass of a body to be measured to the same degree of accuracy as that to which its weight may be determined by the most sensitive balance. The proportionality between gravitational and inertial mass holds in cases, too, in which a diminution of mass is occasioned not by an escape of substance in the old sense, but by an emission of radioactive energy.

The inertial mass of a body has, according to the fundamental law of mechanics, a **universal** significance. It is the inertial mass that regulates the behaviour of the body under the influence of any forces acting on it, of whatever physical nature they may be; the inertial mass of the body is, however, according to the usual view associated only with a special physical field of force, namely, that of gravitation. From this point of view, however, the identity between inertial and gravitational mass remains fully incomprehensible. Due account can be taken of it only by a mechanics which from the outset takes into consideration gravitational as well as inertial mass. This occurs in the case of the mechanics given by the general theory of relativity, in which we assume that **gravitation, just like centrifugal and Coriolis forces, is included in the "pseudo-force" which has its origin in the metrical field**. We shall find actually that the planets pursue the courses mapped out for them by the guiding field, and that we need not have recourse to a special "force of gravitation," as did Newton, to account for the influence which diverts the planets from their paths as prescribed by Galilei's Principle (or Newton's first law of motion). The gravitational forces satisfy the second postulate also; that is, they may be made to vanish at a space-time point if we introduce an appropriate co-ordinate system. A closed box, such as a lift, whose suspension wire has snapped, and which descends without friction in the gravitational field of the earth, is a striking example of such a system of reference. All bodies that are falling freely will appear to be at rest to an observer in the box, and physical events will happen in the box in just the same way as if the box were at rest and there were no gravitational field, in spite of the fact that the gravitational force is acting.

II. The transition from the special to the general theory of relativity, as described in I, is a purely mathematical process. By introducing the metrical groundform (1), we may formulate physical laws so that they remain invariant for arbitrary transformations; this is a possibility that is purely mathematical in essence and denotes no particular peculiarity of these laws. A new physical factor appears only when it is assumed that the metrical structure of the world is not given *a priori*, but that the above quadratic form is related to matter by generally invariant laws. Only this fact justifies us in assigning the name "general theory of relativity" to our reasoning; we are not simply giving it to a theory which has merely borrowed the mathematical form of relativity. The same fact is indispensable if we wish to solve the problem of the relativity of motion; it also enables us to complete the analogy mentioned in I, according to which the metrical field is related to matter in the same way as the electric field to electricity. Only if we accept this fact does the theory briefly quoted at the end of the previous section become possible, according to which **gravitation is a mode of expression of the metrical field**; for we know by experience that the gravitational field is determined (in accordance with Newton's law of attraction) by the distribution of matter. This assumption, rather than the postulate of general invariance, seems to the author to be the real pivot of the general theory of relativity. If we adopt this standpoint we are no longer justified in calling the forces that have their origin in the metrical field pseudo-forces. They then have just as real a meaning as the mechanical forces of the electromagnetic field. Coriolis or centrifugal forces are real force effects, which the gravitational or guiding field exerts on matter. Whereas, in I, we were confronted with the easy problem of extending known physical laws (such as Maxwell's equations) from the special case of a constant metrical fundamental tensor to the general case, we have, in following the ideas set out just above, to discover the **invariant law of gravitation, according to which matter determines the components $\Gamma^{\alpha}_{\beta i}$ of the gravitational field**, and which replaces the Newtonian law of attraction in Einstein's Theory. The well-known laws of the field do not furnish a starting-point for this. Nevertheless Einstein succeeded in solving this problem in a convincing fashion, and in showing that the course of planetary motions may be explained just as well by the new law as by the old one of Newton; indeed, that the only discrepancy which the planetary system discloses towards Newton's Theory, and which has hitherto remained inexplicable, namely, the gradual advance of Mercury's perihelion by 43″ per century, is accounted for accurately by Einstein's theory of gravitation.

Thus this theory, which is one of the greatest examples of the power of speculative thought, presents a solution not only of the problem of the relativity of all motion (the only solution which satisfies the demands of logic), but also of the problem of gravitation (*vide* note 4). We see how cogent arguments added to those in Chapter II bring the ideas of Riemann and Einstein to a successful issue. It may also be asserted that their point of view is the first to give due importance to the circumstance that space and time, in contrast with the material content of the world, are **forms** of phenomena. Only physical phase-quantities can be measured, that is, read off from the behaviour of matter in motion; but we cannot measure the four world-co-ordinates that we assign *a priori* arbitrarily to the world-points so as to be able to represent the phase-quantities extending throughout the world by means of mathematical functions (of four independent variables).

Whereas the potential of the electromagnetic field is built up from the co-efficients of an invariant **linear** differential form of the world-co-ordinates $\phi_i \, dx_i$, the potential of the gravitational field is made up of the co-efficients of an invariant **quadratic** differential form. This fact, which is of fundamental importance, constitutes the form of **Pythagoras' Theorem** to which it has gradually been transformed by the stages outlined above. It does not actually spring from the observation of gravitational phenomena in the true sense (Newton accounted for these observations by introducing a single gravitational potential), but from geometry, from the observations of measurement. Einstein's theory of gravitation is the result of the fusion of two realms of knowledge which have hitherto been developed fully independently of one another; this synthesis may be indicated by the scheme

$$\underbrace{\text{Pythagoras} \quad \text{Newton}}_{\text{Einstein}}$$

To derive the values of the quantities g_{ik} from directly observed phenomena, we use light-signals and point-masses which are moving under no forces, as in the special theory of relativity. Let the world-points be referred to any co-ordinates x_i in some way. The geodetic lines passing through a world-point O, namely,

$$\frac{d^2 x_i}{ds^2} + \begin{Bmatrix} \alpha\beta \\ i \end{Bmatrix} \frac{dx_\alpha}{ds} \frac{dx_\beta}{ds} = 0, \tag{8}$$

$$g_{ik} \frac{dx_i}{ds} \frac{dx_k}{ds} = C = \text{const.}, \tag{9}$$

split up into two classes; (a) those with a **space-like** direction, (b) those with a **time-like** direction ($C < 0$ or $C > 0$ respectively). The latter fill a "double" cone with the common vertex at O and which, at O, separates into two simple cones, of which one opens into the future and the other into the past. The first comprises all world-points that belong to the "active future" of O, the second all world-points that constitute the "passive past" of O. The limiting sheet of the cone is formed by the geodetic null-lines ($C = 0$); the "future" half of the sheet contains all the world-points at which a light-signal emitted from O arrives, or, more generally, the exact initial points of every effect emanating from O. The metrical groundform thus determines in general what world-points are related to one another in effects. If dx_i are the relative co-ordinates of a point O' infinitely near O, then O' will be traversed by a light-signal emitted from O if, and only if, $g_{ik}\, dx_i\, dx_k = 0$. By observing the arrival of light at the points neighbouring to O we can thus determine the ratios of the values of the g_{ik}'s at the point O; and, as for O, so for any other point. It is impossible, however, to derive any further results from the phenomenon of the propagation of light, for it follows from a remark on page 188 that the geodetic null-lines are dependent only on the ratios of the g_{ik}'s.

The optical "direction" picture that an observer ("point-eye" as on page 146) receives, for instance, from the stars in the heavens, is to be constructed as follows. From the world-point O at which the observer is stationed those geodetic null-lines (light-lines) are to be drawn on the backward cone which cuts the world-lines of the stars. The direction of every light-line at O is to be resolved into one component which lies along the direction $\mathbf{e}$ of the world-line of the observer and another $\mathbf{s}$ which is perpendicular to it (the meaning of perpendicular is defined by the metrical structure of the world as given on page 180); $\mathbf{s}$ is the spatial direction of the light-ray. Within the three-dimensional linear manifold of the line-elements at O perpendicular to $\mathbf{e}$, $-ds^2$ is a definitely positive form. The angles (that arise from it when it is taken as the metrical groundform, and which are to be calculated from formula (15), § 11) between the spatial directions $\mathbf{s}$ of the light-rays are those that determine the positions of the stars as perceived by the observer.

The factor of proportionality of the g_{ik}'s which could not be derived from the phenomenon of the transmission of light may be determined from the motion of point-masses which carry a clock with them. For if we assume that—at least for unaccelerated motion under no forces—the time read off from such a clock is the proper-time s, equation (9) clearly makes it possible to apply the unit of measure along the world-line of the motion (cf. Appendix I).

§ 28. Einstein's Fundamental Law of Gravitation

According to the Newtonian Theory the condition (or phase) of matter is characterised by a **scalar**, the mass-density μ; and the gravitational potential is also a scalar Φ: Poisson's equation holds, that is,

$$\Delta\Phi = 4\pi k\mu \tag{10}$$

(Δ = div grad; k = the gravitational constant). This is the law according to which matter determines the gravitational field. But according to the theory of relativity matter can be described rigorously only by a symmetrical **tensor** of the second order T_{ik}, or better still by the corresponding mixed tensor-density $\mathbf{T}_i^k$; in harmony with this the potential of the gravitational field consists of the components of a symmetrical **tensor** g_{ik}. Therefore, in Einstein's Theory we expect equation (10) to be replaced by a system of equations of which the left side consists of differential expressions of the second order in the g_{ik}'s, and the right side of components of the energy-density; this system has to be invariant with respect to arbitrary transformations of the co-ordinates. To find the law of gravitation we shall do best by taking up the thread from Hamilton's Principle formulated at the close of § 26. The *Action* there consisted of three parts: the substance-action of electricity, the field-action of electricity, and the substance-action of mass or gravitation. In it there is lacking a fourth term, the field-action of gravitation, which we have now to find. Before doing this, however, we shall calculate the change in the sum of the first three terms already known, when we leave the potentials ϕ_i of the electromagnetic field and the world-lines of the substance-elements unchanged but subject the g_{ik}'s, **the potentials of the metrical field, to an infinitesimal virtual variation** δ. This is possible only from the point of view of the general theory of relativity.

This causes no change in the substance-action of electricity, but the change in the integrands that occur in the field-action, namely

$$\tfrac{1}{2}\mathbf{S} = \tfrac{1}{4}F_{ik}\mathbf{F}^{ik}$$

is

$$\tfrac{1}{4}\{\sqrt{g}\delta(F_{ik}F^{ik}) + (F_{ik}F^{ik})\delta\sqrt{g}\}.$$

The first summand in the curved bracket here $= \mathbf{F}_{rs}\,\delta F^{rs}$ and hence, since

$$F^{rs} = g^{ri}g^{sk}F_{ik},$$

we immediately get the value

$$2\sqrt{g}F_{ir}F_k^r\,\delta g^{ik}.$$

The second summand, by (58′) § 17,

$$= -\mathbf{S}g_{ik}\,\delta g^{ik}.$$

Thus, finally, we find the variation in the field-action to be

$$= \int \tfrac{1}{2}\mathbf{S}\,\delta g^{ik}\,dx = \int \tfrac{1}{2}\mathbf{S}^{ik}\,\delta g_{ik}\,dx \quad \text{(cf. (59), § 17)}$$

if

$$\mathbf{S}_i^k = \tfrac{1}{2}\mathbf{S}\delta_i^k = F_{ir}\mathbf{F}^{kr} \tag{11}$$

are the components of the energy-density of the electromagnetic field.* It suddenly becomes clear to us now (and only now that we

*The signs are the reverse of those used in Chapter III on account of the change in the sign of the metrical groundform.

have succeeded in calculating the variation of the world's metrical field) what is the origin of the complicated expressions (11) for the energy-momentum density of the electromagnetic field.

We get a corresponding result for the substance-action of the mass; for we have

$$\delta\sqrt{g_{ik}\,dx_i\,dx_k} = \tfrac{1}{2}\,\frac{dx_i\,dx_k\,\delta g_{ik}}{ds} = \tfrac{1}{2}ds\,u^i u^k\,\delta g_{ik},$$

and hence

$$\delta\int\left(dm\int\sqrt{g_{ik}\,dx_i\,dx_k}\right) = \int\tfrac{1}{2}\mu u^i u_k\,\delta g_{ik}\,dx.$$

Hence the total change in the *Action* so far known to us is, for a variation of the metrical field,

$$\int\tfrac{1}{2}\mathbf{T}^{ik}\,\delta g_{ik}\,dx \tag{12}$$

in which $\mathbf{T}_i^k$ denotes the tensor-density of the total energy.

The absent fourth term of the *Action*, namely, the field-action of gravitation, must be an invariant integral, $\int\mathbf{G}\,dx$, of which the integrand $\mathbf{G}$ is composed of the potentials g_{ik} and of the field-components $\left\{\begin{matrix} ik \\ r \end{matrix}\right\}$ of the gravitational field, built up from the g_{ik}'s and their first derivatives. It would seem to us that only under such circumstances do we obtain differential equations of order not higher than the second for our gravitational laws. If the total differential of this function is

$$\delta\mathbf{G} = \tfrac{1}{2}\mathbf{G}^{ik}\,\delta g_{ik} + \tfrac{1}{2}\mathbf{G}^{ik,r}\,\delta g_{ik,r} \qquad (\mathbf{G}^{ki} = \mathbf{G}^{ik} \text{ and } \mathbf{G}^{ki,r} = \mathbf{G}^{ik,r}) \tag{13}$$

we get, for an infinitesimal variation δg_{ik} which disappears for regions beyond a finite limit, by partial integration, that

$$\delta \int \mathbf{G}\, dx = \int \tfrac{1}{2}[\mathbf{G}]^{ik}\, \delta g_{ik}\, dx \tag{14}$$

in which the "Lagrange derivatives" $[\mathbf{G}]^{ik}$, which are symmetrical in i and k, are to be calculated according to the formula

$$[\mathbf{G}] = \mathbf{G}^{ik} - \frac{\partial \mathbf{G}^{ik,r}}{\partial x_r}.$$

The gravitational equations will then actually assume the form which was predicted, namely

$$[\mathbf{G}]_i^k = -\mathbf{T}_i^k. \tag{15}$$

There is no longer any cause for surprise that it happens to be the energy-momentum components that appear as co-efficients when we vary the g_{ik}'s in the first three factors of the *Action* in accordance with (12). Unfortunately a scalar-density $\mathbf{G}$, of the type we wish, does not exist at all; for we can make all the $\begin{Bmatrix} ik \\ r \end{Bmatrix}$'s vanish at any given point by choosing the appropriate co-ordinate system. Yet the scalar R, the curvature defined by Riemann, has made us familiar with an invariant which involves the second derivatives of the g_{ik}'s only **linearly**: it may even be shown that it is the only invariant of this kind (*vide* Appendix II, in which the proof is given). In consequence of this linearity we may use the invariant integral $\int \frac{1}{2} R\sqrt{g}\, dx$ to get the derivatives of the second order by partial integration. We then get

$$\int \tfrac{1}{2} R\sqrt{g}\, dx = \int \mathbf{G}\, dx$$

+ a divergence integral, that is, an integral whose integrand is of the form $\dfrac{\partial \mathbf{w}^i}{\partial x_i}$: **G** here depends only on the g_{ik}'s and their first derivatives. Hence, for variations δg_{ik}, that vanish outside a finite region, we get

$$\delta \int \tfrac{1}{2} R\sqrt{g}\, dx = \delta \int \mathbf{G}\, dx$$

since, according to the principle of partial integration,

$$\int \frac{\partial(\delta \mathbf{w}^i)}{\partial x_i}\, dx = 0.$$

Not $\int \mathbf{G}\, dx$ itself is an invariant, but the variation $\delta \int \mathbf{G}\, dx$, and this is the essential feature of Hamilton's Principle. *We need not, therefore, have fears about introducing* $\int \mathbf{G}\, dx$ *as the* Action *of the gravitational field; and this hypothesis is found to be the only possible one.* We are thus led under compulsion, as it were, to the unique gravitational equations (15). It follows from them that **every kind of energy exerts a gravitational effect**: this is true not only of the energy concentrated in the electrons and atoms, that is of matter in the restricted sense, but also of diffuse field-energy (for the $\mathbf{T}_i^k$'s are the components of the total energy).

Before we carry out the calculations that are necessary if we wish to be able to write down the gravitational equations explicitly, we must first test whether we get analogous results **in the case of Mie's Theory**. The *Action*, $\int \mathbf{L}\, dx$, which occurs in it is an invariant not only for linear, but also for arbitrary transformations. For **L** is composed algebraically (not as a result of tensor analysis) of the components ϕ_i of a co-variant vector (namely, of the electromagnetic potential), of the components F_{ik} of a linear tensor of the second order (namely, of the electromagnetic field), and of the components g_{ik} of the fundamental metrical tensor. We set the total differential $\delta\mathbf{L}$ of this function equal to

$$\tfrac{1}{2}\mathbf{T}^{ik}\,\delta g_{ik} + \delta_0\mathbf{L}, \quad \text{in which } \delta_0\mathbf{L} = \tfrac{1}{2}\mathbf{H}^{ik}\,\delta F_{ik} + \mathbf{s}^i\,\delta\phi_i$$
$$(\mathbf{T}^{ki} = \mathbf{T}^{ik}, \quad \mathbf{H}^{ki} = -\mathbf{H}^{ik}). \tag{16}$$

We then call the tensor-density $\mathbf{T}_i^k$ the energy or matter. By doing this, we affirm once again that the metrical field (with the potentials g_{ik}) is related to matter ($\mathbf{T}^{ik}$) in the same way as the electromagnetic field (with the potentials ϕ_i) is related to the electric current $\mathbf{s}^i$. We are now obliged to prove that the present explanation leads accurately to the expressions given in (64), § 26, for energy and momentum. This will furnish the proof, which was omitted above, of the symmetry of the energy-tensor. To do this we cannot use the method of direct calculation as above in the particular case of Maxwell's Theory, but we must apply the following elegant considerations, the nucleus of which is to be found in Lagrange, but which were discussed with due regard to formal perfection by F. Klein (*vide* note 5).

We subject the world-continuum to an infinitesimal deformation, as a result of which in general the point (x_i) becomes transformed into the point $(\bar{x}_j)$

$$\bar{x}_i = x_i + \epsilon\cdot\xi^i(x_0, x_1, x_2, x_3) \tag{17}$$

(in which ϵ is the constant infinitesimal parameter, all of whose higher powers are to be struck out). We imagine the phase-quantities to follow the deformation so that at its conclusion the new ϕ_i's (we call them $\bar{\phi}_i$) are functions of the co-ordinates of such a kind that, in consequence of (17), the equations

$$\phi_i(x)\,dx_i = \bar{\phi}_i(\bar{x})\,d\bar{x}_i \tag{18}$$

hold; and in the same sense the symmetrical and skew-symmetrical bilinear differential form with the co-efficients g_{ik}, F_{ik}, respectively, remains unchanged. The changes $\bar{\phi}_i(x) - \phi_i(x)$ which the quantities ϕ_i undergo at a fixed world-point (x_i) as a result of the deformation will be denoted by $\delta\phi_i$; δg_{ik} and δF_{ik} have a corresponding meaning.

If we replace the old quantities ϕ_i in the function $\mathbf{L}$ by the $\bar{\phi}_i$ arising from the deformation, we shall suppose the function $\bar{\mathbf{L}} = \mathbf{L} + \delta\mathbf{L}$ to result; the $\delta\mathbf{L}$ in it is given by (16). Furthermore, let $\mathfrak{X}$ be an arbitrary region of the world which, owing to the deformation, becomes $\bar{\mathfrak{X}}$. The deformation causes the *Action* $\int_{\mathfrak{X}} \mathbf{L}\,dx$ to undergo a change $\delta' \int_{\mathfrak{X}} \mathbf{L}\,dx$ which is equal to the difference between the integral $\bar{\mathbf{L}}$ taken over $\mathfrak{X}$ and the integral $\mathbf{L}$ taken over $\bar{\mathfrak{X}}$. The invariance of the *Action* is expressed by the equation

$$\delta' \int_{\mathfrak{X}} \mathbf{L}\,dx = 0. \tag{19}$$

We make a natural division of this difference into two parts: (1) the difference between the integrals of $\bar{\mathbf{L}}$ and $\mathbf{L}$ over $\bar{\mathfrak{X}}$, (2) the difference between the integral of $\mathbf{L}$ over $\bar{\mathfrak{X}}$ and $\mathfrak{X}$. Since $\bar{\mathfrak{X}}$ differs from $\mathfrak{X}$ only by an infinitesimal amount, we may set

$$\delta \int_{\mathfrak{X}} \mathbf{L}\,dx = \int_{\mathfrak{X}} \delta\mathbf{L}\,dx$$

for the first part. On page 165 we found the second part to be

$$\epsilon \int_{\mathfrak{X}} \frac{\partial(\mathbf{L}\xi^i)}{\partial x_i}\,dx.$$

To be able to complete the argument we must next calculate the variations $\delta\phi_i$, δg_{ik}, δF_{ik}. If we set $\bar{\phi}_i(\bar{x}) - \phi_i(x) = \delta'\phi_i$ for a moment, then, owing to (18), we get

$$\delta'\phi_i \cdot dx_i + \epsilon\phi_r \, d\xi^r = 0$$

and hence

$$\delta'\phi_i = -\epsilon \cdot \phi_r \frac{\partial \xi^i}{\partial x^i}.$$

Moreover, since

$$\delta\phi_i = \delta'\phi_i - \left\{\bar{\phi}_i(\bar{x}) - \bar{\phi}_i(x)\right\} = \delta'\phi_i - \epsilon \cdot \frac{\partial \phi}{\partial x_r} \xi^r$$

we get, suppressing the self-evident factor ϵ,

$$-\delta\phi_i = \phi_r \frac{\partial \xi^r}{\partial x_i} + \frac{\partial \phi_i}{\partial x_r} \xi^r. \tag{20}$$

In the same way, we get

$$-\delta g_{ik} = g_{ir} \frac{\partial \xi^r}{\partial x_k} + g_{rk} \frac{\partial \xi^r}{\partial x_i} + \frac{\partial g_{ik}}{\partial x_r} \xi^r, \tag{20$'$}$$

$$-\delta F_{ik} = F_{ir} \frac{\partial \xi^r}{\partial x_k} + F_{rk} \frac{\partial \xi^r}{\partial x_i} + \frac{\partial F_{ik}}{\partial x_r} \xi^r. \tag{20$''$}$$

And, on account of

$$F_{ik} = \frac{\partial \phi_i}{\partial x_k} - \frac{\partial \phi_k}{\partial x_i} \quad \text{we have} \quad \delta F_{ik} = \frac{\partial(\delta\phi_i)}{\partial x_k} - \frac{\partial(\delta\phi_k)}{\partial x_i}, \tag{21}$$

for since the former is an invariant relation, we get from it

$$\bar{F}_{ik}(\bar{x}) = \frac{\partial \bar{\phi}_i(\bar{x})}{\partial \bar{x}_k} - \frac{\partial \bar{\phi}_k(\bar{x})}{\partial \bar{x}_i}, \quad \text{and also } \bar{F}_{ik}(x) = \frac{\partial \bar{\phi}_i(x)}{\partial x_k} - \frac{\partial \bar{\phi}_k(x)}{\partial x_i}.$$

Substitution gives us

$$-\delta \mathbf{L} = (\mathbf{T}_i^k + \mathbf{H}^{rk} F_{ri} + \mathbf{s}^k \phi_i) \frac{\partial \xi}{\partial x_k} + (\tfrac{1}{2} \mathbf{T}^{\alpha\beta} \frac{\partial g_{\alpha\beta}}{\partial x_i} + \cdots +) \xi^i.$$

If we remove the derivatives of ξ^i by partial integration, and use the abbreviation

$$\mathbf{V}_i^k = \mathbf{T}_i^k + F_{ir} \mathbf{H}^{kr} + \phi_i \mathbf{s}^k - \delta_i^k \mathbf{L},$$

we get a formula of the following form

$$-\delta' \int_{\mathfrak{X}} \mathbf{L}\, dx = \int_{\mathfrak{X}} \frac{\partial (\mathbf{V}_i \xi^i)}{\partial x_k}\, dx + \int_{\mathfrak{X}} (\mathbf{t}_i \xi^i)\, dx = 0. \qquad (22)$$

It follows from this that, as we know, by choosing the ξ^i's appropriately, namely, so that they vanish outside a definite region, which we here take to be $\mathfrak{X}$, we must have, at every point,

$$\mathbf{t}_i = 0. \qquad (23)$$

Accordingly, the first summand of (22) is also equal to zero. The identity which comes about in this way is valid for arbitrary quantities ξ^i and for any finite region of integration $\mathfrak{X}$. Hence, since the integral of a continuous function taken over any and every region can vanish only if the function itself $= 0$, we must have

$$\frac{\partial (\mathbf{V}_i^k \xi^i)}{\partial x_k} = \mathbf{V}_i^k \frac{\partial \xi^i}{\partial x_k} + \frac{\partial \mathbf{V}_i^k}{\partial x_k} \xi^i = 0.$$

Now, ξ^i and $\dfrac{\partial \xi^i}{\partial x_k}$ may assume any values at one and the same point. Consequently,

$$\mathbf{V}_i^k = 0 \qquad \left(\frac{\partial \mathbf{V}_i^k}{\partial x_k} = 0 \right).$$

This gives us the desired result

$$\mathbf{T}_i^k = \mathbf{L}\delta_i^k - F_{ir}\mathbf{H}^{kr} - \phi_i \mathbf{s}^k.$$

These considerations simultaneously give us the theorems of conservation of energy and of momentum, which we found by calculation in § 26; they are contained in equations (23). The change in the *Action* of the whole world for an infinitesimal deformation which vanishes outside a finite region of the world is found to be

$$\int \delta\mathbf{L}\,dx = \int \tfrac{1}{2}\mathbf{T}^{ik}\,\delta g_{ik}\,dx + \int \delta_0\mathbf{L}\,dx = 0. \tag{24}$$

In consequence of the equations (21) and of **Hamilton's Principle**, namely

$$\int \delta_0\mathbf{L}\,dx = 0, \tag{25}$$

which is here valid, the second part (in Maxwell's equations) disappears. But the first part, as we have already calculated, is

$$-\int\left(\mathbf{T}_i^k\frac{\partial\xi^i}{\partial x_k} + \tfrac{1}{2}\frac{\partial g_{\alpha\beta}}{\partial x_i}\mathbf{T}^{\alpha\beta}\xi^i\right)dx = \int\left(\frac{\partial\mathbf{T}_i^k}{\partial x_k} - \tfrac{1}{2}\frac{\partial g_{\alpha\beta}}{\partial x_i}\mathbf{T}^{\alpha\beta}\right)\xi^i\,dx.$$

Thus, **as a result of the laws of the electromagnetic field, we get the mechanical equations**

$$\frac{\partial\mathbf{T}_i^k}{\partial x_k} - \tfrac{1}{2}\frac{\partial g_{\alpha\beta}}{\partial x_i}\mathbf{T}^{\alpha\beta} = 0. \tag{26}$$

(On account of the presence of the additional term due to gravitation these equations can no longer in the general theory of relativity be fitly termed theorems of conservation. The question whether proper theorems of conservation may actually be set up will be discussed in § 33.)

The Hamiltonian Principle which has been **supplemented by the *Action* of the gravitational field**, namely

$$\delta \int (\mathbf{L} + \mathbf{G})\, dx = 0, \tag{27}$$

and in which the electromagnetic and the **gravitational** condition (phase) of the field may be subjected independently of one another to virtual infinitesimal variations gives rise to the gravitational equations (15) in addition to the electromagnetic laws. If we apply the process above, which ended in (26), to $\mathbf{G}$ instead of to $\mathbf{L}$—here, too, we have, for the variation δ caused by a deformation of the world-continuum which vanishes outside a finite region, that

$$\delta \int \mathbf{G}\, dx = \delta \int \tfrac{1}{2} R \sqrt{g}\, dx = 0$$

—we arrive at **mathematical identities** analogous to (26), namely

$$\frac{\partial [\mathbf{G}]_i^k}{\partial x_k} - \frac{1}{2} \frac{\partial g_{\alpha\beta}}{\partial x_i} [\mathbf{G}]^{\alpha\beta} = 0.$$

The fact that $\mathbf{G}$ contains the derivatives of the g_{ik}'s as well as the g_{ik}'s themselves is of no account. Accordingly, *the mechanical equations* (26) *are just as much a consequence of the gravitational equations* (15) *as of the electromagnetic laws of the field.*

The wonderful relationships, which here reveal themselves, may be formulated in the following way independently of the question whether Mie's theory of electrodynamics is valid or not. The phase (or condition) of a physical system is described relatively to a co-ordinate system by means of certain variable space-time phase-quantities ϕ (these were our ϕ_i's above). Besides these, we have also to take account of the **metrical field** in which the system is embedded and which is characterised by its potentials g_{ik}. The uniformity underlying the phenomena occurring in the system is expressed by an invariant integral $\int \mathbf{L}\, dx$; in it, the scalar-density **L** is a function of the ϕ's and of their derivatives of the first and if need be, of the second order, and also a function of the g_{ik}'s, but the latter quantities alone and not their derivatives occur in **L**. We form the total differential of the function **L** by writing down explicitly only that part which contains the differentials δg_{ik}, namely,

$$\delta \mathbf{L} = \tfrac{1}{2}\mathbf{T}^{ik}\delta g_{ik} + \delta_0 \mathbf{L}.$$

$\mathbf{T}_i^k$ is then the tensor-density of the **energy** (identical with **matter**) associated with the physical state or phase of the system. The determination of its components is thus reduced once and for all to a determination of Hamilton's Function **L**. *The general theory of relativity alone, which allows the process of variation to be applied to the metrical structure of the world, leads to a true definition of energy.* The phase-laws emerge from the "partial" principle of action in which only the phase-quantities ϕ are to be subjected to variation; just as many equations arise from it as there are quantities ϕ. The additional ten gravitational equations (15) for the ten potentials g_{ik} result if we enlarge the partial principle of action to the total one (27), in which the g_{ik}'s are also to be subjected to variation. The **mechanical equations** (26) are a consequence of the phase-laws as well as of the gravitational laws; they may, indeed, be termed the eliminant of the latter. Hence, in the system of phase and gravitational laws, there are four superfluous equations. The general solution must, in fact, contain four arbitrary functions, since the equations, in virtue of their invariant character, leave the co-ordinate system of the x_i's indeterminate; hence, arbitrary continuous transformations of these co-ordinates derived from **one** solution of the equations always give rise to new solutions in their turn. (These solutions, however, represent the same objective course of the world.) The old subdivision into geometry, mechanics, and physics must be replaced in Einstein's Theory by the separation into physical phases and metrical or gravitational fields.

For the sake of completeness we shall once again revert to the Hamiltonian Principle used in the theory of Lorentz and Maxwell. Variation applied to the ϕ_i's gives the electromagnetic laws, but applied to the g_{ik}'s the gravitational laws. Since the *Action* is an invariant, the infinitesimal change which an infinitesimal deformation of the world-continuum calls up in it $= 0$; this deformation is to affect the electromagnetic and the gravitational field as well as the world-lines of the substance-elements. This change consists of three summands, namely, of the changes which are caused in turn by the variation of the electromagnetic field, of the gravitational field, and of the substance-paths. The first two parts are zero as a consequence of the electromagnetic and the gravitational laws; hence the third part also vanishes and we see that the mechanical equations are a result of the two groups of laws mentioned just above. Recapitulating our former calculations we may derive this result by taking the following steps. From the gravitational laws there follow (26), i.e.

$$\mu U_i + u_i M = -\left\{\frac{\partial \mathbf{S}_i^k}{\partial x_k} - \frac{1}{2}\frac{\partial g_{\alpha\beta}}{\partial x_i}\mathbf{S}^{\alpha\beta}\right\}, \tag{28}$$

in which $\mathbf{S}_i^k$ is the tensor-density of the electromagnetic energy of field, namely, of

$$U_i = \frac{du_i}{ds} - \frac{1}{2}\frac{\partial g_{\alpha\beta}}{\partial x_i} u^\alpha u^\beta,$$

and M is the left-hand member of the equation of continuity for matter, namely

$$M = \frac{\partial(\mu u^i)}{\partial x_i}.$$

As a result of Maxwell's equations the right-hand member of (28)

$$= \mathbf{p}_i = -F_{ik}\mathbf{s}^k \qquad (\mathbf{s}^i = \rho u^i).$$

If we then multiply (28) by u^i and sum up with respect to i, we get $M = 0$; in this way we have arrived at the equation of continuity for matter and also at the mechanical equations in their usual form.

After having gained a full survey of how the gravitational laws of Einstein are to be arranged into the scheme of the remaining physical laws, we are still faced with the task of working out the explicit expression for the $[\mathbf{G}]_i^k$'s (*vide* note 6). The virtual change

$$\delta\Gamma_{ik}^r = \delta\begin{Bmatrix} ik \\ r \end{Bmatrix} = \gamma_{ik}^r$$

of the components of the affine relationship is, as we know (page 168), a tensor. If we use a geodetic co-ordinate system at a certain point, then we get directly from the formula for R^{ik} ((60), § 17) that

$$\delta R_{ik} = \frac{\partial \gamma_{ik}^r}{\partial x_r} - \frac{\partial \gamma_{ir}^r}{\partial x_k}$$

and

$$g^{ik}\,\delta R_{ik} = g^{ik}\,\frac{\partial \gamma^r_{ik}}{\partial x_r} - g^{ir}\,\frac{\partial \gamma^k_{ik}}{\partial x_r}.$$

If we set

$$g^{ik}\gamma^r_{ik} - g^{ir}\gamma^k_{ik} = w^r$$

we get

$$g^{ik}\,\delta R_{ik} = \frac{\partial w^r}{\partial x_r},$$

or, for any arbitrary co-ordinate system,

$$\delta R = R_{ik}\,\delta g^{ik} + \frac{1}{\sqrt{g}}\,\frac{\partial(\sqrt{g}w^r)}{\partial x_r}.$$

The divergence disappears in the integration and hence, since by definition we are to have

$$\delta\int R\sqrt{g}\,dx = \int[\mathbf{G}]^{ik}\,\delta g_{ik}\,dx = -\int[\mathbf{G}]_{ik}\,\delta g^{ik}\,dx$$

and since the R_{ik}'s are symmetrical in Riemann's space, we get

$$[\mathbf{G}]_{ik} = \sqrt{g}(\tfrac{1}{2}g_{ik}R - R_{ik}) = \tfrac{1}{2}g_{ik}\mathbf{R} - \mathbf{R}_{ik},$$
$$[\mathbf{G}]^k_i = \tfrac{1}{2}\delta^k_i\mathbf{R} - \mathbf{R}^k_i.$$

Therefore the gravitational laws are

$$\boxed{\mathbf{R}^k_i - \tfrac{1}{2}\delta^k_i\mathbf{R} = \mathbf{T}^k_i} \tag{29}$$

Here, of course (exactly as was done for the unit of charge in electromagnetic equations), the unit of mass has been suitably chosen. If we retain the units of the c.g.s. system, a universal constant $8\pi\kappa$

will have to be added as a factor to the right-hand side. It might still appear doubtful now at the outset whether κ is positive or negative, and whether the right-hand side of equation (29) should not be of opposite sign. We shall find, however, in the next paragraph that, in virtue of the fact that masses attract one another and do not repel, κ is actually positive.

It is of mathematical importance to notice that **the exact gravitational laws are not linear**; although they are linear in the derivatives of the field-components $\begin{Bmatrix} ik \\ r \end{Bmatrix}$, they are not linear in the field-components themselves. If we contract equations (29), that is, set $k = i$, and sum with respect to i, we get $-\mathbf{R} = \mathbf{T} = \mathbf{T}_i^i$; hence, in place of (29) we may also write

$$\mathbf{R}_i^k = \mathbf{T}_i^k - \tfrac{1}{2}\delta_i^k\mathbf{T}. \tag{30}$$

In the first paper in which Einstein set up the gravitational equations without following on from Hamilton's Principle, the term $-\frac{1}{2}\delta_i^k\mathbf{T}$ was missing on the right-hand side; he recognised only later that it is required as a result of the energy-momentum-theorem (*vide* note 7). The whole series of relations here described and which is subject to Hamilton's Principle, has become manifest in further works by H. A. Lorentz, Hilbert, Einstein, Klein, and the author (*vide* note 8).

In the sequel we shall find it desirable to know the value of $\mathbf{G}$. To convert

$$\int R\sqrt{g}\,dx \quad \text{into} \quad 2\int \mathbf{G}\,dx$$

by means of partial integration (that is, by detaching a divergence), we must set

$$\sqrt{g}g^{ik}\frac{\partial}{\partial x_r}\begin{Bmatrix} ik \\ r \end{Bmatrix} = \frac{\partial}{\partial x_r}\left(\sqrt{g}g^{ik}\begin{Bmatrix} ik \\ r \end{Bmatrix}\right) - \begin{Bmatrix} ik \\ r \end{Bmatrix}\frac{\partial}{\partial x_r}(\sqrt{g}g^{ik}),$$
$$\sqrt{g}g^{ik}\frac{\partial}{\partial x_k}\begin{Bmatrix} ir \\ r \end{Bmatrix} = \frac{\partial}{\partial x_k}\left(\sqrt{g}g^{ik}\begin{Bmatrix} ir \\ r \end{Bmatrix}\right) - \begin{Bmatrix} ir \\ r \end{Bmatrix}\frac{\partial}{\partial x_k}(\sqrt{g}g^{ik}).$$

Thus we get

$$2\mathbf{G} = \begin{Bmatrix} is \\ s \end{Bmatrix}\frac{\partial}{\partial x_k}(\sqrt{g}g^{ik}) - \begin{Bmatrix} ik \\ r \end{Bmatrix}\frac{\partial}{\partial x_r}(\sqrt{g}g^{ik}) + \left(\begin{Bmatrix} ik \\ r \end{Bmatrix}\begin{Bmatrix} rs \\ s \end{Bmatrix} - \begin{Bmatrix} ir \\ s \end{Bmatrix}\begin{Bmatrix} ks \\ r \end{Bmatrix}\right)\sqrt{g}g^{ik}.$$

By (57′), (57″) of § 17, however, the first two terms on the right, if we omit the factor $\sqrt{g}$,

$$= -\begin{Bmatrix} is \\ s \end{Bmatrix}\begin{Bmatrix} kr \\ i \end{Bmatrix}g^{kr} + 2\begin{Bmatrix} ik \\ r \end{Bmatrix}\begin{Bmatrix} rs \\ i \end{Bmatrix}g^{sk} - \begin{Bmatrix} ik \\ r \end{Bmatrix}\begin{Bmatrix} rs \\ s \end{Bmatrix}g^{ik}$$
$$= \left(-\begin{Bmatrix} rs \\ s \end{Bmatrix}\begin{Bmatrix} ik \\ r \end{Bmatrix} + 2\begin{Bmatrix} sk \\ r \end{Bmatrix}\begin{Bmatrix} ri \\ s \end{Bmatrix} - \begin{Bmatrix} ik \\ r \end{Bmatrix}\begin{Bmatrix} rs \\ s \end{Bmatrix}\right)g^{ik}$$
$$= 2g^{ik}\left(\begin{Bmatrix} ir \\ s \end{Bmatrix}\begin{Bmatrix} ks \\ r \end{Bmatrix} - \begin{Bmatrix} ik \\ r \end{Bmatrix}\begin{Bmatrix} rs \\ s \end{Bmatrix}\right).$$

Hence we finally arrive at

$$\frac{1}{\sqrt{g}}\mathbf{G} = \tfrac{1}{2}g^{ik}\left(\begin{Bmatrix} ir \\ s \end{Bmatrix}\begin{Bmatrix} ks \\ r \end{Bmatrix} - \begin{Bmatrix} ik \\ r \end{Bmatrix}\begin{Bmatrix} rs \\ s \end{Bmatrix}\right). \tag{31}$$

This completes our development of the foundations of Einstein's Theory of Gravitation. We must now inquire whether observation confirms this theory which has been built up on purely speculative grounds, and above all, whether the motions of the planets can be explained just as well (or better) by it as by Newton's law of attraction. §§ 29–32 treat of the solution of the gravitational equations. The discussion of the general theory will not be resumed till § 33.

§ 29. The Stationary Gravitational Field—Comparison with Experiment

To establish the relationship of Einstein's laws with the results of observations of the planetary system, we shall first specialise them for the case of a stationary gravitational field (*vide* note 9). The latter is characterised by the circumstance that, if we use appropriate co-ordinates, the world resolves into space and time, so that for the metrical form

$$ds^2 = f^2\,dt^2 - d\sigma^2, \qquad d\sigma^2 = \sum_{i,k=1}^{3} \gamma_{ik}\,dx_i\,dx_k,$$

we get

$$g_{00} = f^2; \quad g_{0i} = g_{i0} = 0; \quad g_{ik} = -\gamma_{ik} \qquad (i, k = 1, 2, 3),$$

and also that the co-efficients f and γ^{ik} occurring in it depend only on the space-co-ordinates x_1, x_2, x_3, and not on the time $t = x_0$. $d\sigma^2$ is a positive definite quadratic differential form which determines the metrical nature of the space having co-ordinates x_1, x_2, x_3; f is obviously the velocity of light. The measure t of time is fully determined (when the unit of time has been chosen) by the postulates that have been set up, whereas the space co-ordinates x_1, x_2, x_3 are fixed only to the extent of an arbitrary continuous transformation of these co-ordinates among themselves. In the statical case, therefore, the metrics of the world gives, besides the measure-determination of the space, also a scalar field f in space.

If we denote the Christoffel 3-indices symbol, relating to the ternary form $d\sigma^2$, by an appended $*$, and if the index letters i, k, l assume only the values $1, 2, 3$ in turn, then it easily follows from definition that

$$\begin{Bmatrix} ik \\ l \end{Bmatrix} = \begin{Bmatrix} ik \\ l \end{Bmatrix}^*,$$

$$\begin{Bmatrix} ik \\ 0 \end{Bmatrix} = 0, \qquad \begin{Bmatrix} 0i \\ k \end{Bmatrix} = 0, \qquad \begin{Bmatrix} 00 \\ 0 \end{Bmatrix} = 0,$$

$$\begin{Bmatrix} i0 \\ 0 \end{Bmatrix} = \frac{f_i}{f}, \qquad \begin{Bmatrix} 00 \\ 0 \end{Bmatrix} = ff^i.$$

In the above, $f_i = \dfrac{\partial f}{\partial x_i}$ are co-variant components of the three-dimensional gradient, and $f^i = \gamma^{ik} f_k$ are the corresponding contra-variant components, whereas $\sqrt{\gamma} f^i = \mathfrak{f}^i$ are the components of a contra-variant vector-density in space. For the determinant γ of the γ_{ik}'s we have $\sqrt{g} = f\sqrt{\gamma}$. If we further set

$$f_{ik} = \frac{\partial f_i}{\partial x_k} - \begin{Bmatrix} ik \\ r \end{Bmatrix}^* f_r = \frac{\partial^2 f}{\partial x_i\, \partial x_k} - \begin{Bmatrix} ik \\ r \end{Bmatrix}^* \frac{\partial f}{\partial x_r}$$

(the summation letter r also assumes only the three values $1, 2, 3$), and if we also set

$$\Delta f = \frac{\partial \mathfrak{f}}{\partial x_i} \qquad \Delta f = \sqrt{\gamma} \cdot f_i^i),$$

we arrive by an easy calculation at the following relations between the components R_{ik} and P_{ik} of the curvature tensor of the second order which belongs to the quadratic groundform ds^2 for $d\sigma^2$, respectively

$$R_{ik} = \mathsf{P}_{ik} - \frac{f_{ik}}{f},$$

$$R_{i0} = R_{0i} = 0,$$

$$R_{00} = f \cdot \frac{\Delta f}{\sqrt{\gamma}} \qquad (\mathbf{R}_0^0 = \Delta f).$$

For statical matter which is non-coherent (i.e. of which the parts do not act on one another by means of stresses), $\mathbf{T}_0^0 = \mu$ is the only component of the energy-density tensor that is not zero; hence $\mathbf{T} = \mu$. Matter at rest produces a statical gravitational field. Among the gravitational equations (30) the only one that is of interest to us is the $\binom{0}{0}$th: it gives us

$$\Delta f = \tfrac{1}{2}\mu \tag{32}$$

or, if we insert the constant factor of proportionality $8\pi\kappa$, we get

$$\Delta f = 4\pi\kappa\mu. \tag{32'}$$

If we assume that, for an appropriate choice of the space-coordinates x_1, x_2, x_3, ds^2 differs only by an infinitesimal amount from

$$c^2\,dt^2 - (dx_1^2 + dx_2^2 + dx_3^2) \tag{33}$$

—the masses producing the gravitational field must be infinitely small if this is to be true—we get, by setting

$$f = c + \frac{\Phi}{c}, \tag{34}$$

that

$$\Delta\Phi = \frac{\partial^2\Phi}{\partial x_1^2} + \frac{\partial^2\Phi}{\partial x_2^2} + \frac{\partial^2\Phi}{\partial x_3^2} = 4\pi\kappa c\mu, \tag{10}$$

and μ is c-times the mass-density in the ordinary units. We find that actually, according to all our geometric observations, this assumption is very approximately true for the planetary system.

Since the masses of the planets are very small compared with the mass of the sun which produces the field and is to be considered at rest, we may treat the former as "test-bodies" that are embedded in the gravitational field of the sun. The motion of each of them is then given by a geodetic world-line in this statical gravitational field, if we neglect the disturbances due to the influence of the planets on one another. The motion thus satisfies the principle of variation

$$\delta \int ds = 0,$$

the ends of the portion of world-line remaining fixed. For the case of rest, this gives us

$$\delta \int \sqrt{f^2 - v^2}\, dt = 0,$$

in which

$$v^2 = \left(\frac{d\sigma}{dt}\right)^2 = \sum_{i,k=1}^{3} \gamma_{ik} \frac{dx_i}{dt} \frac{dx_k}{dt}$$

is the square of the velocity. This is a principle of variation of the same form as that of classical mechanics; the "Lagrange Function" in this case is

$$L = \sqrt{f^2 - v^2}.$$

If we make the same approximation as just above and notice that in an infinitely weak gravitational field the velocities that occur will also be infinitely small (in comparison with c), we get

$$\sqrt{f^2 - v^2} = \sqrt{c^2 - 2\Phi - v^2} = c + \frac{1}{c}(\Phi - \tfrac{1}{2}v^2),$$

and since we may now set

$$v^2 = \sum_{i,k=1}^{3} \left(\frac{dx_i}{dt}\right)^2 = \sum_i \dot{x}_i^2,$$

we arrive at

$$\delta \int \left\{ \tfrac{1}{2} \sum_i \dot{x}_i^2 - \Phi \right\} dt = 0;$$

that is, the planet of mass m moves according to the laws of classical mechanics, if we assume that a force with the potential $m\Phi$ acts in it. **In this way we have linked up the theory with that of Newton**: Φ is the Newtonian potential that satisfies Poisson's equation (10), and $\mathsf{k} = c^2\kappa$ is the gravitational constant of Newton. From the well-known numerical value of the Newtonian constant k, we get for $8\pi\kappa$ the numerical value

$$8\pi\kappa = \frac{8\pi\mathsf{k}}{c^2} = 1.87 \cdot 10^{-27} \mathrm{cm} \cdot \mathrm{gr}^{-1}.$$

The deviation of the metrical groundform from that of Euclid (33) is thus considerable enough to make the geodetic world-lines differ from rectilinear uniform motion by the amount actually shown by planetary motion—although the geometry which is valid in space and is founded on $d\sigma^2$ differs only very little from Euclidean geometry as far as the dimensions of the planetary system are concerned. (The sum of the angles in a geodetic triangle of these dimensions differs very very slightly from 180°.) The chief cause of this is that the radius of the earth's orbit amounts to about eight light-minutes whereas the time of revolution of the world in its orbit is a whole year!

We shall pursue the exact theory of the motion of a point-mass and of light-rays in a statical gravitational field a little further (*vide* note 10). According to § 17 the geodetic world-lines may be characterised by the two principles of variation

$$\delta \int \sqrt{Q}\, ds = 0 \quad \text{or} \quad \delta \int Q\, ds = 0, \quad \text{in which } Q = g_{ik}\,\frac{dx_i}{ds}\,\frac{dx_k}{ds}. \tag{35}$$

The second of these takes for granted that the parameter s has been chosen suitably. The second alone is of account for the "null-lines" which satisfy the condition $Q = 0$ and depict the progress of a light-signal. The variation must be performed in such a way that the ends of the piece of world-line under consideration remain unchanged. If we subject only $x_0 = t$ to variation, we get in the statical case

$$\delta \int Q\, ds = \left[2f^2\,\frac{dx_0}{ds}\,\delta x_0\right] - 2\int \frac{d}{ds}\left(f^2\,\frac{dx_0}{ds}\right)\delta x_0\, ds. \tag{36}$$

Thus we find that

$$f^2\,\frac{dx_0}{ds} = \text{const.} \quad \text{holds.}$$

If, for the present, we keep our attention fixed on the case of the light-ray, we can, by choosing the unit of measure of the parameter s appropriately (s is standardised by the principle of variation itself except for an arbitrary unit of measure), make the constant which occurs on the right equal to unity. If we now carry out the variation more generally by varying the spatial path of the ray whilst keeping the ends fixed but dropping the subsidiary condition imposed by time, namely, that $\delta x_0 = 0$ for the ends, then, as is evident from (36), the principle becomes

$$\delta \int Q\,ds = 2[\delta t] = 2\delta \int dt.$$

If the path after variation is, in particular, traversed with the velocity of light just as the original path, then for the varied world-line, too, we have

$$Q = 0, \qquad d\sigma = f\,dt,$$

and we get

$$\delta \int dt = \delta \int \frac{d\sigma}{f} = 0. \tag{37}$$

This equation fixes only the spatial position of the light-ray; it is nothing other than **Fermat's principle of the shortest path**. In the last formulation time has been eliminated entirely; it is valid for any arbitrary portion of the path of the light-ray if the latter alters its position by an infinitely small amount, its ends being kept fixed.

If, for a statical field of gravitation, we use any space-co-ordinates x_1, x_2, x_3, we may construct a graphical representation of a Euclidean space by representing the point whose co-ordinates are x_1, x_2, x_3 by means of a point whose Cartesian co-ordinates are x_1, x_2, x_3. If we mark the position of two stars S_1, S_2 which are at rest and also an observer B, who is at rest, in this picture-space, then the angle at which the stars appear to the observer is not equal to the angle between the straight lines BS_1, BS_2 connecting the stars with the observer; we must connect B with S_1, S_2 by means of the curved lines of shortest path resulting from (37) and then, by means of an auxiliary construction, transform the angle which these two lines make with one another at B from Euclidean measure to that of Riemann determined by the metrical groundform $d\sigma^2$ (cf. formula (15), § 11). The angles which have been calculated in this way are those which determine the actually observed position of the stars to one another, and which are read off on the divided circle of the observing instrument. Whereas B, S_1, S_2 retain their positions in space, this angle S_1BS_2 may change, if great masses happen to get into proximity of the path of the rays. It is in this sense that we may talk of **light-rays being curved as a result of the gravitational field**. But the rays are not, as we assumed in § 12 to get at general results, geodetic lines in space with the metrical groundform $d\sigma^2$; they do not make the integral $\int d\sigma$ but $\int \frac{d\sigma}{f}$ assume a limiting value. The bending of light-rays occur, in particular, in the gravitational field of the sun. If for our graphical representation we use co-ordinates x_1, x_2, x_3, for which the Euclidean formula $d\sigma^2 = dx_1^2+dx_2^2+dx_3^2$ holds at infinity, then numerical calculation for the case of a light-ray passing by close to the sun shows that it must be diverted from its path to the extent of 1.74 seconds (*vide* § 31). This entails a displacement of the positions of the stars in the apparent immediate neighbourhood of the sun, which should certainly be measurable. These positions of the stars can be observed, of course, only during a total eclipse of the sun. The stars which come into consideration must be sufficiently bright, as numerous as possible, and sufficiently close to the sun to lead to a measurable effect, and yet sufficiently far removed to avoid being masked by the brilliance of the corona. The most favourable day for such an observation is the 29th May, and it was a piece of great good fortune that a total eclipse of the sun occurred on the 29th May, 1919. Two English expeditions were dispatched to the zone in which the total eclipse was observable, one to Sobral in North Brazil, the other to the Island of Principe in the Gulf of Guinea, for the express purpose of ascertaining the presence or absence of the Einstein displacement. The effect was found to be present to the amount predicted; the final results of the measurements were $1.98'' \pm 0.12''$ for Sobral, $1.61'' \pm 0.30''$ for Principe (*vide* note 11).

Another optical effect which should present itself, according to Einstein's theory of gravitation, in the statical field and which, under favourable conditions, may just be observable, arises from the relationship

$$ds = f\,dt$$

holding between the cosmic time dt and the proper-time ds at a fixed point in space. If two sodium atoms at rest are objectively fully alike, then the events that give rise to the light-waves of the D-line in each must have the same frequency, as measured in **proper-time**. Hence, if f has the values f_1, f_2, respectively at the points at which the atoms are situated, then between f_1, f_2 and the frequencies ν_1, ν_2 in cosmic time, there will exist the relationship

$$\frac{\nu_1}{f_1} = \frac{\nu_2}{f_2}.$$

But the light-waves emitted by an atom will have, of course, the same frequency, measured in **cosmic** time, at all points in space (for, in a **static** metrical field, Maxwell's equations have a solution in which time is represented by the factor $e^{i\nu t}$, ν being an arbitrary **constant** frequency). Consequently, if we compare the sodium D-line produced in a spectroscope by the light sent from a star of great mass with the same line sent by an earth-source into the same spectroscope, there should be a slight displacement of the former line towards the red as compared with the latter, since f has a slightly smaller value in the neighbourhood of great masses than at a great distance from them. The ratio in which the frequency is reduced, has according to our approximate formula (34) the value $1-\frac{\kappa m_0}{r}$ at the distance r from a mass m_0. At the surface of the sun this amounts to a displacement of .008 Angströms for a line in the blue corresponding to the wave-length 4000 Å. This effect lies just within the limits of observability. Superimposed on this, there are the disturbances due to the Doppler effect, the uncertainty of the means used for comparison on the earth, certain irregular fluctuations in the sun's lines the causes of which have been explained only partly, and finally, the mutual disturbances of the densely packed lines of the sun owing to the overlapping of their intensities (which, under certain circumstances, causes two lines to merge into one with a single maximum of intensity). If all these factors are taken into consideration, the observations that have so far been made, seem to confirm the displacement towards the red to the amount stated (*vide* note 12). This question cannot, however, yet be considered as having been definitely answered.

A third possibility of controlling the theory by means of experiment is this. According to Einstein, Newton's theory of the planets is only a first approximation. The question suggests itself whether the divergence between Einstein's Theory and the latter are sufficiently great to be detected by the means at our disposal. It is clear that the chances for this are most favourable for the planet Mercury which is nearest the sun. In actual fact, after Einstein had carried the approximation a step further, and after Schwarzschild (*vide* note 13) had determined accurately the radially symmetrical field of gravitation produced by a mass at rest and also the path of a point-mass of infinitesimal mass, both found that the **elliptical orbit of Mercury should undergo a slow rotation in the same direction as the orbit is traversed** (over and above the disturbances produced by the remaining planets), **amounting to 43″ per century**. Since the time of Leverrier an effect of this magnitude has been known among the secular disturbances of Mercury's perihelion, which could not be accounted for by the usual causes of disturbance. Manifold hypotheses have been proposed to remove this discrepancy between theory and observation (*vide* note 14). We shall revert to the rigorous solution given by Schwarzschild in § 31.

Thus we see that, however great is the revolution produced in our ideas of space and time by Einstein's theory of gravitation, the actual deviations from the old theory are exceedingly small in our field of observation. Those which are measurable have been confirmed up to now. The chief support of the theory is to be found less in that lent by observation hitherto than in its inherent logical consistency, in which it far transcends that of classical mechanics, and also in the fact that it solves the perplexing problem of gravitation and of the relativity of motion at one stroke in a manner highly satisfying to our reason.

Using the same method as for the light-ray, we may set up for the motion of a point-mass in a statical gravitational field a "minimum" principle affecting only the path in space, corresponding to Fermat's principle of the shortest path. If s is the parameter of proper-time, then

$$Q = 1, \quad \text{and} \quad f^2 \frac{dt}{ds} = \text{const.} = \frac{1}{E} \tag{38}$$

is the energy-integral. We now apply the first of the two principles of variation (35) and generalise it as above by varying the spatial path quite arbitrarily while keeping the ends, $x_0 = t$, fixed. We get

$$\delta \int \sqrt{Q}\, ds = \left[\frac{1}{E}\, \delta t\right] = \delta \int \frac{dt}{E}. \tag{39}$$

To eliminate the proper-time we divide the first of the equations (38) by the square of the second; the result is

$$\frac{1}{f^4}\left\{f^2 - \left(\frac{d\sigma}{dt}\right)^2\right\} = E^2 \qquad d\sigma = f^2\sqrt{U}\, dt, \tag{40}$$

in which

$$U = \frac{1}{f^2} - E^2.$$

(40) is the law of velocity according to which the point-mass traverses its path. If we perform the variation so that the varied path is traversed according to the same law with the same constant E, it follows from (39) that

$$\delta \int \frac{dt}{E} = \delta \int \sqrt{f^2 - \left(\frac{d\sigma}{dt}\right)^2}\, dt = \delta \int E f^2\, dt \quad \text{i.e. } \delta \int f^2 U\, dt = 0$$

or, finally, by expressing dt in terms of the spatial element of arc $d\sigma$, and thus eliminating the time entirely, we get

$$\delta \int \sqrt{U}\, d\sigma = 0.$$

The path of the point-mass having been determined in this way, we get as a relation giving the time of the motion in this path, from (40), that

$$dt = \frac{d\sigma}{f^2\sqrt{U}}.$$

For $E = 0$, we again get the laws for the light-ray.

§ 30. Gravitational Waves

By assuming that the generating energy-field $\mathbf{T}_i^k$ is infinitely weak, Einstein has succeeded in integrating the gravitational equations generally (*vide* note 15). The g_{ik}'s will, under these circumstances, if the co-ordinates are suitably chosen, differ from the $\overset{0}{g}_{ik}$'s by only infinitesimal amounts γ_{ik}. We then regard the world as "Euclidean," having the metrical groundform

$$\overset{0}{g}_{ik}\, dx_i\, dx_k \tag{41}$$

and the γ_{ik}'s as the components of a symmetrical tensor-field of the second order in this world. The operations that are to be performed in the sequel will always be based on the metrical groundform (41). For the present we are again dealing with the special theory of relativity. We shall consider the co-ordinate system which is chosen to be a "normal" one, so that $\overset{0}{g}_{ik} = 0$ for $i \neq k$ and

$$g_{00} = 1, \qquad \overset{0}{g}_{11} = \overset{0}{g}_{22} = \overset{0}{g}_{33} = -1.$$

x_0 is the time, x_1, x_2, x_3 are Cartesian space-co-ordinates; the velocity of light is taken equal to unity.

We introduce the quantities

$$\psi_i^k = \gamma_i^k - \gamma\delta_i^k \qquad (\gamma = \tfrac{1}{2}\gamma_i^i),$$

and we next assert that we may without loss of generality set

$$\frac{\partial \psi_i^k}{\partial x_k} = 0. \tag{42}$$

For, if this is not so initially, we may, by an infinitesimal change, alter the co-ordinate system so that (42) holds. The transformation formulæ that lead to a new co-ordinate system $\bar{x}$, namely,

$$\bar{x}_i = x_i + \xi(x_0, x_1, x_2, x_3)$$

contain the unknown functions ξ^i, which are of the same order of infinitesimals as the γ's. We get new co-efficients $\bar{g}_{ik}$ for which, according to earlier formulæ, we must have

$$g_{ik}(x) - \bar{g}_{ik}(x) = g_{ir}\frac{\partial \xi^r}{\partial x_k} + g_{kr}\frac{\partial \xi^r}{\partial x_i} + \frac{\partial g_{ik}}{\partial x_r}\xi^r$$

so that, here, we have

$$\gamma_{ik}(x) - \bar{\gamma}_{ik}(x) = \frac{\partial \xi_i}{\partial x_k} + \frac{\partial \xi_k}{\partial x_i}, \qquad \gamma(x) - \bar{\gamma}(x) = \frac{\partial \xi^i}{\partial x_i} = \Xi,$$

and we finally get

$$\frac{\partial \gamma_i^k}{\partial x_k} - \frac{\partial \bar{\gamma}_i^k}{\partial x_k} = \nabla \xi_i + \frac{\partial \Xi}{\partial x_i}, \qquad \frac{\partial \gamma}{\partial x_i} - \frac{\partial \bar{\gamma}}{\partial x_i} = \frac{\partial \Xi}{\partial x_i},$$

in which ∇ denotes, for an arbitrary function, the differential operator

$$\nabla f = \frac{\partial}{\partial x_i}\left(\overset{0}{g}_{ik}\frac{\partial f}{\partial x_k}\right) = \frac{\partial^2 f}{\partial x_0^2} - \left(\frac{\partial^2 f}{\partial x_1^2} + \frac{\partial^2 f}{\partial x_2^2} + \frac{\partial^2 f}{\partial x_3^2}\right).$$

The desired condition will therefore be fulfilled in the new co-ordinate system if the ξ^i's are determined from the equations

$$\nabla\xi^i = \frac{\partial\psi_i^k}{\partial x_k},$$

which may be solved by means of retarded potentials (cf. Chapter III, page 244). If the linear Lorentz transformations are discarded, the co-ordinate system is defined not only to the first order of small quantities but also to the second. It is very remarkable that such an invariant normalisation is possible.

We now calculate the components R_{ik} of curvature. As the field-quantities $\begin{Bmatrix} ik \\ r \end{Bmatrix}$ are infinitesimal, we get, by confining ourselves to terms of the first order

$$R_{ik} = \frac{\partial}{\partial x_r}\begin{Bmatrix} ik \\ r \end{Bmatrix} - \frac{\partial}{\partial x_k}\begin{Bmatrix} ir \\ r \end{Bmatrix}.$$

Now,

$$\begin{bmatrix} ik \\ r \end{bmatrix} = \tfrac{1}{2}\left(\frac{\partial\gamma_{ir}}{\partial x_k} + \frac{\partial\gamma_{kr}}{\partial x_i} - \frac{\partial\gamma_{ik}}{\partial x_r}\right),$$

hence

$$\begin{Bmatrix} ik \\ r \end{Bmatrix} = \tfrac{1}{2}\left(\frac{\partial\gamma_i^r}{\partial x_k} + \frac{\partial\gamma_k^r}{\partial x_i} - \overset{0}{g}_{rs}\frac{\partial\gamma_{ik}}{\partial x_s}\right).$$

Taking into account equations (42) or

$$\frac{\partial\gamma_i^k}{\partial x_k} = \frac{\partial\gamma}{\partial x_i},$$

we get

$$\frac{\partial}{\partial x_r}\begin{Bmatrix} ik \\ r \end{Bmatrix} = \frac{\partial^2\gamma}{\partial x_i\,\partial x_k} - \tfrac{1}{2}\nabla\gamma_{ik}.$$

In the same way we obtain

$$\frac{\partial}{\partial x_k}\begin{Bmatrix} ir \\ r \end{Bmatrix} = \frac{\partial^2 \gamma}{\partial x_i\,\partial x_k}.$$

The result is

$$R_{ik} = -\tfrac{1}{2}\nabla\gamma_{ik}.$$

Consequently, $R = -\nabla\gamma$ and

$$R_i^k - \tfrac{1}{2}\delta_i^k R = -\tfrac{1}{2}\nabla\psi_i^k.$$

The gravitational equations are, however,

$$\tfrac{1}{2}\nabla\psi_i^k = -T_i^k, \tag{43}$$

and may be directly integrated with the help of retarded potentials (cf. page 244). Using the same notation, we get

$$\psi_i^k = -\int \frac{T_i^k(t-r)}{2\pi r}\,dV.$$

Accordingly, *every change in the distribution of matter produces a gravitational effect which is propagated in space with the velocity of light.* Oscillating masses produce gravitational waves. Nowhere in the Nature accessible to us do mass-oscillations of sufficient power occur to allow the resulting gravitational waves to be observed.

Equations (43) correspond fully to the electromagnetic equations

$$\nabla\phi^i = s^i$$

and, just as the potentials ϕ^i of the electric field had to satisfy the secondary condition

$$\frac{\partial\phi^i}{\partial x_i} = 0$$

because the current s^i fulfils the condition

$$\frac{\partial s^i}{\partial x_i} = 0,$$

so we had here to introduce the secondary conditions (42) for the system of gravitational potentials ψ_i^k, because they hold for the matter-tensor

$$\frac{\partial T_i^k}{\partial x_k} = 0.$$

Plane gravitational waves may exist: they are propagated in space free from matter: we get them by making the same supposition as in optics, i.e. by setting

$$\psi_i^k = a_i^k \cdot e^{(\alpha_0 x_0 + \alpha_1 x_1 + \alpha_2 x_2 + \alpha_3 x_3)\sqrt{-1}}.$$

The a_i^k's and the α_i's are constants; the latter satisfy the condition $\alpha_i \alpha^i = 0$. Moreover, $\alpha_0 = \nu$ is the frequency of the vibration and $\alpha_1 x_1 + \alpha_2 x_2 + \alpha_3 x_3 =$ const. are the planes of constant phase. The differential equations $\nabla \psi_i^k = 0$ are satisfied identically. The secondary conditions (42) require that

$$a_i^k \alpha_k = 0. \tag{44}$$

If the x_1-axis is the direction of propagation of the wave, we have

$$\alpha_2 = \alpha_3 = 0, \qquad -\alpha_1 = \alpha_0 = \nu,$$

and equations (44) state that

$$a_i^0 = a_i^1 \quad \text{or} \quad a_{0i} = -a_{1i}. \tag{45}$$

Accordingly, it is sufficient to specify the space part of the constant symmetrical tensor a, namely,

$$\begin{Vmatrix} a_{11} & a_{12} & a_{13} \\ a_{21} & a_{22} & a_{23} \\ a_{31} & a_{32} & a_{33} \end{Vmatrix}$$

since the a's with the index 0 are determined from these by (45); the space part, however, is subject to no limitation. In its turn it splits up into the three summands in the direction of propagation of the waves:

$$\begin{Vmatrix} a_{11} & 0 & 0 \\ 0 & 0 & 0 \\ 0 & 0 & 0 \end{Vmatrix} + \begin{Vmatrix} 0 & a_{12} & a_{13} \\ a_{21} & 0 & 0 \\ a_{31} & 0 & 0 \end{Vmatrix} + \begin{Vmatrix} 0 & 0 & 0 \\ 0 & a_{22} & a_{23} \\ 0 & a_{32} & a_{33} \end{Vmatrix}.$$

The tensor-vibration may hence be resolved into three independent components: a longitudinal-longitudinal, a longitudinal-transverse, and a transverse-transverse wave.

H. Thirring has made two interesting applications of integration based on the method of approximation used here for the gravitational equations (*vide* note 16). With its help he has investigated the influence of the rotation of a large, heavy, hollow sphere on the motion of point-masses situated near the centre of the sphere. He discovered, as was to be expected, a force effect of the same kind as centrifugal force. In addition to this a second force appears which seeks to drag the body into the equatorial plane according to the same law as that according to which centrifugal force seeks to drive it away from the axis. Secondly (in conjunction with J. Lense), he has studied the influence of the rotation of a central body on its planets or moons, respectively. In the case of the fifth moon of Jupiter, the disturbance caused attains an amount that may make it possible to compare theory with observation.

Now that we have considered in §§ 29, 30 the approximate integration of the gravitational equations that occur if only linear terms are taken into account, we shall next endeavour to arrive at rigorous solutions: our attention will, however, be confined to statical gravitation.

§ 31. Rigorous Solution of the Problem of One Body*

For a statical gravitational field we have

$$ds^2 = f^2\, dx_0^2 - d\sigma^2$$

in which $d\sigma^2$ is a definitely positive quadratic form in the three-space variables x_1, x_2, x_3; the velocity of light f is likewise dependent only on these. The field is **radially symmetrical** if, for a proper choice of the space-co-ordinates, f and $d\sigma^2$ are invariant with respect to linear orthogonal transformations of these co-ordinates. If this is to be the case, f must be a function of the distance

$$r = \sqrt{x_1^2 + x_2^2 + x_3^2},$$

from the centre, but $d\sigma^2$ must have the form

$$\lambda(dx_1^2 + dx_2^2 + dx_3^2) + l(x_1\, dx_1 + x_2\, dx_2 + x_3\, dx_3)^2 \tag{46}$$

in which λ and l are likewise functions of r alone. Without disturbing this normal form we may subject the space-co-ordinates to a further transformation which consists in replacing x_1, x_2, x_3

* *Vide* note 17.

by τx_1, τx_2, τx_3, the factor of proportionality τ being an arbitrary function of the distance r. By choosing λ appropriately we may clearly succeed in getting $\lambda = 1$; let us suppose this to have been done. Then, using the notation of § 29, we have

$$\gamma_{ik} = -g_{ik} = \delta_i^k + l \cdot x_i x_k \qquad (i, k = 1, 2, 3).$$

We shall next define this radially symmetrical field so that it satisfies the homogeneous gravitational equations which hold wherever there is no matter, that is, wherever the energy-density $\mathbf{T}_i^k$ vanishes. These equations are all included in the principle of variation

$$\delta \int \mathbf{G}\, dx = 0.$$

The gravitational field, which we are seeking, **is that which is produced by statical masses which are distributed about the centre with radial symmetry.** If the accent signify differentiation with respect to r, we get

$$\frac{\partial \gamma_{ik}}{\partial x_\alpha} = l' \frac{x_\alpha}{r} x_i x_k + l(\delta_i^\alpha x_k + \delta_k^\alpha x_i),$$

and hence

$$-\begin{bmatrix} ik \\ \alpha \end{bmatrix} = \tfrac{1}{2} \frac{x_\alpha}{r}\, l' x_i x_k + l \delta_i^k x_\alpha \qquad (i, k, \alpha = 1, 2, 3).$$

Since it follows from

$$x_\alpha = \sum_{\beta=1}^{3} \gamma_{\alpha\beta} x^\beta$$

that

$$x_\alpha = \frac{1}{h^2} x_\alpha \quad \text{and} \quad h^2 = 1 + l r^2,$$

as may be verified by direct substitution, we must have

$$\left\{\begin{matrix} ik \\ \alpha \end{matrix}\right\} = \tfrac{1}{2}\,\frac{x_\alpha}{r}\,\frac{l' x_i x_k + 2lr\delta_i^k)}{h^2}.$$

It is sufficient to carry out the calculation of $\mathbf{G}$ for the point $x_1 = r$, $x_2 = 0$, $x_3 = 0$. At this point, we get for the three-indices symbols just calculated:

$$\left\{\begin{matrix} 11 \\ 1 \end{matrix}\right\} = \frac{h'}{h} \quad \text{and} \quad \left\{\begin{matrix} 22 \\ 1 \end{matrix}\right\} = \left\{\begin{matrix} 33 \\ 1 \end{matrix}\right\} = \frac{lr}{h^2},$$

whereas the remaining ones are equal to zero. Of the three-indices symbols containing 0, we find by § 29 that

$$\left\{\begin{matrix} 10 \\ 0 \end{matrix}\right\} = \left\{\begin{matrix} 01 \\ 0 \end{matrix}\right\} = \frac{f'}{f} \quad \text{and} \quad \left\{\begin{matrix} 00 \\ 1 \end{matrix}\right\} = \frac{ff'}{h^2},$$

whereas all the others = 0. Of the g_{ik}'s all those situated in the main diagonal ($i = k$) are equal, respectively, to

$$f^2, \quad -h^2, \quad -1, \quad -1$$

whereas the lateral ones all vanish. Hence definition (31) of $\mathbf{G}$ gives us

$$-\frac{2}{\sqrt{g}}\mathbf{G} =$$

$$\begin{array}{r|l}
\dfrac{1}{f^2} & \left\{\begin{matrix} 00 \\ 1 \end{matrix}\right\}\left(\left\{\begin{matrix} 10 \\ 0 \end{matrix}\right\} + \left\{\begin{matrix} 11 \\ 1 \end{matrix}\right\}\right) - 2\left\{\begin{matrix} 01 \\ 0 \end{matrix}\right\}\left\{\begin{matrix} 00 \\ 1 \end{matrix}\right\} \\
-\dfrac{1}{h^2} & \left\{\begin{matrix} 11 \\ 1 \end{matrix}\right\}\left(\left\{\begin{matrix} 10 \\ 0 \end{matrix}\right\} + \left\{\begin{matrix} 11 \\ 1 \end{matrix}\right\}\right) - \left\{\begin{matrix} 10 \\ 0 \end{matrix}\right\}\left\{\begin{matrix} 10 \\ 0 \end{matrix}\right\} - \left\{\begin{matrix} 11 \\ 1 \end{matrix}\right\}\left\{\begin{matrix} 11 \\ 1 \end{matrix}\right\} \\
-1 & \left\{\begin{matrix} 22 \\ 1 \end{matrix}\right\}\left(\left\{\begin{matrix} 10 \\ 0 \end{matrix}\right\} + \left\{\begin{matrix} 11 \\ 1 \end{matrix}\right\}\right) \\
-1 & \left\{\begin{matrix} 33 \\ 1 \end{matrix}\right\}\left(\left\{\begin{matrix} 10 \\ 0 \end{matrix}\right\} + \left\{\begin{matrix} 11 \\ 1 \end{matrix}\right\}\right).
\end{array}$$

The terms in the first and second row taken together lead to

$$\left(\begin{Bmatrix}11\\1\end{Bmatrix}-\begin{Bmatrix}10\\0\end{Bmatrix}\right)\left(\frac{1}{f^2}\begin{Bmatrix}00\\1\end{Bmatrix}-\frac{1}{h^2}\begin{Bmatrix}10\\0\end{Bmatrix}\right).$$

The second factor in this product, however, is equal to zero. Since, by (57) § 17

$$\sum_{i=0}^{3}\begin{Bmatrix}1i\\i\end{Bmatrix}=\frac{\Delta'}{\Delta}\qquad(\Delta=\sqrt{g}=hf),$$

the sum of the terms in the third and fourth row is equal to

$$-\frac{2lr}{h^2}\cdot\frac{\Delta'}{\Delta}.$$

If we wish to take the world-integral **G** over a fixed interval with respect to the time x_0, and over a shell enclosed by two spherical surfaces with respect to space, then, since the element of integration is

$$dx=dx_0\cdot d\Omega\cdot r^2\,dr\qquad(d\Omega=\text{solid angle}),$$

the equation of variation that is to be solved is

$$\delta\int\mathbf{G}r^2\,dr=0.$$

Hence, if we set

$$\frac{lr^3}{h^2}=\frac{lr^3}{1+lr^2}=\left(1-\frac{1}{h^2}\right)r=w,$$

we get

$$\delta\int w\Delta'\,dr=0$$

in which Δ and w may be regarded as the two functions that may be varied arbitrarily.

By varying w, we get

$$\Delta' = 0, \qquad \Delta = \text{const.}$$

and hence, if we choose the unit of time suitably

$$\Delta = hf = 1.$$

Partial integration gives

$$\int w\Delta' \, dr = [w, \Delta] - \int \Delta w' \, dr.$$

Hence, if we vary Δ, we arrive at

$$w' = 0, \qquad w = \text{const.} = 2m.$$

Finally, from the definition of w and $\Delta = 1$, we get

$$\boxed{f^2 = 1 - \frac{2m}{r}, \qquad h^2 = \frac{1}{f^2}}$$

This completes the solution of the problem. The unit of time has been chosen so that the velocity of light at infinity $= 1$. For distances r, which are great compared with m, the Newtonian value of the potential holds in the sense that the quantity m_0, introduced by the equation $m = \kappa m_0$ occurs as the **field-producing mass** in it; we call m the **gravitational radius** of the matter causing the disturbance of the field. Since $4\pi m$ is the flux of the spatial vector-density $\mathbf{f}^i$ through an arbitrary sphere enclosing the masses, we get, from (32′), for discrete or non-coherent mass

$$m_0 = \int \mu \, dx_1 \, dx_2 \, dx_3.$$

Since f^2 cannot become negative, it is clear from this that, if we use the co-ordinates here introduced for the region of space devoid of matter, r must be $> 2m$. Further light is shed on this by the special case of a sphere of liquid which is to be discussed in § 32, and for which the gravitational field *inside* the mass, too, will be determined. We may apply the solution found to the gravitational field of the sum external to itself if we neglect the effect due to the planets and the distant stars. The gravitational radius is about 1.47 kilometres for the sun's mass, and only 5 millimetres for the earth.

The motion of a planet (supposed infinitesimal in comparison with the sun's mass) is represented by a geodetic world-line. Of its four equations

$$\frac{d^2 x_i}{ds^2} + \left\{ \begin{matrix} \alpha\beta \\ i \end{matrix} \right\} \frac{dx_\alpha}{ds} \frac{dx_\beta}{ds} = 0,$$

the one corresponding to the index $i = 0$ gives, for the statical gravitational field, the energy-integral

$$f^2 \frac{dx_0}{ds} = \text{const.}$$

as we saw above; or, since,

$$\left(f \frac{dx_0}{ds} \right)^2 = 1 + \left(\frac{d\sigma}{ds} \right)^2,$$

we get

$$f^2 \left[1 + \left(\frac{d\sigma}{ds} \right)^2 \right] = \text{const.}$$

In the case of a radially symmetrical field the equations corresponding to the indices $i = 1, 2, 3$ give the proportion

$$\frac{d^2 x_1}{ds^2} : \frac{d^2 x_2}{ds^2} : \frac{d^2 x_3}{ds^2} = x_1 : x_2 : x_3$$

(this is readily seen from the three-indices symbols that are written down). And from them, there results, in the ordinary way, the three equations which express the Law of Areas

$$\ldots\ldots\ldots\ldots\ldots\ldots, \qquad x_1 \frac{dx_2}{ds} - x_2 \frac{dx_1}{ds} = \text{const.}$$

This theorem differs from the similar one derived in Newton's Theory, in that the differentiations are made, not according to cosmic time, but according to the proper-time s of the planet. On account of the Law of Areas the motion takes place in a plane that we may choose as our co-ordinate plane $x_3 = 0$. If we introduce polar co-ordinates into it, namely

$$x_1 = r \cos \phi, \qquad x_2 = r \sin \phi,$$

the integral of the area is

$$r^2 \frac{d\phi}{ds} = \text{const.} = b. \tag{47}$$

The energy-integral, however, since

$$dx_1^2 + dx_2^2 = dr^2 + r^2\, d\phi^2, \qquad x_1\, dx_1 + x_2\, dx_2 = r\, dr,$$
$$d\sigma^2 = (dr^2 + r^2\, d\phi^2) + l(r\, dr)^2 = h^2\, dr^2 + r^2\, d\phi^2,$$

becomes

$$f^2\left\{1+h^2\left(\frac{dr}{ds}\right)^2+r^2\left(\frac{d\phi}{ds}\right)^2\right\}=\text{const.}$$

Since $fh = 1$, we get, by substituting for f^2 its value, that

$$-\frac{2m}{r}+\left(\frac{dr}{ds}\right)^2+r(r-2m)\left(\frac{d\phi}{ds}\right)^2=-E=\text{const.} \tag{48}$$

Compared with the energy-equation of Newton's Theory this equation differs from it only in having $r-2m$ in place of r in the last term of the left-hand side.

The succeeding steps are the same as those of Newton's Theory. We substitute $\dfrac{d\phi}{ds}$ from (47) into (48), getting

$$\left(\frac{dr}{ds}\right)^2=\frac{2m}{r}-E-\frac{b^2(r-2m)}{r^3},$$

or, using the reciprocal distance $\rho=\dfrac{1}{r}$ in place of r,

$$\left(\frac{d\rho}{\rho^2\,ds}\right)^2=2m\rho-E-b^2\rho^2(1-2m\rho).$$

To arrive at the orbit of the planet we eliminate the proper-time by dividing this equation by the square of (47), thus

$$\left(\frac{d\rho}{d\phi}\right)^2=\frac{2m}{b^2}\rho-\frac{E}{b^2}-\rho^2+2m\rho^3.$$

In Newton's Theory the last term on the right is absent. Taking into account the numerical conditions that are presented in the case of planets, we find that the polynomial of the third degree in ρ on the right has three positive roots $\rho_0 > \rho_1 > \rho_2$ and hence

$$= 2m(\rho_0 - \rho)(\rho_1 - \rho)(\rho - \rho_2);$$

ρ assumes values ranging between ρ_1 and ρ_2. The root ρ_0 is very great in comparison with the remaining two. As in Newton's Theory, we set

$$\frac{1}{\rho_1} = a(1-e), \qquad \frac{1}{\rho_2} = a(1+e),$$

and call a the semi-major axis and e the eccentricity. We then get

$$\rho_1 + \rho_2 = \frac{2}{a(1-e^2)}.$$

If we compare the co-efficients of ρ^2 with one another, we find that

$$\rho_0 + \rho_1 + \rho_2 = \frac{1}{2m}.$$

ϕ is expressed in terms of ρ by an elliptic integral of the first kind and hence, conversely, ρ is an elliptic function of ϕ. The motion is of precisely the same type as that executed by the spherical pendulum. To arrive at simple formulæ of approximation, we make the same substitution as that used to determine the Kepler orbit in the Newtonian Theory, namely

$$\rho - \frac{\rho_1 + \rho_2}{2} + \frac{\rho_1 - \rho_2}{2} \cos\theta.$$

Then

$$\phi = \int \frac{d\theta}{\sqrt{2m\left(\rho_0 - \frac{\rho_1+\rho_2}{2} - \frac{\rho_1-\rho_2}{2}\cos\theta\right)}}. \tag{49}$$

The perihelion is characterised by the values $\theta = 0, 2\pi, \ldots$. The increase of the azimuth ϕ after a full revolution from perihelion to perihelion is furnished by the above integral, taken between the limits 0 and 2π. With easily sufficient accuracy this increase may be set

$$= \frac{2\pi}{\sqrt{2m\left(\rho_0 - \frac{\rho_1 + \rho_2}{2}\right)}}.$$

We find, however, that

$$\rho_0 + \frac{\rho_1 + \rho_2}{2} = (\rho_0 + \rho_1 + \rho_2) - \tfrac{3}{2}(\rho_1 + \rho_2) = \frac{1}{2m} - \frac{3}{a(1-e^2)}.$$

Consequently the above increase (of azimuth)

$$= \frac{2\pi}{\sqrt{1 - \frac{6m}{a(1-e^2)}}} \sim 2\pi \left\{1 + \frac{3m}{a(1-e^2)}\right\},$$

and **the advance of the perihelion per revolution**

$$= \frac{6\pi m}{a(1-e^2)}.$$

In addition, m, the gravitational radius of the sun may be expressed according to Kepler's third law, in terms of the time of revolution T of the planet and the semi-major axis a, thus

$$m = \frac{4\pi^2 a^3}{c^2 T^2}.$$

Using the most delicate means at their disposal, astronomers have hitherto been able to establish the existence of this advance of the perihelion only in the case of Mercury, the planet nearest the sun (*vide* note 18).

Formula (49) also gives the deflection α of the path of a ray of light. If $\theta_0 = \frac{\pi}{2} + \epsilon$ is the angle θ for which $\rho = 0$, then the value of the integral, taken between $-\theta_0$ and $+\theta_0 = \pi + \alpha$. Now in the present case

$$2m(\rho_0 - \rho)(\rho_1 - \rho)(\rho - \rho_2) = \frac{1}{b^2} - \rho^2 + 2m\rho^3.$$

The values of ρ fluctuate between 0 and ρ_2. Moreover, $\frac{1}{\rho_1} = r$ is the nearest distance to which the light-ray approaches the centre of mass O, whilst b is the distance of the two asymptotes of the light-ray from O (for in the case of any curve, this distance is given by the value of $\frac{d\phi}{d\rho}$ for $\rho = 0$). Now,

$$2m(\rho_0 + \rho_1 + \rho_2) = 1$$

is accurately true. If $\frac{m}{b}$ is a small fraction, we get to a first degree of approximation that

$$m\rho_1 = -m\rho_2 = \frac{m}{b}, \qquad \frac{m}{2}(\rho_1 + \rho_2) = \left(\frac{m}{b}\right)^2, \qquad \epsilon = \frac{m}{b},$$

$$\alpha = \int_{-\theta_0}^{\theta_0} (1 + \frac{m}{b}\cos\theta)\, d\theta - \pi = 2\epsilon + \frac{2m}{b} \quad \text{and hence} \quad \boxed{\alpha = \frac{4m}{b}}$$

If we calculate the path of the light-ray according to Newton's Theory, taking into account the gravitation of light, that is, considering it as the path of a body that has the velocity c at infinity, then if we set

$$\frac{1}{b^2} + \frac{2m}{b^2}\rho - \rho^2 = (\rho_1 - \rho)(\rho - \rho_2)$$

in which $\rho_1 > 0$, $\rho_2 < 0$ and set

$$\cos\theta_0 = -\frac{\rho_1 + \rho_2}{\rho_1 - \rho_2},$$

we get

$$\pi + \alpha = 2\theta_0, \qquad \alpha \sim \frac{2m}{b}.$$

Thus Newton's law of attraction leads to a deflection which is only half as great as that predicted by Einstein. The observations made at Sobral and Principe decide the question definitely in favour of Einstein (*vide* note 19).

§ 32. Additional Rigorous Solutions of the Statical Problem of Gravitation

In a Euclidean space with Cartesian co-ordinates x_1, x_2, x_3, the equation of a surface of revolution having as its axis of rotation the x_3-axis is

$$x_3 = F(r), \qquad r = \sqrt{x_1^2 + x_2^2}.$$

On it, the square of the distance $d\sigma$ between two infinitely near points is

$$\begin{aligned} d\sigma^2 &= (dx_1^2 + dx_2^2) + \big(F'(r)\big)^2\, dr^2 \\ &= (dx_1^2 + dx_2^2) + \left(\frac{F'(r)}{r}\right)^2 (x_1\, dx_1 + x_2\, dx_2)^2. \end{aligned}$$

In a radially symmetrical statical gravitational field we have for a plane ($x_3 = 0$) passing through the centre

$$d\sigma^2 = (dx_1^2 + dx_2^2) + l(x_1\, dx_1 + x_2\, dx_2)^2$$

in which

$$l = \frac{h^2 - 1}{r^2} = \frac{2m}{r^2(r - 2m)}.$$

The two formulæ are identical if we set

$$F'(r) = \sqrt{\frac{2m}{r - 2m}}, \qquad F(r) = \sqrt{8m(r - 2m)}.$$

The geometry which holds on this plane is therefore the same as that which holds in Euclidean space on the surface of revolution of a parabola

$$z = \sqrt{8m(r - 2m)}$$

(*vide* note 20).

A **charged sphere**, besides calling up a radially symmetrical gravitational field, calls up a similar electrostatic field. Since both fields influence one another mutually, they may be determined only conjointly and simultaneously (*vide* note 21). If we use the ordinary units of the c.g.s. system (and not those of Heaviside which dispose of the factor 4π in another way and which we have generally used in the foregoing) for electricity as well as for the other quantities, then in the region devoid of masses and charges the integral becomes

$$\int \left\{ w\Delta' - \kappa \frac{\Phi'^2 r^2}{\Delta} \right\} dr.$$

It assumes a stationary value for the condition of equilibrium. The notation is the same as above, Φ denoting the electrostatic potential. The square of the numerical value of the field is used as a basis for the function of *Action* of the electric field, in accordance with the classical theory. Variation of w gives, just as in the case of no charges,

$$\Delta' = 0, \qquad \Delta = \text{const.} = c.$$

But variation of Φ leads to

$$\frac{d}{dr}\left(\frac{r^2\Phi'}{\Delta}\right) = 0 \quad \text{and hence} \quad \Phi = \frac{e_0}{r}.$$

For the electrostatic potential we therefore get the same formula as when gravitation is disregarded. The constant e_0 is the electric charge which excites the field. If, finally, Δ be varied, we get

$$w' - \kappa\frac{\Phi'^2 r^2}{\Delta^2} = 0$$

and hence

$$w = 2m - \frac{\kappa}{c^2}\frac{e_0^2}{r}, \qquad \frac{1}{h^2} = \left(\frac{f}{c}\right)^2 = 1 - \frac{2\kappa m_0}{r} + \frac{\kappa}{c^2}\frac{e_0^2}{r^2}$$

in which m_0 denotes the mass which produces the gravitational field. In f^2 there occurs, as we see, in addition to the term depending on the mass, an electrical term which decreases more rapidly as r increases. We call $m = \kappa m_0$ the gravitational radius of the mass m_0, and $\frac{\sqrt{\kappa}}{c}e_0 = e$ the gravitational radius of the charge e_0. Our formula leads to **a view of the structure of the electron which diverges essentially from the one commonly**

accepted. A finite radius has been attributed to the electron; this has been found to be necessary, if one is to avoid coming to the conclusion that the electrostatic field it produces has infinite total energy, and hence an infinitely great inertial mass. If the inertial mass of the electron is derived from its field-energy alone, then its radius is of the order of magnitude

$$a = \frac{e_0^2}{m_0 c^2}.$$

But in our formula a finite mass m_0 (producing the gravitational field) occurs quite independently of the smallness of the value of r for which the formula is regarded as valid; how are these results to be reconciled? According to Faraday's view the charge enclosed by a surface Ω is nothing more than the flux of the electrical field through Ω. Analogously to this it will be found in the next paragraph that the true meaning of the conception of mass, both as field-producing mass and as inertial or gravitational mass, is expressed by a field-flux. If we are to regard the statical solution here given as valid for all space, the flux of the electrical field through any sphere is $4\pi e_0$ at the centre. On the other hand the mass which is enclosed by a sphere of radius r, assumes the value

$$m_0 - \tfrac{1}{2}\frac{e_0^2}{c^2 r}$$

which is dependent on the value of r. The mass is consequently distributed continuously. The density of mass coincides, of course, with the density of energy. The "initial level" at the centre, from which the mass is to be calculated, is not equal to 0 but to $-\infty$. Therefore the mass m_0 of the electron cannot be determined from this level at all, but signifies the "ultimate level" at an infinitely great distance. a now signifies the radius of the sphere which encloses the mass zero. Contrary to Mie's view **matter** now appears **as a real singularity of the field**. In the general theory of relativity, however, space is no longer assumed to be Euclidean, and hence we are not compelled to ascribe to it the relationships of Euclidean space. It is quite possible that it has other limits besides infinity, and, in particular, that its relationships are like those of a Euclidean space which contains punctures (cf. § 34). We may, therefore, claim for the ideas here developed—according to which there is no connection between the total mass of the electron and the potential of the field it produces, and in which there is no longer a meaning in talking of a cohesive pressure holding the electron together—equal rights as for those of Mie. An unsatisfactory feature of the present theory is that the field is to be entirely free of charge, whereas the mass (= energy) is to permeate the whole of the field with a density that diminishes continuously.

It is to be noted that $a : e = e : m$ or, that $e = \sqrt{am}$. In the case of the electron the quotient $\frac{e}{m}$ is a number of the order of magnitude 10^{20}, $\frac{a}{m}$ of the order 10^{40}; that is, the electric repulsion which two electrons (separated by a great distance) exert upon one another is 10^{40} times as great as that which they exert in virtue of gravitation. The circumstance that in an electron an integral number of this kind occurs which is of an order of magnitude varying greatly from unity makes the thesis contained in Mie's Theory, namely, that all pure figures determined from the measures of the electron must be derivable as mathematical constants from the exact physical laws, rather doubtful: on the other hand, we regard with equal scepticism the belief that the structure of the world is founded on certain pure figures of accidental numerical value.

The gravitational field that is present in the interior of **massive bodies** is, according to Einstein's Theory, determined only when the dynamical constitution of the bodies are fully known; since the mechanical conditions are included in the gravitational equations, the conditions of equilibrium are given for the statical case. The simplest conditions that offer themselves for consideration are given when we deal with bodies that are composed of a **homogeneous incompressible fluid**. The energy-tensor of a fluid on which no volume forces are acting is given according to § 25, by

$$T_{ik} = \mu^* u^i u_k - p g_{ik}$$

in which the u_i's are co-variant components of the world-direction of the matter, the scalar p denotes the pressure, and μ^* is determined from the constant density μ_0 by means of the equation $\mu^* = \mu_0 + p$. We introduce the quantities

$$\mu^* u_i = v_i$$

as independent variables, and set

$$L = \frac{1}{\sqrt{g}}\mathbf{L} = \mu_0 - \sqrt{v_i v^i}.$$

Then, if we vary only the g^{ik}'s, not the v_i's,

$$d\mathbf{L} = -\tfrac{1}{2}\mathbf{T}_{ik}\,\delta g_{ik}.$$

Consequently, by referring these equations to this kind of variation, we may epitomise them in the formula

$$\delta \int (\mathbf{L} + \mathbf{G})\,dx = 0.$$

It must carefully be noted, however, that, if the v_i's are varied as independent variables in this principle, it does **not** lead to the correct hydrodynamical equations (instead, we should get $\dfrac{v^i}{\sqrt{v_i v^i}} = 0$, which leads to nowhere). But these conservation theorems of energy and momentum, are already included in the gravitational equations.

In the statical case, $v_1 = v_2 = v_3 = 0$, and all quantities are independent of the time. We set $v_0 = v$ and apply the symbol of variation δ just as in § 28 to denote a change that is produced by an infinitesimal deformation (in this case a pure spatial deformation). Then

$$\delta\mathbf{L} = \tfrac{1}{2}\mathbf{T}^{ik}\,\delta g_{ik} - h\,\delta v \qquad \left(h = \frac{\Delta}{f}\right)$$

in which δv denotes nothing more than the difference of v at two points in space that are generated from one another as a result of the displacement. By now arguing backwards from the conclusion which gave us the energy-momentum theorem in § 28, we infer from this theorem, namely

$$\int \mathbf{T}^{ik}\,\delta g_{ik}\,dx = 0,$$

and from the equation

$$\int \delta\mathbf{L}\,dx = 0,$$

which expresses the invariant character of the world-integral of **L**, that $\delta v = 0$. This signifies that, **in a connected space filled with fluid,** v **has a constant value**. The theorem of energy is true identically, and the law of momentum is expressed most simply by this fact. A single mass of fluid in equilibrium will be radially symmetrical in respect of the distribution of its mass and its field. In this special case we must make the same assumption for ds^2, involving the three unknown functions λ, l, f, as at the beginning of § 31. If we start by setting $\lambda = 1$, we lose the equation which is derived by varying λ. A full substitute for it is clearly given by the equation that asserts the invariance of the *Action* during an infinitesimal spatial displacement in radial directions, that is, the theorem of momentum : $v =$ const. The problem of variation that has now to be solved is given by

$$\delta \int \left\{ \Delta' w + r^2 \mu_0 \Delta - r^2 v h \right\} dr = 0$$

in which Δ and h are to undergo variation, whereas

$$w = \left(1 - \frac{1}{h^2}\right) r.$$

Let us begin by varying Δ; we get

$$w' - \mu_0 r^2 = 0 \quad \text{and} \quad w = \frac{\mu_0}{3} r^3,$$

that is

$$\boxed{\frac{1}{h^2} = 1 - \frac{\mu_0}{3} r^2} \tag{50}$$

Let the spherical mass of fluid have a radius $r = r_0$. It is obvious that r_0 must remain

$$< a = \sqrt{\frac{3}{\mu_0}}.$$

The energy and the mass are expressed in the rational units given by the theory of gravitation. For a sphere of water, for example, this upper limit of the radius works out to

$$\sqrt{\frac{3}{8\pi\kappa}} = 4 \cdot 10^8 \text{ km.} = 22 \text{ light-minutes.}$$

Outside the sphere our earlier formulæ are valid, in particular

$$\frac{1}{h^2} = 1 - \frac{2m}{r}, \qquad \Delta = 1.$$

The boundary conditions require that h and f have continuous values in passing over the spherical surface, and that the pressure p vanish at the surface. From the continuity of h we get for the gravitational radius m of the sphere of fluid

$$m = \frac{\mu_0 r_0^3}{6}.$$

The inequality, which holds between r_0 and μ_0, shows that the radius r_0 must be greater than $2m$. Hence, if we start from infinity, then, before we get to the singular sphere $r = 2m$ mentioned above, we reach the fluid, within which other laws hold. If we now adopt the gramme as our unit, we must replace μ_0 by $8\pi\kappa\mu_0$, whereas $m = \kappa m_0$, if m_0 denotes the gravitating mass. We then find that

$$m_0 = \mu_0 \frac{4\pi r_0^3}{3}.$$

Since

$$v = \mu^* f = \frac{\mu^* \Delta}{h}$$

is a constant, and assumes the value $\frac{\mu_0}{h_0}$ at the surface of the sphere, in which h_0 denotes the value of h there as given by (50), we see that in the whole interior

$$v = (\mu_0 + p)f = \frac{\mu_0}{h_0}. \tag{51}$$

Variation of h leads to

$$-\frac{2\Delta'}{h^3} + rv = 0.$$

Since it follows from (50) that

$$\frac{h'}{h^3} = \frac{\mu_0}{3} r,$$

we get immediately

$$\Delta = \frac{3v}{2\mu_0} h + \text{const.}$$

Further, if we use the value of the constant v given by (51), and calculate the value of the integration constant that occurs, by using the boundary condition $\Delta = 1$ at the surface of the sphere, then

$$\Delta = \frac{3h - h_0}{2h_0}, \qquad \boxed{f = \frac{3h - h_0}{2hh_0}}$$

Finally, we get from (51)

$$\boxed{p = \mu_0 \cdot \frac{h_0 - h}{3h - h_0}}$$

These results determine the metrical groundform of space

$$d\sigma^2 = (dx_1^2 + dx_2^2 + dx_3^2) + \frac{(x_1\,dx_1 + x_2\,dx_2 + x_3\,dx_3)}{a^2 - r^2}, \tag{52}$$

the gravitational potential or the velocity of light f, and the pressure-field p.

If we introduce a superfluous co-ordinate

$$x_4 = \sqrt{a^2 - r^2}$$

into space, then

$$x_1^2 + x_2^2 + x_3^2 + x_4^2 = a^2 \tag{53}$$

and hence

$$x_1\,dx_1 + x_2\,dx_2 + x_3\,dx_3 + x_4\,dx_4 = 0.$$

(52) then becomes

$$d\sigma^2 = dx_1^2 + dx_2^2 + dx_3^2 + dx_4^2.$$

In the whole interior of the fluid sphere spatial spherical geometry is valid, namely, that which is true on the "sphere" (53) *in four-dimensional Euclidean space with Cartesian co-ordinates x_i.* The fluid covers a cap-shaped portion of the sphere. The pressure in it is a linear fractional function of the "vertical height," $z = x_4$ on the sphere:

$$\frac{p}{\mu_0} = \frac{z - z_0}{3z_0 - z}.$$

Further, it is shown by this formula that, since the pressure p may not pass, on a sphere of latitude, $z =$ const., from positive to negative values through infinity, $3z_0$ must be $> a$, and the upper limit a found above for the radius of the fluid sphere must be correspondingly reduced to $\frac{2a\sqrt{2}}{3}$.

These results for a sphere of fluid were first obtained by Schwarzschild (*vide* note 22). After the most important cases of radially symmetrical statical gravitational fields had been solved, the author succeeded in solving the more general problem of the **cylindrically symmetrical statical field** (*vide* note 23). We shall here just mention briefly the simplest results of this investigation. Let us consider first **uncharged masses** and a gravitational field in space free from matter. It then follows from the gravitational equations, if certain space-co-ordinates r, θ, z (so-called canonical **cylindrical co-ordinates**) are used, that

$$ds^2 = f^2\,dt^2 - d\sigma^2, \qquad d\sigma^2 = h(dr^2 + dz^2) + \frac{r^2\,d\theta^2}{f^2}.$$

θ is an angle whose modulus is 2π; that is, corresponding to values of θ that differ by integral multiples of 2π there is only one point. On the axis of rotation $r = 0$. Also, h and f are functions of r and z. We shall plot real space in terms of a Euclidean space, in which r, θ, z are cylindrical co-ordinates. The canonical co-ordinate system is uniquely defined except for a displacement in the direction of the axis of rotation $z' = z + \text{const.}$ When $h = f = 1$, $d\sigma^2$ is identical with the metrical groundform of the Euclidean picture-space (used for the plotting). The gravitational problem may be solved just as easily on this theory as on that of Newton, if the distribution of the matter is known in terms of canonical co-ordinates. For if we transfer these masses into our picture-space, that is, if we make the mass contained in a portion of each space equal to the mass contained in the corresponding portion of the picture-space, and if ψ is then the Newtonian potential of this mass-distribution in the Euclidean picture-space, the simple formula

$$f = e^{\psi/c^2} \tag{54}$$

holds. The second still unknown function h may also be determined by the solution of an ordinary Poisson equation (referring to the meridian plane $\theta = 0$). In the case of **charged bodies**, too, the canonical co-ordinate system exists. If we assume that the masses are negligible in comparison with the charges, that is, that for an arbitrary portion of space the gravitational radius of the electric charges contained in it is much greater than the gravitational radius of the masses contained in it, and if ϕ denotes the electrostatic potential (calculated according to the classical theory) of the transposed charges in the canonical picture-space, then f and the electrostatic potential Φ in real space are given by the formulæ

$$\Phi = \frac{c}{\sqrt{\kappa}} \tan\left(\frac{\sqrt{\kappa}}{c}\phi\right), \qquad f = \frac{1}{\cos\left(\frac{\sqrt{\kappa}}{c}\phi\right)}. \tag{54'}$$

It is not quite easy to subordinate the radially symmetrical case to this more general theory: it becomes necessary to carry out a rather complicated transformation of the space-co-ordinates, into which we shall not enter here.

Just as the laws of Mie's electrodynamics are non-linear, so also **Einstein's laws of gravitation**. This non-linearity is not perceptible in those measurements that are accessible to direct observation, because, in them, the non-linear terms are quite negligible in comparison with the linear ones. It is as a result of this that the **principle of superposition** is found to be confirmed by the interplay of forces in the visible world. Only, perhaps, for the unusual occurrences within the atom, of which we have as yet no clear picture, does this non-linearity come into consideration. Non-linear differential equations involve, in comparison with linear equations, particularly as regards singularities, extremely intricate, unexpected, and, at the present, quite uncontrollable conditions. The suggestion immediately arises that these two circumstances, the remarkable behaviour of non-linear differential equations and the peculiarities of intra-atomic occurrences, are to be related to one another. Equations (54) and (54′) offer a beautiful and simple example of how the principle of superposition becomes modified in the strict theory of gravitation: the field-potentials f and Φ depend in the one case on the exponential function of the quantity ψ, and in the other on a trigonometrical function of the quantity ϕ, these quantities being those which satisfy the principle of superposition. At the same time, however, these equations demonstrate clearly that the non-linearity of the gravitational equations will be of no assistance whatever for explaining the occurrences within the atom or the constitution of the electron. For the differences between ϕ and Φ become appreciable only when $\dfrac{\sqrt{\kappa}}{c}\phi$ assumes values that are comparable with 1. But even in the interior of the electron this case arises only for spheres whose radius corresponds to the order of gravitational radius

$$e = \frac{\sqrt{\kappa}}{c} e_0 \sim 10^{-33} \text{ cms.}$$

for the charge e_0 of the electron.

It is obvious that the statical differential equations of gravitation cannot uniquely determine the solutions, but that boundary conditions at infinity, or conditions of symmetry such as the postulate of radial symmetry must be added. The solutions which we found were those for which the metrical groundform converges, at spatial infinity, to

$$dx_0^2 - (dx_1^2 + dx_2^2 + dx_3^2),$$

the expression which is a characteristic of the special theory of relativity.

A further series of elegant investigations into problems of statical gravitation have been initiated by Levi-Civita (*vide* note 24). The Italian mathematicians have studied, besides the statical case, also the "stationary" one, which is characterised by the circumstance that all the g_{ik}'s are independent of the time-co-ordinate x_0, whereas the "lateral" co-efficients g_{01}, g_{02}, g_{03} need not vanish (*vide* note 25): an example of this is given by the field that surrounds a body which is in stationary rotation.

§ 33. Gravitational Energy. The Theorems of Conservation

An **isolated system** sweeps out in the course of its history a "world-canal"; we assume that outside this canal the stream-density $\mathbf{s}^i$ vanishes (if not entirely, at least to such a degree that the following argument retains its validity). It follows from the equation of continuity

$$\frac{\partial \mathbf{s}^i}{\partial x_i} = 0 \tag{55}$$

that the flux of the vector-density $\mathbf{s}^i$ has the same value e through every three-dimensional "plane" across the canal. To fix the sign of e, we shall agree to take for its direction that leading from the past into the future. The invariant e is the **charge** of our system. If the co-ordinate system fulfils the conditions that every "plane" $x_0 =$ const. intersects the canal in a finite region and that these planes, arranged according to increasing values of x_0, follow one another in the order, past $\rightarrow$ future, then we may calculate e by means of the equation

$$\int \mathbf{s}^0 \, dx_1 \, dx_2 \, dx_3 = e$$

in which the integration is taken over any arbitrary plane of the family $x_0 =$ const. This integral $e = e(x_0)$ is accordingly independent of the "time" x_0, as is readily seen, too, from (55) if we integrate it with respect to the "space-co-ordinates" x_1, x_2, x_3. What has been stated above is valid in virtue of the equation of continuity alone; the idea of substance and the convention to which it leads in Lorentz's Theory, namely, $\mathbf{s}^i = \rho u^i$ do not come into question in this case.

Does a similar **theorem of conservation** hold true for **energy and momentum**? This can certainly not be decided from the equation (26) of § 28, since the latter contains the additional term which is a characteristic of the theory of gravitation. **It is possible**, however, to write this addition term, too, in the form of a divergence. We choose a definite co-ordinate system and subject the world-continuum to an infinitesimal **deformation** in the true sense, that is, we choose constants for the deformation components ξ^i in § 28. Then, of course, for any finite region $\mathfrak{X}$

$$\delta' \int_{\mathfrak{X}} \mathbf{G} \, dx = 0$$

(this is true for **every** function of the g_{ik}'s and their derivatives: it has nothing to do with properties of invariance; δ' denotes, as in § 28, the variation effected by the displacement). Hence, the displacement gives us

$$\int_{\mathfrak{X}} \frac{\partial(\mathbf{G}\xi^k)}{\partial x_k}\, dx + \int_{\mathfrak{X}} \delta\mathbf{G}\, dx = 0.$$

If, as earlier, we set

$$\delta\mathbf{G} = \tfrac{1}{2}\mathbf{G}^{\alpha\beta}\,\delta g_{\alpha\beta} + \tfrac{1}{2}\mathbf{G}^{\alpha\beta,k}\,\delta g_{\alpha\beta,k}, \tag{13}$$

then partial integration gives

$$2\int_{\mathfrak{X}} \delta\mathbf{G}\, dx = \int_{\mathfrak{X}} \frac{\partial(\mathbf{G}^{\alpha\beta,k})\,\delta g_{\alpha\beta,k}}{\partial x_k} + \int_{\mathfrak{X}} [\mathbf{G}]^{\alpha\beta}\,\delta g_{\alpha\beta}\, dx.$$

Now, in this case, since the ξ's are constants,

$$\delta g_{\alpha\beta} = -\frac{\partial g_{\alpha\beta}}{\partial x_i}\,\xi^i.$$

If we introduce the quantities

$$\mathbf{G}\delta_i^k - \tfrac{1}{2}\mathbf{G}^{\alpha\beta,k}\,\frac{\partial g_{\alpha\beta}}{\partial x_i} = \mathbf{t}_i^k$$

then, by the preceding relation, we get the equation

$$\int_{\mathfrak{X}} \left\{ \frac{\partial \mathbf{t}_i^k}{\partial x_k} - \tfrac{1}{2}[\mathbf{G}]^{\alpha\beta}\,\frac{\partial g_{\alpha\beta}}{\partial x_i} \right\} \xi^i\, dx = 0.$$

Since this holds for any arbitrary region $\mathfrak{X}$, the integrand must be equal to zero. In it the ξ^i's denote arbitrary constant numbers; hence we get four identities:

$$\tfrac{1}{2}[\mathbf{G}]^{\alpha\beta}\,\frac{\partial g_{\alpha\beta}}{\partial x_i} = \frac{\partial \mathbf{t}_i^k}{\partial x_k}.$$

The left-hand side, by the gravitational equations,

$$= -\tfrac{1}{2}\mathbf{T}^{\alpha\beta}\frac{\partial g_{\alpha\beta}}{\partial x_i}$$

and, accordingly, the mechanical equations (26) become

$$\frac{\partial \mathbf{U}_i^k}{\partial x_k} = 0, \qquad \text{where } \mathbf{U}_i^k = \mathbf{T}_i^k + \mathbf{t}_i^k. \tag{56}$$

It is thus shown that if we regard the $\mathbf{t}_i^k$'s, which are dependent only on the potentials and the field-components of gravitation, as the components of **the energy-density of the gravitational field**, we get pure divergence equations for **all** energy associated with "physical state or phase" and "gravitation" (*vide* note 26).

And yet, physically, it seems devoid of sense to introduce the $\mathbf{t}_i^k$'s as energy-components of the gravitational field, for these quantities **neither form a tensor nor are they symmetrical**. In actual fact, if we choose an appropriate co-ordinate system, we may make all the $\mathbf{t}_i^k$'s at one point vanish; it is only necessary to choose a geodetic co-ordinate system. And, on the other hand, if we use a curvilinear co-ordinate system in a "Euclidean" world totally devoid of gravitation, we get $\mathbf{t}_i^k$'s that are all different from zero, although the existence of gravitational energy in this case can hardly come into question. Hence, although the differential relations (56) have no real physical meaning, we can derive from them, by **integrating over an isolated system**, an invariant theorem of conservation (*vide* note 27).

During motion an isolated system with its accompanying gravitational field sweeps out a canal in the "world". Beyond the canal, in the empty surroundings of the system, we shall assume that the tensor-density $\mathbf{T}_i^k$ and the gravitational field vanish. We may then use co-ordinates x_0 $(= t)$, x_1, x_2, x_3, such that the metrical groundform assumes constant co-efficients outside the canal, and in particular assumes the form

$$dt^2 - (dx_1^2 + dx_2^2 + dx_3^2).$$

Hence, outside the canal, the co-ordinates are fixed except for a linear (Lorentz) transformation, and the $\mathbf{t}_i^k$'s vanish there. We assume that each of the "planes" $t =$ const. has only a finite portion of section in common with the canal. If we integrate the equations (56) with respect to x_1, x_2, x_3 over such a plane, we find that the quantities

$$J_i = \int \mathbf{U}_i^0 \, dx_1 \, dx_2 \, dx_3$$

are independent of the time; that is $\dfrac{dJ_i}{dt} = 0$. We call J_0 the **energy**, and J_1, J_2, J_3 the **momentum co-ordinates** of the system.

These quantities have a significance which is independent of the co-ordinate system. We affirm, firstly, that they retain their value if the co-ordinate system is changed anywhere **within the canal**. Let $\bar{x}_i$ be the new co-ordinates, identical with the old ones for the region outside the canal. We mark out two "surfaces"

$$x_0 = \text{const.} = a \quad \text{and} \quad \bar{x}_0 = \text{const.} = \bar{a} \qquad (\bar{a} \neq a)$$

which do not intersect in the canal (for this it suffices to choose a and $\bar{a}$ sufficiently different from one another). We can then construct a third co-ordinate system x_i^* which is identical with the x_i's in the neighbourhood of the first surface, identical with the $\bar{x}_i$ in that of the second system, and is identical with both outside the canal. If we give expression to the fact that the energy-momentum components J_i^* in this system assume the same values for $x_0^* = a$ and $x_0 = \bar{a}$, then we get the result which we enunciated, namely, $J_i = \bar{J}_i$.

Consequently, the behaviour of the J_i's need be investigated only in the case of **linear** transformations of the co-ordinates. With respect to such, however, the conception of a tensor with components that are constant (that is, independent of position) is invariant. We make use of an arbitrary vector p^i of this type, and form $\mathbf{U}^k = \mathbf{U}_i^k p^i$, and deduce from (56) that

$$\frac{\partial \mathbf{U}^k}{\partial x_k} = 0.$$

By applying the same reasoning as was used above in the case of the electric current, it follows from this that

$$\int \mathbf{U}^0 \, dx_1 \, dx_2 \, dx_3 = J_i p^i$$

is an invariant with respect to linear transformations. **Accordingly, the J_i's are the components of a constant co-variant vector in the "Euclidean" surroundings of the system**; this energy-momentum vector is uniquely determined by the phase (or state) of the physical system. The direction of this vector determines generally the direction in which the canal traverses the surrounding world (a purely descriptive datum that can be expressed in an exact form accessible to mathematical analysis only with great difficulty). The invariant

$$\sqrt{J_0^2 - J_1^2 - J_2^2 - J_3^2}$$

is the **mass** of the system.

In the statical case $J_1 = J_2 = J_3 = 0$, whereas J_0 is equal to the space-integral of $\mathbf{R}_0^0 - (\frac{1}{2}\mathbf{R} - \mathbf{G})$. According to § 29 and § 28 (page 359), respectively,

$$\mathbf{R}_0^0 = \frac{\partial \mathbf{f}^i}{\partial x_i}, \quad \text{and in general,}$$

$$\tfrac{1}{2}\mathbf{R} - \mathbf{G} = \tfrac{1}{2}\frac{\partial}{\partial x_i}\sqrt{g}\left(g^{\alpha\beta}\begin{Bmatrix}\alpha\beta\\ i\end{Bmatrix} - g^{i\alpha}\begin{Bmatrix}\alpha\beta\\ \beta\end{Bmatrix}\right),$$

and hence, in the notation of § 29 and § 31, the mass J_0 is equal to the flux of the (spurious) spatial vector-density

$$\mathbf{m}_i = \tfrac{1}{2} f\sqrt{g}\left(\gamma^{\alpha\beta}\begin{Bmatrix}\alpha\beta\\ i\end{Bmatrix} - \gamma^{i\alpha}\begin{Bmatrix}\alpha\beta\\ \beta\end{Bmatrix}\right) \quad (i, \alpha, \beta = 1, 2, 3), \quad (57)$$

which has yet to be multiplied by $\dfrac{1}{8\pi\kappa}$ if we use the ordinary units. Since at a great distance from the system the solution of the field laws, which was found in § 31, is always valid, and for which $\mathbf{m}^i$ is a radial current of intensity

$$\frac{1 - f^2}{8\pi\kappa r} = \frac{m_0}{4\pi r^2},$$

we get that *the energy, J_0, or the inertial mass of the system, is equal to the mass m_0, which is characteristic of the gravitational field generated by the system* (*vide* note 28). On the other hand it is to be remarked parenthetically that the physics based on the notion of substance leads to the space-integral of μ/f for the value of the mass, whereas, in reality, for incoherent matter $J_0 = m_0 =$ the space-integral of μ; this is a definite indication of how radically erroneous is the whole idea of substance.

§ 34. Concerning the Inter-connection of the World as a Whole

The general theory of relativity leaves it quite undecided whether the world-points may be represented by the values of four co-ordinates x_i in a singly reversible continuous manner or not. It merely assumes that the **neighbourhood** of every world-point admits of a singly reversible continuous representation in a region of the four-dimensional "number-space" (whereby "point of the four-dimensional number-space" is to signify any number-quadruple); it makes no assumptions at the outset about the inter-connection of the world. When, in the theory of surfaces, we start with a parametric representation of the surface to be investigated, we are referring only to a piece of the surface, not to the whole surface, which in general can by no means be represented uniquely and continuously on the Euclidean plane or by a plane region. Those properties of surfaces that persist during all one-to-one continuous transformations form the subject-matter of *analysis situs* (the analysis of position); connectivity, for example, is a property of *analysis situs*. Every surface that is generated from the sphere by continuous deformation does not, from the point of view of *analysis situs*, differ from the sphere, but does differ from an anchor-ring, for instance. For on the anchor-ring there exist closed lines, which do not divide it into several regions, whereas such lines are not to be found on the sphere. From the geometry which is valid on a sphere, we derived "spherical geometry" (which, following Riemann, we set up in contrast with the geometry of Bolyai-Lobatschefsky) by identifying two diametrically opposite points of the sphere. The resulting surface **F** is from the point of view of *analysis situs* likewise different from the sphere, in virtue of which property it is called one-sided. If we imagine on a surface a small wheel in continual rotation in the one direction to be moved along this surface during the rotation, the centre of the wheel describing a closed curve, then we should expect that when the wheel has returned to its initial position it would rotate in the same direction as at the commencement of its motion. If this is the case, then whatever curve the centre of the wheel may have described on the surface, the latter is called **two-sided**; in the reverse case, it is called **one-sided**. The existence of one-sided surfaces was first pointed out by Möbius. The surface **F** mentioned above is one-sided, whereas the sphere is, of course, two-sided. This is obvious if the centre of the wheel be made to describe a great circle; on the sphere the **whole** circle must be traversed if this path is to be closed, whereas on **F** only the half need be covered. Quite analogously to the case of two-dimensional manifolds, four-dimensional ones may be endowed with diverse properties with regard to *analysis situs*. But in every four-dimensional manifold the neighbourhood of a point may, of course, be represented in a continuous manner by four co-ordinates in such a way that different co-ordinate quadruples always correspond to different points of this neighbourhood. The use of the four world-co-ordinates is to be interpreted in just this way.

Every world-point is the origin of the double-cone of the active future and the passive past. Whereas in the special theory of relativity these two portions are separated by an intervening region, it is certainly possible in the present case for the cone of the active future to overlap with that of the passive past; so that, in principle, it is possible to experience events now that will in part be an effect of my future resolves and actions. Moreover, it is not impossible for a world-line (in particular, that of my body), although it has a time-like direction at every point, to return to the neighbourhood of a point which it has already once passed through. The result would be a spectral image of the world more fearful than anything the weird fantasy of E. T. A. Hoffmann has ever conjured up. In actual fact the very considerable fluctuations of the g_{ik}'s that would be necessary to produce this effect do not occur in the region of world in which we live. Nevertheless there is a certain amount of interest in speculating on these possibilities inasmuch as they shed light on the philosophical problem of cosmic and phenomenal time. Although paradoxes of this kind appear, nowhere do we find any real contradiction to the facts directly presented to us in experience.

We saw in § 26 that, apart from the consideration of gravitation, the fundamental electrodynamic laws (of Mie) have a form such as is demanded by the **principle of causality**. The time-derivatives of the phase-quantities are expressed in terms of these quantities themselves and their spatial differential co-efficients. These facts persist when we introduce gravitation and thereby increase the table of phase-quantities ϕ_i, F_{ik}, by the g_{ik}'s and the $\left\{ {ik \atop r} \right\}$'s. But on account of the general invariance of physical laws we must formulate our statements so that, from the values of the phase-quantities for one moment, all those assertions concerning them, **which have an invariant character**, follow as a consequence of physical laws; moreover, it must be noted that this statement does not refer to the world as a whole but only to a portion which can be represented by four co-ordinates. Following Hilbert (*vide* note 29) we proceed thus. In the neighbourhood of the world-point O we introduce 4 co-ordinates x_i such that, at O

itself,

$$ds^2 = dx_0^2 - (dx_1^2 + dx_2^2 + dx_3^2).$$

In the three-dimensional space $x_0 = 0$ surrounding O we may mark off a region $\mathbf{R}$, such that, in it, $-ds^2$ remains definitely positive. Through every point of this region we draw the geodetic world-line which is orthogonal to that region, and which has a time-like direction. These lines will cover singly a certain four-dimensional neighbourhood of O. We now introduce new co-ordinates which will coincide with the previous ones in the three-dimensional space $\mathbf{R}$, for we shall now assign the co-ordinates x_0, x_1, x_2, x_3 to the point P at which we arrive, if we go from the point $P_0 = (x_1, x_2, x_3)$ in $\mathbf{R}$ along the orthogonal geodetic line passing through it, so far that the proper-time of the arc traversed, P_0P, is equal to x_0. This system of co-ordinates was introduced into the theory of surfaces by Gauss. Since $ds^2 = dx_0^2$ on each of the geodetic lines, we must get identically for all four co-ordinates in this co-ordinate system:

$$g_{00} = 1. \tag{58}$$

Since the lines are orthogonal to the three-dimensional space $x_0 = 0$, we get for $x_0 = 0$

$$g_{01} = g_{02} = g_{03} = 0. \tag{59}$$

Moreover, since the lines that are obtained when x_1, x_2, x_3 are kept constant and x_0 is varied are geodetic, it follows (from the equation of geodetic lines) that

$$\begin{Bmatrix} 00 \\ i \end{Bmatrix} = 0 \qquad (i = 0, 1, 2, 3),$$

and hence also that

$$\begin{bmatrix} 00 \\ i \end{bmatrix} = 0.$$

Taking (58) into consideration, we get from the latter

$$\frac{\partial g_0}{\partial x_0} = 0 \qquad (i = 1, 2, 3)$$

and, on account of (59), we have consequently not only for $x_0 = 0$ but also identically for the four co-ordinates that

$$g_{0i} = 0 \qquad (i = 1, 2, 3). \tag{60}$$

The following picture presents itself to us: a family of geodetic lines with time-like direction which covers a certain world-region singly and completely (without gaps); also, a similar uniparametric family of three-dimensional spaces x_0 = const. According to (60) these two families are everywhere orthogonal to one another, and all portions of arc cut off from the geodetic lines by two of the "parallel" spaces x_0 = const. have the same proper-time. If we use this particular co-ordinate system, then

$$\frac{\partial g_{ik}}{\partial x_0} = -2\begin{Bmatrix} ik \\ 0 \end{Bmatrix} \qquad (i, k = 1, 2, 3)$$

and the gravitational equations enable us to express the derivatives

$$\frac{\partial}{\partial x_0}\begin{Bmatrix} ik \\ 0 \end{Bmatrix} \qquad (i, k = 1, 2, 3)$$

not only in terms of the ϕ_i's and their derivatives, but also in terms of the g_{ik}'s, their derivatives (of the first and second order) with respect to x_1, x_2, x_3, and the $\begin{Bmatrix} ik \\ 0 \end{Bmatrix}$'s themselves. Hence, by regarding the twelve quantities,

$$g_{ik}, \quad \begin{Bmatrix} ik \\ 0 \end{Bmatrix} \qquad (i, k = 1, 2, 3)$$

together with the electromagnetic quantities, as the unknowns, we arrive at the required result (x_0 playing the part of time). The cone of the passive past starting from the point O' with a positive x_0 co-ordinate will cut a certain portion $\mathbf{R}'$ out of $\mathbf{R}$, which, with the sheet of the cone, will mark off a finite region of the world $\mathbf{G}$ (namely, a conical cap with its vertex at O'). If our assertion that the geodetic null-lines denote the initial points of all action is rigorously true, then the values of the above twelve quantities as well as the electromagnetic potentials ϕ_i and the field-quantities F_{ik} in the three-dimensional region of space $\mathbf{R}'$ determine fully the values of the two latter quantities in the world-region $\mathbf{G}$. This has hitherto not been proved. *In any case, we see that the differential equations of the field contain the physical laws of nature in their complete form*, and that there cannot be a further limitation due to boundary conditions at spatial infinity, for example.

Einstein, arguing from cosmological considerations of the inter-connection of the world as a whole (*vide* note 30) came to the conclusion that the world is finite in space. Just as in the Newtonian theory of gravitation the law of contiguous action expressed in Poisson's equation entails the Newtonian law of attraction only if the condition that the gravitational potential vanishes at infinity is superimposed, so Einstein in his theory seeks to supplement the differential equations by introducing boundary conditions at spatial infinity. To overcome the difficulty of formulating conditions of a general invariant character, which are in agreement with astronomical facts, he finds himself constrained to assume that the world is closed with respect to space; for in this case the boundary conditions are absent. In consequence of the above remarks the author cannot admit the cogency of this deduction, since the differential equations in themselves, without boundary conditions, contain the physical laws of nature in an unabbreviated form excluding every ambiguity. So much more weight is accordingly to be attached to another consideration which arises from the question: How does it come about that our stellar system with the relative velocities of the stars, which are extraordinarily small in comparison with that of light, persists and maintains itself and has not, even ages ago, dispersed itself into infinite space? This system presents exactly the same view as that which a molecule in a gas in equilibrium offers to an observer of correspondingly small dimensions. In a gas, too, the individual molecules are not at rest but the small velocities, according to Maxwell's law of distribution, occur much more often than the large ones, and the distribution of the molecules over the volume of the gas is, on the average, uniform, so that perceptible differences of density occur very seldom. If this analogy is legitimate, we could interpret the state of the stellar system and its gravitational field according to the same **statistical principles** that tell us that an isolated volume of gas is almost always in equilibrium. This would, however, be possible only if the **uniform distribution of stars at rest in a static gravitational field, as an ideal state of equilibrium**, is reconcilable with the laws of gravitation. In a statical field of gravitation the world-line of a point-mass at rest, that is, a line on which x_1, x_2, x_3 remain constant and x_0 alone varies, is a geodetic line if

$$\begin{Bmatrix} 00 \\ i \end{Bmatrix} = 0, \qquad (i = 1, 2, 3),$$

and hence

$$\begin{bmatrix} 00 \\ i \end{bmatrix} = 0, \qquad \frac{\partial g_{00}}{\partial x_i} = 0.$$

Therefore, a distribution of mass at rest is possible only if

$$\sqrt{g_{00}} = f = \text{const.} = 1.$$

The equation

$$\Delta f = \tfrac{1}{2}\mu \qquad (\mu = \text{density of mass}) \tag{32}$$

then shows, however, that the ideal state of equilibrium under consideration **is incompatible** with the laws of gravitation, as hitherto assumed.

In deriving the gravitational equations in § 28, however, we committed a sin of omission. R is not the only invariant dependent on the g_{ik}'s and their first and second differential co-efficients, and which is linear in the latter; for the most general invariant of this description has the form $\alpha R + \beta$, in which α and β are numerical constants. Consequently we may generalise the laws of gravitation by replacing R by $R+\lambda$ (and $\mathbf{G}$ by $\mathbf{G}+\frac{1}{2}\lambda\sqrt{g}$), in which λ denotes a universal constant. If it is not equal to 0, as we have hitherto assumed, we may take it equal to 1; by this means not only has the unit of time been reduced by the principle of relativity to the unit of length, and the unit of mass by the law of gravitation to the same unit, but the unit of length itself is fixed absolutely. With these modifications the gravitational equations for statical non-coherent matter ($\mathbf{T}_0^0 = \mu = \mu_0\sqrt{g}$, all other components of the tensor-density $\mathbf{T}$ being equal to zero) give, if we use the equation $f = 1$ and the notation of § 29:

$$\lambda = \mu_0 \quad \text{[in place of (32)]}$$

and

$$P_{ik} - \lambda\gamma_{ik} = 0 \qquad (i, k = 1, 2, 3). \tag{61}$$

Hence this ideal state of equilibrium is possible under these circumstances if the mass is distributed with the density λ. The space must then be homogeneous metrically; and indeed the equations (61) are then actually satisfied for a spherical space of radius $a = \sqrt{2/\lambda}$. Thus, in space, we may introduce four co-ordinates, connected by

$$x_1^2 + x_2^2 + x_3^2 + x_4^2 = a^2, \tag{62}$$

for which we get

$$d\sigma^2 = dx_1^2 + dx_2^2 + dx_3^2 + dx_4^2.$$

From this we conclude that space is closed and hence finite. If this were not the case, it would scarcely be possible to imagine how a state of statistical equilibrium could come about. If the world is closed, spatially, it becomes possible for an observer to see several pictures of one and the same star. These depict the star at epochs separated by enormous intervals of time (during which light travels once entirely round the world). We have yet to inquire whether the points of space correspond singly and reversibly to the value-quadruples x_i which satisfy the condition (62), or whether two value-systems

$$(x_1, x_2, x_3, x_4) \quad \text{and} \quad (-x_1, -x_2, -x_3, -x_4)$$

correspond to the same point. From the point of view of *analysis situs* these two possibilities are different even if both spaces are two-sided. According as the one or the other holds, the total mass of the world in grammes would be

$$\frac{\pi a}{2\kappa} \quad \text{or} \quad \frac{\pi a}{4\kappa}, \quad \text{respectively.}$$

Thus our interpretation demands that the total mass that happens to be present in the world bear a definite relation to the universal constant $\lambda = \frac{2}{a^2}$ which occurs in the law of action, this obviously makes great demands on our credulity.

The radially symmetrical solutions of the modified homogeneous equations of gravitation that would correspond to a world empty of mass are derivable by means of the principle of variation (*vide* § 31 for the notation)

$$\delta \int (2w\Delta' + \lambda\Delta r^2)\, dr = 0.$$

The variation of w gives, as earlier, $\Delta = 1$. On the other hand, variation of Δ gives

$$w' = \frac{\lambda}{2} r^2. \tag{63}$$

It we demand regularity at $r = 0$, it follows from (63) that

$$w = \frac{\lambda}{6} r^3$$

$$\text{and} \quad \frac{1}{h^2} = f^2 = 1 - \frac{\lambda}{6} r^2. \tag{64}$$

The space may be represented congruently on a "sphere"

$$x_1^2 + x_2^2 + x_3^2 + x_4^2 = 3a^2 \tag{65}$$

of radius $a\sqrt{3}$ in four-dimensional Euclidean space (whereby one of the two poles on the sphere, whose first three co-ordinates, x_1, x_2, x_3 each $= 0$, corresponds to the centre in our case). The world is a cylinder erected on this sphere in the direction of a fifth co-ordinate axis t. But since on the "greatest sphere" $x_4 = 0$, which may be designated as the equator or the space-horizon for that centre, f becomes zero, and hence the metrical groundform of the world becomes singular, we see that the possibility of a stationary empty world is contrary to the physical laws that are here regarded as valid. There must at least be masses at the horizon. The calculation may be performed most readily if (merely to orient ourselves on the question) we assume an incompressible fluid to be present there. According to § 32 the problem of variation that is to be solved is (if we use the same notation and add the λ term)

$$\delta \int \left\{ \Delta' w + \left(\mu_0 + \frac{\lambda}{2} \right) r^2 \Delta - r^2 v h \right\} dr = 0.$$

In comparison with the earlier expression we note that the only change consists in the constant μ_0 being replaced by $\mu_0 + \dfrac{\lambda}{2}$. As earlier, it follows that

$$w' - \left(\mu_0 + \frac{\lambda}{2} \right) r^2 = 0, \qquad w = -2M + \frac{2\mu_0 + \lambda}{6} r^3,$$

$$\frac{1}{h^2} = 1 + \frac{2M}{r} - \frac{2\mu_0 + \lambda}{6} r^2. \tag{66}$$

If the fluid is situated between the two meridians $x_4 = \text{const.}$, which have a radius r_0 $(< a\sqrt{3})$, then continuity of argument with (64) demands that the constant

$$M = \frac{\mu_0}{6} r_0^3.$$

To the first order $\frac{1}{h^2}$ becomes equal to zero for a value $r = b$ between r_0 and $a\sqrt{3}$. Hence the space may still be represented on the sphere (65), but this representation is no longer congruent for the zone occupied by fluid. The equation for Δ (page 397) now yields a value of f that does not vanish at the equator. The boundary condition of vanishing pressure gives a transcendental relation between μ_0 and r_0, from which it follows that, if the mass-horizon is to be taken arbitrarily small, then the fluid that comes into question must have a correspondingly great density, namely, such that the total mass does not become less than a certain positive limit (*vide* note 31).

The general solution of (63) is

$$\frac{1}{h^2} = f^2 = 1 - \frac{2m}{r} - \frac{\lambda}{6} r^2 \qquad (m = \text{const.}).$$

It corresponds to the case in which a spherical mass is situated at the centre. The world can be empty of mass only in a zone $r_0 \leq r \leq r_1$, in which this f^2 is positive; a mass-horizon is again necessary. Similarly, if the central mass is charged electrically; for in this case, too, $\Delta = 1$. In the expression for $\frac{1}{h^2} = f^2$ the electrical term $+\frac{e^2}{r^2}$ has to be added, and the electrostatic potential $= \frac{e}{r}$.

Perhaps in pursuing the above reflections we have yielded too readily to the allurement of an imaginary flight into the region of masslessness. Yet these considerations help to make clear what the new views of space and time bring within the realm of **possibility**. The assumption on which they are based is at any rate the simplest on which it becomes explicable that, in the world as actually presented to us, statical conditions obtain as a whole, so far as the electromagnetic and the gravitational field is concerned, and that just those solutions of the statical equations are valid which vanish at infinity or, respectively, converge towards Euclidean metrics. For on the sphere these equations will have a unique solution (boundary conditions do not enter into the question as they are replaced by the postulate of regularity over the whole of the closed configuration). If we make the constant λ arbitrarily small, the spherical solution converges to that which satisfies at infinity the boundary conditions mentioned for the infinite world which results when we pass to the limit.

A metrically homogeneous world is obtained most simply if, in a five-dimensional space with the metrical groundform $ds^2 = -\Omega(dx)$, ($-\Omega$ denotes a non-degenerate quadratic form with constant co-efficients), we examine the four-dimensional "conic-section" defined by the equation $\Omega(x) = \dfrac{6}{\lambda}$. Thus this basis gives us a solution of the Einstein equations of gravitation, modified by the λ term, for the case of no mass. If, as must be the case, the resulting metrical groundform of the world is to have one positive and three negative dimensions, we must take for Ω a form with four positive dimensions and one negative, thus

$$\Omega(x) = x_1^2 + x_2^2 + x_3^2 + x_4^2 - x_5^2.$$

By means of a simple substitution this solution may easily be transformed into the one found above for the statical case. For if we set

$$x_4 = z \cosh t, \qquad x_5 = z \sinh t,$$

we get

$$x_1^2 + x_2^2 + x_3^2 + z^2 = \frac{6}{\lambda}, \qquad -ds^2 = (dx_1^2 + dx_2^2 + dx_3^2 + dz^2) - z^2\,dt^2.$$

These "new" z, t co-ordinates, however, enable only the "wedge-shaped" section $x_4^2 - x_5^2 > 0$ to be represented. At the "edge" of the wedge (at which $x_4 = 0$ simultaneously with $x_5 = 0$), t becomes indeterminate. This edge, which appears as a two-dimensional configuration in the original co-ordinates is, therefore, three-dimensional in the new co-ordinates; it is the cylinder erected in the direction of the t-axis over the equator $z = 0$ of the sphere (65). The question arises whether it is the first or the second co-ordinate system that serves to represent the whole world in a regular manner. In the former case the world would not be static as a whole, and the absence of matter in it would be in agreement with physical laws; de Sitter argues from this assumption (*vide* note 32). In the latter case we have a static world that cannot exist without a mass-horizon; this assumption, which we have treated more fully, is favoured by Einstein.

§ 35. The Metrical Structure of the World as the Origin of Electromagnetic Phenomena*

We now aim at a final synthesis. To be able to characterise the physical state of the world at a certain point of it by means of numbers we must not only refer the neighbourhood of this point to

* *Vide* note 33.

a co-ordinate system but we must also fix on certain units of measure. We wish to achieve just as fundamental a point of view with regard to this second circumstance as is secured for the first one, namely, the arbitrariness of the co-ordinate system, by the Einstein Theory that was described in the preceding paragraph. This idea, when applied to geometry and the conception of distance (in Chapter II) after the step from Euclidean to Riemann geometry had been taken, effected the final entrance into the realm of infinitesimal geometry. Removing every vestige of ideas of "action at a distance," let us assume that world-geometry is of this kind; we then find that the metrical structure of the world, besides being dependent on the quadratic form (1), is also dependent on a linear differential form $\phi_i \, dx_i$.

Just as the step which led from the special to the general theory of relativity, so this extension affects immediately only the world-geometrical foundation of physics. Newtonian mechanics, as also the special theory of relativity, assumed that uniform translation is a unique state of motion of a set of vector axes, and hence that the position of the axes at one moment determines their position in all other moments. But this is incompatible with the intuitive principle of the **relativity of motion**. This principle could be satisfied, if facts are not to be violated drastically, only by maintaining the conception of **infinitesimal** parallel displacement of a vector set of axes; but we found ourselves obliged to regard the affine relationship, which determines this displacement, as something physically real that depends physically on the states of matter ("**guiding field**"). The properties of *gravitation* known from experience, particularly the equality of inertial and gravitational mass, teach us, finally, that gravitation is already contained in the guiding field besides inertia. And thus the general theory of relativity gained a significance which extended beyond its original important bearing on **world-geometry** to a significance which is specifically *physical.* The same certainty that characterises the relativity of motion accompanies the principle of the **relativity of magnitude**. We must not let our courage fail in maintaining this principle, according to which the size of a body at one moment does not determine its size at another, in spite of the existence of rigid bodies.* But, unless we are to come into violent conflict with fundamental facts, this principle cannot be maintained without retaining the conception of *infinitesimal* congruent transformation; that is, we shall have to assign to the world besides its *measure-determination* at every point also a *metrical relationship.* Now this is not to be regarded as revealing a "geometrical" property which belongs to the world as a form of phenomena, but as being a phase-field having physical reality. Hence, as the fact of the propagation of action and of the existence of rigid bodies leads us to found the affine relationship on the *metrical* character of the world which lies a grade lower, it immediately suggests itself to us, not only to identify the co-efficients of the quadratic groundform $g_{ik}\, dx_i\, dx_k$ with the potentials of the gravitational field, but also to identify **the co-efficients of the linear groundform $\phi_i\, dx_i$ with the electromagnetic potentials**. The electromagnetic field and the electromagnetic forces are then derived from the metrical structure of the world or the *metrics,* as we may call it. No other truly essential actions of forces are,

*It must be recalled in this connection that the spatial direction-picture which a point-eye with a given world-line receives at every moment from a given region of the world, depends only on the ratios of the g_{ik}'s, inasmuch as this is true of the geodetic null-lines which are the determining factors in the propagation of light.

however, known to us besides those of gravitation and electromagnetic actions; for all the others statistical physics presents some reasonable argument which traces them back to the above two by the method of mean values. We thus arrive at the inference: **The world is a $(3+1)$-dimensional metrical manifold; all physical field-phenomena are expressions of the metrics of the world.** (Whereas the old view was that the four-dimensional metrical continuum is the scene of physical phenomena; the physical essentialities themselves are, however, things that exist "in" this world, and we must accept them in type and number in the form in which experience gives us cognition of them: nothing further is to be "comprehended" of them.) We shall use the phrase "state of the world-æther" as synonymous with the word "metrical structure," in order to call attention to the character of reality appertaining to metrical structure; but we must beware of letting this expression tempt us to form misleading pictures. In this terminology the fundamental theorem of infinitesimal geometry states that the guiding field, and hence also gravitation, is determined by the state of the æther. The antithesis of "physical state" and "gravitation" which was enunciated in § 28 and was expressed in very clear terms by the division of Hamilton's Function into two parts, is overcome in the new view, which is uniform and logical in itself. Descartes' dream of a purely geometrical physics seems to be attaining fulfilment in a manner of which he could certainly have had no presentiment. The quantities of intensity are sharply distinguished from those of magnitude.

The linear groundform $\phi_i \, dx_i$ is determined except for an additive total differential, but the tensor of distance-curvature

$$f_{ik} = \frac{\partial \phi_i}{\partial x_k} - \frac{\partial \phi_k}{\partial x_i}$$

which is derived from it, is free of arbitrariness. According to Maxwell's Theory the same result obtains for the electromagnetic potential. The electromagnetic field-tensor, which we denoted earlier by F_{ik}, is now to be identified with the distance-curvature f_{ik}. If our view of the nature of electricity is true, then the first system of Maxwell's equations

$$\frac{\partial f_{ik}}{\partial x_l} + \frac{\partial f_{kl}}{\partial x_i} + \frac{\partial f_{li}}{\partial x_k} = 0 \tag{67}$$

is an intrinsic law, the validity of which is wholly independent of whatever physical laws govern the series of values that the physical phase-quantities actually run through. In a four-dimensional metrical manifold the simplest integral invariant that exists at all is

$$\int \mathbf{l}\,dx = \tfrac{1}{4}\int f_{ik}\mathbf{f}^{ik}\,dx \tag{68}$$

and it is just this one, in the form of *Action*, on which Maxwell's Theory is founded! We have accordingly a good right to claim that the whole fund of experience which is crystallised in Maxwell's Theory weighs in favour of the world-metrical nature of electricity. And since it is impossible to construct an integral invariant at all of such a simple structure in manifolds of more or less than four dimensions the new point of view does not only lead to a deeper understanding of Maxwell's Theory but the fact that the world is four-dimensional, which has hitherto always been accepted as merely "accidental," becomes intelligible through it. In the linear groundform $\phi_i\,dx_i$ there is an arbitrary factor in the form of an additive total differential, but there is not a factor of proportionality; the quantity *Action* is a pure number. But this is only as it should be, if the theory is to be in agreement with that atomistic structure of the world which, according to the most recent results (Quantum Theory), carries the greatest weight.

The **statical case** occurs when the co-ordinate system and the calibration may be chosen so that the linear groundform becomes equal to $\phi\, dx_0$ and the quadratic groundform becomes equal to

$$f^2\, dx_0^2 - d\sigma^2,$$

whereby ϕ and f are not dependent on the time x_0, but only on the space-co-ordinates x_1, x_2, x_3, whilst $d\sigma^2$ is a definitely positive quadratic differential form in the three space-variables. This particular form of the groundform (if we disregard quite particular cases) remains unaffected by a transformation of co-ordinates and a re-calibration only if x_0 undergoes a linear transformation of its own, and if the space-co-ordinates are likewise transformed only among themselves, whilst the calibration ratio must be a constant. Hence, in the statical case, we have a three-dimensional Riemann space with the groundform $d\sigma^2$ and two scalar fields in it: the electrostatic potential ϕ, and the gravitational potential or the velocity of light f. The length-unit and the time-unit (centimetre, second) are to be chosen as arbitrary units; $d\sigma^2$ has dimensions cm^2, f has dimensions $\text{cm} \cdot \text{sec}^{-1}$, and ϕ has sec^{-1}. Thus, as far as one may speak of a space at all in the general theory of relativity (namely, in the statical case), it appears as a **Riemann** space, and not as one of the more general type, in which the transference of distances is found to be non-integrable.

We have the case of the special theory of relativity again, if the co-ordinates and the calibration may be chosen so that

$$ds^2 = dx_0^2 - (dx_1^2 + dx_2^2 + dx_3^2).$$

If x_i, $\bar{x}_i$ denote two co-ordinate systems for which this normal form for ds^2 may be obtained, then the transition from x_i to $\bar{x}_i$ is a conformal transformation, that is, we find

$$dx_0^2 - (dx_1^2 + dx_2^2 + dx_3^2),$$

except for a factor of proportionality, is equal to

$$d\bar{x}_0^2 - (d\bar{x}_1^2 + d\bar{x}_2^2 + d\bar{x}_3^2).$$

The conformal transformations of the four-dimensional Minkowski world coincide with spherical transformations (*vide* note 34), that is, with those transformations which convert every "sphere" of the world again into a sphere. A sphere is represented by a linear homogeneous equation between the homogeneous "hexaspherical" co-ordinates

$$u_0 : u_1 : u_2 : u_3 : u_4 : u_5 = x_0 : x_1 : x_2 : x_3 : \frac{(x,x)+1}{2} : \frac{(x,x)-1}{2},$$

where

$$(x,x) = x_0^2 - (x_1^2 + x_2^2 + x_3^2).$$

They are bound by the condition

$$u_0^2 - u_1^2 - u_2^2 - u_3^2 - u_4^2 + u_5^2 = 0.$$

The spherical transformations therefore express themselves as those linear homogeneous transformations of the u_i's which leave this condition, as expressed in the equation, invariant. Maxwell's equations of the æther, in the form in which they hold in the special theory of relativity, are therefore invariant not only with respect to the 10-parameter group of the linear Lorentz transformations but also indeed with respect to the more comprehensive 15-parameter group of spherical transformations (*vide* note 35).

To test whether the new hypothesis about the nature of the electromagnetic field is able to account for phenomena, we must work out its implications. We choose as our initial physical law a Hamilton principle which states that the change in the *Action* $\int \mathbf{W}\,dx$ for every infinitely small variation of the metrical structure of the world that vanishes outside a finite region is zero. The *Action* is an invariant, and hence $\mathbf{W}$ is a scalar-density (in the true sense) which is derived from the metrical structure. Mie, Hilbert, and Einstein assumed the *Action* to be an invariant with respect to transformations of the co-ordinates. We have here to add the further limitation that it must also be invariant with respect to the process of re-calibration, in which ϕ_i, g_{ik} are replaced by

$$\phi_i - \frac{1}{\lambda}\frac{\partial\lambda}{\partial x_i} \quad \text{and} \quad \lambda g_{ik}, \quad \text{respectively,} \tag{69}$$

in which λ is an arbitrary positive function of position. We assume that $\mathbf{W}$ is an expression of the second order, that is, built up, on the one hand, of the g_{ik}'s and their derivatives of the first and second order, on the other hand, of the ϕ_i's and their derivatives of the first order. The simplest example is given by Maxwell's *density of action* $\mathbf{l}$. But we shall here carry out a general investigation without binding ourselves to any particular form of $\mathbf{W}$ at the beginning. According to Klein's method, used in § 28 (and which will only now be applied with full effect), we shall here deduce certain mathematical identities, which are valid for every scalar-density $\mathbf{W}$ which has its origin in the metrical structure.

I. If we assign to the quantities ϕ_i, g_{ik}, which describe the metrical structure relative to a system of reference, infinitely small increments $\delta\phi_i$, δg_{ik}, and if $\mathfrak{X}$ denote a finite region of the world, then the effect of partial integration is to separate the integral of the corresponding change $\delta\mathbf{W}$ in $\mathbf{W}$ over the region $\mathfrak{X}$ into two parts: (a) a divergence integral and (b) an integral whose integrand is only a linear combination of $\delta\phi_i$ and δg_{ik}, thus

$$\int_{\mathfrak{X}} \delta\mathbf{W}\,dx = \int_{\mathfrak{X}} \frac{\partial(\delta\mathbf{v}^k)}{\partial x_k}\,dx + \int_{\mathfrak{X}} (\mathbf{w}^i\,\delta\phi_i + \tfrac{1}{2}\mathbf{W}^{ik}\,\delta g_{ik})\,dx \qquad (70)$$

whereby $\mathbf{W}^{ki} = \mathbf{W}^{ik}$.

The $\mathbf{w}^i$'s are components of a contra-variant vector-density, but the $\mathbf{W}_i^k$'s are the components of a mixed tensor-density of the second order (in the true sense). The $\delta\mathbf{v}^k$'s are linear combinations of

$$\delta\phi_\alpha, \qquad \delta g_{\alpha\beta} \quad \text{and} \quad \delta g_{\alpha\beta,i} \qquad \left[\delta g_{\alpha\beta,i} = \frac{\partial g_{\alpha\beta}}{\partial x_i}\right].$$

We indicate this by the formula

$$\delta\mathbf{v}^k = (k,\alpha)\,\delta\phi_\alpha + (k,\alpha,\beta)\,\delta g_{\alpha\beta} + (k,i,\alpha,\beta)\,\delta g_{\alpha\beta,i}.$$

The $\delta\mathbf{v}^k$'s are defined uniquely by equation (70) only if the normalising condition that the co-efficients (k,i,α,β) be symmetrical in the indices k and i is added. In the normalisation the $\delta\mathbf{v}^k$'s are components of a vector-density (in the true sense), if the $\delta\phi_i$'s are regarded as the components of a co-variant vector of weight zero and the δg_{ik}'s as the components of a tensor of weight unity. (There is, of course, no objection to applying another normalisation in place of this one, provided that it is invariant in the same sense.)

First of all, we express that $\int_{\mathfrak{X}} \mathbf{W}\, dx$ is a calibration invariant, that is, that it does not alter when the calibration of the world is altered infinitesimally. If the calibration ratio between the altered and the original calibration is $\lambda = 1 + \pi$, π is an infinitesimal scalar-field which characterises the event and which may be assigned arbitrarily. As a result of this process, the fundamental quantities assume, according to (69), the following increments:

$$\delta g_{ik} = \pi g_{ik}, \qquad \delta \phi_i = -\frac{\partial \pi}{\partial x_i}. \tag{71}$$

If we substitute these values in $\delta \mathbf{v}^k$, let the following expressions result:

$$\mathbf{s}^k(\pi) = \pi \cdot \mathbf{s}^k + \frac{\partial \pi}{\partial x_\alpha} \cdot \mathbf{h}^{k\alpha}. \tag{72}$$

They are the components of a vector-density which depends on the scalar-field π in a linear-differential manner. It further follows from this, that, since the $\frac{\partial \pi}{\partial x_\alpha}$'s are the components of a co-variant vector-field which is derived from the scalar-field, $\mathbf{s}^k$ is a vector-density, and $\mathbf{h}^{k\alpha}$ is a contra-variant tensor-density of the second order. The variation (70) of the integral of *Action* must vanish on account of its calibration invariance; that is, we have

$$\int_{\mathfrak{X}} \frac{\partial \mathbf{s}^k(\pi)}{\partial x_k}\, dx + \int_{\mathfrak{X}} \left(-\mathbf{w}^i \frac{\partial \pi}{\partial x_i} + \tfrac{1}{2} \mathbf{W}_i^i \pi \right) dx = 0.$$

If we transform the first term of the second integral by means of partial integration, we may write, instead of the preceding equation,

$$\int_{\mathfrak{X}} \frac{\partial \left(\mathbf{s}^k(\pi) - \pi \mathbf{w}^k \right)}{\partial x_k}\, dx + \int_{\mathfrak{X}} \pi \left(\frac{\partial \mathbf{w}^i}{\partial x_i} + \tfrac{1}{2} \mathbf{W}_i^i \right) dx = 0. \tag{73}$$

This immediately gives the identity

$$\frac{\partial \mathbf{w}^i}{\partial x_i} + \tfrac{1}{2}\mathbf{W}_i^i = 0 \tag{74}$$

in the manner familiar in the calculus of variations. If the function of position on the left were different from 0 at a point x_i, say positive, then it would be possible to mark off a neighbourhood $\mathfrak{X}$ of this point so small that this function would be positive at every point within $\mathfrak{X}$. If we choose this region for $\mathfrak{X}$ in (73), but choose for π a function which vanishes for points outside $\mathfrak{X}$ but is > 0 throughout $\mathfrak{X}$, then the first integral vanishes, but the second is found to be positive—which contradicts equation (73). Now that this has been ascertained, we see that (73) gives

$$\int_{\mathfrak{X}} \frac{\partial\big(\mathbf{s}^k(\pi) - \pi\mathbf{w}^k\big)}{\partial x_k}\, dx = 0.$$

For a given scalar-field π it holds for every finite region $\mathfrak{X}$, and consequently we must have

$$\frac{\partial\big(\mathbf{s}^k(\pi) - \pi\mathbf{w}^k\big)}{\partial x_k} = 0. \tag{75}$$

If we substitute (72) in this, and observe that, for a particular point, arbitrary values may be assigned to π, $\dfrac{\partial \pi}{\partial x}$, $\dfrac{\partial^2 \pi}{\partial x_i\, \partial x_k}$, then this single formula resolves into the identities:

$$\frac{\partial \mathbf{s}^k}{\partial x_k} = \frac{\partial \mathbf{w}^k}{\partial x_k}; \qquad \mathbf{s}^i + \frac{\partial \mathbf{h}^{\alpha i}}{\partial x_\alpha} = \mathbf{w}^i; \qquad \mathbf{h}^{\alpha\beta} + \mathbf{h}^{\beta\alpha} = 0. \tag{$75_{1,2,3}$}$$

According to the third identity, $\mathbf{h}^{ik}$ is a linear tensor-density of the second order. In view of the skew-symmetry of $\mathbf{h}$ the first is a result of the second, since

$$\frac{\partial^2 \mathbf{h}^{\alpha\beta}}{\partial x_\alpha \, \partial x_\beta} = 0.$$

II. We subject the world-continuum to an infinitesimal deformation, in which each point undergoes a displacement whose components are ξ^i; let the metrical structure accompany the deformation without being changed. Let δ signify the change occasioned by the deformation in a quantity, if we remain at the same space-time point, δ' the change in the same quantity if we share in the displacement of the space-time point. Then, by (20), (21′), (71)

$$\left.\begin{aligned} -\delta\phi_i &= \left(\phi_r \frac{\partial \xi^r}{\partial x_i} + \frac{\partial \phi_i}{\partial x_r}\xi^r\right) + \frac{\partial \pi}{\partial x_i}, \\ -\delta g_{ik} &= \left(g_{ir}\frac{\partial \xi^r}{\partial x_k} + g_{kr}\frac{\partial \xi^r}{\partial x_i} + \frac{\partial g_{ik}}{\partial x_r}\xi^r\right) - \pi g_{ik}, \end{aligned}\right\} \tag{76}$$

in which π denotes an infinitesimal scalar-field that has still been left arbitrary by our conventions. The invariance of the *Action* with respect to transformation of co-ordinates and change of calibration is expressed in the formula which relates to this variation:

$$\delta' \int_{\mathfrak{X}} \mathbf{W}\, dx = \int_{\mathfrak{X}} \left\{ \frac{\partial(\mathbf{W}\xi^k)}{\partial x_k} + \delta\mathbf{W} \right\} dx = 0. \tag{77}$$

If we wish to express the invariance with respect to the co-ordinates alone we must make $\pi = 0$; but the resulting formulæ of variation (76) have not then an invariant character. This convention, in fact, signifies that the deformation is to make the two groundforms vary in such a way that the measure l of a line-element remains unchanged, that is, $\delta' l = 0$. This equation does not, however, express the process of congruent transference of a distance, but indicates that

$$\delta' l = -l(\phi_i\, \delta' x_i) = -l(\phi_i \xi^i).$$

Accordingly, in (76) we must choose π not equal to zero but equal to $-(\phi_i \xi^i)$ if we are to arrive at invariant formulæ, namely,

$$\left.\begin{aligned} -\delta\phi_i &= f_{ir}\xi^r, \\ -\delta g_{ik} &= \left(g_{ir}\frac{\partial \xi^r}{\partial x_k} + g_{kr}\frac{\partial \xi^r}{\partial x_i}\right) + \left(\frac{\partial g_{ik}}{\partial x_r} + g_{ik}\phi_r\right)\xi^r. \end{aligned}\right\} \qquad (78)$$

The change in the two groundforms which it represents is one that makes *the metrical structure appear carried along unchanged by the deformation and every line-element to be transferred congruently.* The invariant character is easily recognised analytically, too; particularly in the case of the second equation (78), if we introduce the mixed tensor

$$\frac{\partial \xi^i}{\partial x_k} + \Gamma^i_{kr}\xi^r = \xi^i_k.$$

The equation then becomes

$$-\delta g_{ik} = \xi_{ik} + \xi_{ki}.$$

Now that the calibration invariance has been applied in I, we may in the case of (76) restrict ourselves to the choice of π, which was discussed just above, and which we found to be alone possible from the point of view of invariance.

For the variation (78) let

$$\mathbf{W}\xi^k + \delta\mathbf{v}^k = \mathbf{S}^k(\xi).$$

$\mathbf{S}^k(\xi)$ is a vector-density which depends in a linear differential manner on the arbitrary vector-field ξ^i. We write in an explicit form

$$\mathbf{S}^k(\xi) = \mathbf{S}_i^k \xi^i + \overline{\mathbf{H}}_i^{k\alpha} \frac{\partial \xi^i}{\partial x_\alpha} + \tfrac{1}{2}\mathbf{H}_i^{k\alpha\beta} \frac{\partial^2 \xi^i}{\partial x_\alpha \, \partial x_\beta}$$

(the last co-efficient is, of course, symmetrical in the indices α, β). The fact that $\mathbf{S}^k(\xi)$ is a vector-density dependent on the vector-field ξ^i expresses most simply and most fully the character of invariance possessed by the co-efficients which occur in the expression for $\mathbf{S}^k(\xi)$; in particular, it follows from this that the $\mathbf{S}_i^k$'s are not components of a mixed tensor-density of the second order: we call them the components of a "pseudo-tensor-density". If we insert in (77) the expressions (70) and (78), we get an integral, whose integrand is

$$\frac{\partial \mathbf{S}^k(\xi)}{\partial x_k} - \xi^i \left\{ f_{ki}\mathbf{w}^k + \tfrac{1}{2}\left(\frac{\partial g_{\alpha\beta}}{\partial x_i} + g_{\alpha\beta}\phi_i\right)\mathbf{W}^{\alpha\beta}\right\} \mathbf{W}_i^k \frac{\partial \xi^i}{\partial x_k}.$$

On account of

$$\frac{\partial g_{\alpha\beta}}{\partial x_i} + g_{\alpha\beta}\phi_i = \Gamma_{\alpha,\beta i} + \Gamma_{\beta,\alpha i}$$

and of the symmetry of $\mathbf{W}^{\alpha\beta}$ we find

$$\tfrac{1}{2}\left(\frac{\partial g_{\alpha\beta}}{\partial x_i} + g_{\alpha\beta}\phi_i\right)\mathbf{W}^{\alpha\beta} = \Gamma_{\alpha,\beta i}\mathbf{W}^{\alpha\beta} = \Gamma_{\beta i}^{\alpha}\mathbf{W}_{\alpha}^{\beta}.$$

If we apply partial integration to the last member of the integrand, we get

$$\int_{\mathfrak{X}} \frac{\partial\left(\mathbf{S}^k(\xi) - \mathbf{W}_i^k \xi^i\right)}{\partial x_k}\, dx + \int_{\mathfrak{X}} [\ldots]_i \xi^i \, dx = 0.$$

According to the method of inference used above we get from this the identities:

$$[\dots]_i, \quad \text{that is,} \quad \left(\frac{\partial \mathbf{W}_i^k}{\partial x_k} - \Gamma_\beta^\alpha \mathbf{W}_\alpha^\beta\right) + f_{ik}\mathbf{w}^k = 0 \tag{79}$$

and

$$\frac{\partial\big(\mathbf{S}^k(\xi) - \mathbf{W}_i^k \xi^i\big)}{\partial x_k} = 0. \tag{80}$$

The latter resolves into the following four identities:

$$\left.\begin{aligned} \frac{\partial \mathbf{S}_i^k}{\partial x_k} = \frac{\partial \mathbf{W}_i^k}{\partial x_k}; \qquad & \mathbf{S}_i^k + \frac{\partial \overline{\mathbf{H}}_i^{\alpha k}}{\partial x_\alpha} = \mathbf{W}_i^k; \\ (\overline{\mathbf{H}}_i^{\alpha\beta} + \overline{\mathbf{H}}_i^{\beta\alpha}) + \frac{\partial \mathbf{H}_i^{\gamma\alpha\beta}}{\partial x_\gamma} = 0; \quad & \mathbf{H}_i^{\alpha\beta\gamma} + \mathbf{H}_i^{\beta\gamma\alpha} + \mathbf{H}_i^{\gamma\alpha\beta} = 0. \end{aligned}\right\} \quad (80_{1,2,3,4})$$

If from $({}_4)$ we replace in $({}_3)$

$$\overline{\mathbf{H}}_i^{\gamma\alpha\beta} \quad \text{by} \quad -\mathbf{H}_i^{\alpha\beta\gamma} - \mathbf{H}_i^{\beta\alpha\gamma},$$

we get that

$$\overline{\mathbf{H}}_i^{\alpha\beta} - \frac{\partial \mathbf{H}_i^{\alpha\beta\gamma}}{\partial x_\gamma} = \mathbf{H}_i^{\alpha\beta}$$

is skew-symmetrical in the indices α, β. If we introduce $\mathbf{H}_i^{\alpha\beta}$ in place of $\overline{\mathbf{H}}_i^{\alpha\beta}$ we see that $({}_3)$ and $({}_4)$ are merely statements regarding symmetry, but $({}_2)$ becomes

$$\mathbf{S}_i^k + \frac{\partial \mathbf{H}_i^{\alpha k}}{\partial x_\alpha} + \frac{\partial^2 \mathbf{H}_i^{\alpha\beta k}}{\partial x_\alpha \, \partial x_\beta} = \mathbf{W}_i^k. \tag{81}$$

($_1$) follows from this because, on account of the conditions of symmetry

$$\frac{\partial^2 \mathbf{H}_i^{\alpha\beta}}{\partial x_\alpha \, \partial x_\beta} = 0, \quad \text{we get} \quad \frac{\partial^3 \mathbf{H}_i^{\alpha\beta\gamma}}{\partial x_\alpha \, \partial x_\beta \, \partial x_\gamma} = 0.$$

Example.—In the case of Maxwell's Action-density we have, as is immediately obvious

$$\delta \mathbf{v}^k = \mathbf{f}^{ik} \, \delta\phi_i.$$

Consequently

$$\mathbf{s}^i = 0, \;\; \mathbf{h}^{ik} = \mathbf{f}^{ik}; \;\; \mathbf{S}_i^k = \mathbf{l}\delta_i^k - f^{i\alpha}\mathbf{f}^{k\alpha}, \quad \text{and the quantities } \mathbf{H} = 0.$$

Hence our identities lead to

$$\mathbf{w}^i = \frac{\partial \mathbf{f}^{\alpha i}}{\partial x_\alpha} \qquad \frac{\partial \mathbf{w}^i}{\partial x_i} = 0, \qquad \mathbf{W}_i^i = 0,$$

$$\mathbf{W}_i^k = \mathbf{S}_i^k \qquad \left(\frac{\partial \mathbf{S}_i^k}{\partial x_k} - \tfrac{1}{2} \frac{\partial g_{\alpha\beta}}{\partial x_i} \, \mathbf{S}^{\alpha\beta} \right) + f_{i\alpha} \frac{\partial \mathbf{f}^{\beta\alpha}}{\partial x_\beta} = 0.$$

We arrived at the last two formulæ by calculation earlier, the former on page 343, the latter on page 247; the latter was found to express the desired connection between Maxwell's tensor-density $\mathbf{S}_i^k$ of the field-energy and the ponderomotive force.

Field Laws and Theorems of Conservation.—If, in (70), we take for δ an arbitrary variation which vanishes outside a finite region, and for $\mathfrak{X}$ we take the whole world or a region such that, outside it, $\delta = 0$, we get

$$\int \delta \mathbf{W} \, dx = \int (\mathbf{w}^i \, \delta\phi^i + \tfrac{1}{2} \mathbf{W}^{ik} \, \delta g_{ik}) \, dx.$$

If $\int \mathbf{W}\, dx$ is the *Action*, we see from this that the following invariant laws are contained in Hamilton's Principle:

$$\mathbf{w}^i = 0, \qquad \mathbf{W}_i^k = 0.$$

Of these, we have to call the former the electromagnetic laws, the latter the gravitational laws. Between the left-hand sides of these equations there are five identities, which have been stated in (74) and (79). Thus there are among the field-equations five superfluous ones corresponding to the transition (dependent on five arbitrary functions) from one system of reference to another.

According to (75_2) the electromagnetic laws have the following form:

$$\frac{\partial \mathbf{h}^{ik}}{\partial x_k} = \mathbf{s}^i \quad \text{[and (67)]} \tag{82}$$

in full agreement with Maxwell's Theory; $\mathbf{s}^i$ is the density of the 4-current, and the linear tensor-density of the second order $\mathbf{h}^{ik}$ is the electromagnetic density of field. Without specialising the *Action* at all we can read off the whole structure of Maxwell's Theory from the calibration invariance alone. The particular form of Hamilton's function $\mathbf{W}$ affects only the formulæ which state that current and field-density are determined by the phase-quantities ϕ_i, g_{ik} of the æther. In the case of Maxwell's Theory in the restricted sense ($\mathbf{W} = \mathbf{l}$), which is valid only in empty space, we get $\mathbf{h}^{ik} = \mathbf{f}^{ik}$, $\mathbf{s}^i = 0$, which is as it should be.

Just as the $\mathbf{s}^i$'s constitute the density of the 4-current, so the scheme of $\mathbf{S}_i^k$'s is to be interpreted as the pseudo-tensor-density of the energy. In the simplest case, $\mathbf{W} = \mathbf{l}$, this explanation becomes identical with that of Maxwell. According to (75_1) and (80_1) **the**

theorems of conservation

$$\frac{\partial \mathbf{s}^i}{\partial x_i} = 0, \qquad \frac{\partial \mathbf{S}_i^k}{\partial x_k} = 0$$

are generally valid; and, indeed, they follow in two ways from the field laws. For $\frac{\partial \mathbf{s}^i}{\partial x_i}$ is not only identically equal to $\frac{\partial \mathbf{w}^i}{\partial x_i}$, but also to $-\frac{1}{2}\mathbf{W}_i^i$, and $\frac{\partial \mathbf{S}_i^k}{\partial x_k}$ is not only identically equal to $\frac{\partial \mathbf{W}_i^k}{\partial x_k}$, but also to $\Gamma_{i\beta}^{\alpha}\mathbf{W}_{\alpha}^{\beta} - f_{ik}\mathbf{w}^k$. The form of the gravitational equations is given by (81). The field laws and their accompanying laws of conservation may, by (75) and (80), be summarised conveniently in the two equations

$$\frac{\partial \mathbf{s}^i(\pi)}{\partial x_i} = 0, \qquad \frac{\partial \mathbf{S}^i(\xi)}{\partial x_i} = 0.$$

Attention has already been directed above to the intimate connection between the laws of conservation of the energy-momentum and the co-ordinate-invariance. To these four laws there is to be added the law of conservation of electricity, and, corresponding to it, there must, logically, be a property of invariance which will introduce a fifth arbitrary function; the calibration-invariance here appears as such. Earlier we derived the law of conservation of energy-momentum from the co-ordinate-invariance only owing to the fact that Hamilton's function consists of two parts, the *Action*-function of the gravitational field and that of the "physical phase"; each part had to be treated differently, and the component results had to be combined appropriately (§ 33). If those quantities, which are derived from $\mathbf{W}\xi^k + \delta\mathbf{v}^k$ by taking the variation of the fundamental quantities from (76) for the case $\pi = 0$, instead of from (78), are distinguished by a prefixed asterisk, then, in consequence of the co-ordinate-invariance, the "theorems of conservation" $\frac{\partial^* \mathbf{S}_i^k}{\partial x_k} = 0$ are generally valid. But the $^*\mathbf{S}_i^k$'s are not the energy-momentum components of the twofold action-function which have been used as a basis since § 28. For the gravitational component ($\mathbf{W} = \mathbf{G}$) we defined the energy by means of $^*\mathbf{S}_i^k$ (§ 33), but for the electromagnetic component ($\mathbf{W} = \mathbf{L}$, § 28) we introduced $\mathbf{W}_i^k$ as the energy components. This second component $\mathbf{L}$ contains only the g_{ik}'s themselves, not their derivatives; for a quantity of this kind we have, by (80_2), $\mathbf{W}_i^k = \mathbf{S}_i^k$. Hence (**if we use the transformations which the fundamental quantities undergo during an infinitesimal alteration of the calibration**), we can adapt the two different definitions of energy to one another although we cannot reconcile them entirely. These discrepancies are removed only here since it is the new theory which first furnishes us with an explanation of the current $\mathbf{s}^i$, of the electromagnetic density of field $\mathbf{h}^{ik}$, and of the **energy** $\mathbf{S}_i^k$, which is no longer bound by the assumption that the *Action* is composed of two parts, of which the one does not contain the ϕ_i's and their derivatives, and the other does not contain the derivatives of the g^{ik}'s. The virtual deformation of the world-continuum which leads to the definition of $\mathbf{S}_i^k$ must, accordingly, carry along the metrical structure and the line-elements "unchanged" in **our** sense and not in that of **Einstein**. The laws of conservation of the $\mathbf{s}^i$'s and the $\mathbf{S}_i^k$'s are then likewise not bound by an assumption concerning the composition of the *Action*. Thus, after the total energy had been introduced in § 33, we have once again passed beyond the stand taken in § 28 to a point of view which gives a more compact survey of the whole. What is done by Einstein's theory of gravitation with respect to the equality of inertial and gravitational matter, namely, that it recognises their identity as necessary but not as a consequence of an undiscovered law of physical nature, is accomplished by the present theory with respect to the facts that find expression in the structure of Maxwell's equations and the laws of conservation. Just as is the case in § 33 in which we integrate over the cross-section of a canal of the system, so we find here that, as a result of the laws of conservation, if the $\mathbf{s}^i$'s and $\mathbf{S}_i^k$'s vanish outside the canal, the system has a constant charge e and a constant energy-momentum J. Both may be represented, by Maxwell's equations (82) and the gravitational equations (81), as the flux of a certain spatial field through a surface Ω that encloses the system. If we regard this representation as a definition, the integral theorems of conservation hold, even if the field has a real singularity within the canal of the system. To prove this, let us replace this field within the canal in any arbitrary way (preserving, of course, a continuous connection with the region outside it) by a regular field, and let us define the $\mathbf{s}^i$'s and the $\mathbf{S}_i^k$'s by the equations (82), (81) (in which the right-hand sides are to be replaced by zero) in terms of the quantities $\mathbf{h}$ and $\mathbf{H}$ belonging to the altered field. The integrals of these fictitious quantities $\mathbf{s}^0$ and $\mathbf{S}_i^0$, which are to be taken over the cross-section of the canal (the interior of Ω), are constant; on the other hand, they coincide with the fluxes mentioned above over the surface Ω, since on Ω the imagined field coincides with the real one.

§ 36. Application of the Simplest Principle of Action. The Fundamental Equations of Mechanics

We have now to show that if we uphold our new theory it is possible to make an assumption about **W** which, as far as the results that have been confirmed in experience are concerned, agrees with Einstein's Theory. The simplest assumption* for purposes of calculation (I do not insist that it is realised in nature) is:

$$\mathbf{W} = -\tfrac{1}{4}F^2\sqrt{g} + \alpha\mathbf{l}. \tag{83}$$

The quantity *Action* is thus to be composed of the volume, measured in terms of the radius of curvature of the world as unit of length (cf. (62), § 17) and of Maxwell's action of the electromagnetic field; the positive constant α is a pure number. It follows that

$$\delta\mathbf{W} = -\tfrac{1}{2}F\delta(F\sqrt{g}) + \tfrac{1}{4}F^2\delta\sqrt{g} + \alpha\,\delta\mathbf{l}.$$

We assume that $-F$ is positive; the calibration may then be uniquely determined by the postulate $F = -1$; thus

$$\delta\mathbf{W} = \text{the variation of } \tfrac{1}{2}F\sqrt{g} + \tfrac{1}{4}\sqrt{g} + \alpha\mathbf{l}.$$

If we use the formula (61), § 17 for F, and omit the divergence

$$\delta\frac{(\partial\sqrt{g}\phi^i)}{\partial x_i}$$

which vanishes when we integrate over the world, and if, by means of partial integration, we convert the world-integral of $\delta(\frac{1}{2}R\sqrt{g})$

* *Vide* note 36.

into the integral of $\delta\mathbf{G}$ (§ 28), then our principle of action takes the form

$$\delta \int \mathbf{V}\,dx = 0, \text{ and we get } \mathbf{V} = \mathbf{G} + \alpha\mathbf{l} + \tfrac{1}{4}\sqrt{g}\{1 - 3(\phi_i\phi^i)\}. \quad (84)$$

This normalisation denotes that we are measuring with cosmic measuring rods. If, in addition, we choose the co-ordinates x_i so that points of the world whose co-ordinates differ by amounts of the order of magnitude 1, are separated by cosmic distances, then we may assume that the g_{ik}'s and the ϕ_i's are of the order of magnitude 1. (It is, of course, a fact that the potentials vary perceptibly by amounts that are extraordinarily small in comparison with cosmic distances.) By means of the substitution $x_i = \epsilon x'_i$ we introduce co-ordinates of the order of magnitude in general use (that is having dimensions comparable with those of the human body); ϵ is a very small constant. The g_{ik}'s do not change during this transformation, if we simultaneously perform the re-calibration which multiplies ds^2 by $\frac{1}{\epsilon^2}$. In the new system of reference we then have

$$g'_{ik} = g_{ik}, \qquad \phi'_i = \phi_i; \qquad F' = -\epsilon^2.$$

$\frac{1}{\epsilon}$ is accordingly, in our ordinary measures, the radius of curvature of the world. If g_{ik}, ϕ_i retain their old significance, but if we take x_i to represent the co-ordinates previously denoted by x'_i, and if Γ^r_{ik} are the components of the affine relationship corresponding to these co-ordinates, then

$$\mathbf{V} = (\mathbf{G} + \alpha\mathbf{l}) + \frac{\epsilon^2}{4}\sqrt{g}\{1 - 3(\phi_i\phi^i)\},$$

$$\Gamma^r_{ik} = \begin{Bmatrix} ik \\ r \end{Bmatrix} + \tfrac{1}{2}\epsilon^2(\delta^r_i\phi_k + \delta^r_k\phi_i - g_{ik}\phi^r).$$

Thus, by neglecting the exceedingly small cosmological terms, we arrive exactly at the classical Maxwell-Einstein theory of electricity and gravitation. To make the expression correspond exactly with that of § 34 we must set $\frac{\epsilon^2}{2} = \lambda$. Hence our theory necessarily gives us Einstein's cosmological term $\frac{1}{2}\lambda\sqrt{g}$. The uniform distribution of electrically neutral matter at rest over the whole of (spherical) space is thus a state of equilibrium which is compatible with our law. But, whereas in Einstein's Theory (cf. § 34) there must be a pre-established harmony between the universal physical constant λ that occurs in it, and the total mass of the earth (because each of these quantities in themselves already determine the curvature of the world), here (where λ **denotes** merely the curvature), we have that the mass present in the world **determines** the curvature. It seems to the author that just this is what makes Einstein's cosmology physically possible. In the case in which a physical field is present, Einstein's cosmological term must be supplemented by the further term $-\frac{3}{2}\lambda\sqrt{g}(\phi_i\phi^i)$; and in the components Γ^r_{ik} of the gravitational field, too, a cosmological term that is dependent on the electromagnetic potentials occurs. Our theory is founded on a definite unit of electricity; let it be e in ordinary electrostatic units. Since, in (84), if we use these units, $\frac{2\kappa}{c^2}$ occurs in place of α, we have

$$\frac{2e^2\kappa}{c^2} = \frac{\alpha}{-F}, \qquad \frac{e\sqrt{\kappa}}{c} = \frac{1}{\epsilon}\sqrt{\frac{\alpha}{2}}:$$

our unit is that quantity of electricity whose gravitational radius is $\sqrt{\frac{\alpha}{2}}$ times the radius of curvature of the world. It is, therefore, like the quantum of action **l**, of cosmic dimensions. The cosmological factor which Einstein added to his theory later is part of ours from the very beginning.

Variation of the ϕ_i's gives us Maxwell's equations

$$\frac{\partial \mathbf{f}^{ik}}{\partial x_k} = \mathbf{s}^i$$

and, in this case, we have simply

$$\mathbf{s}^i = -\frac{3\lambda}{\alpha}\,\phi_i\sqrt{g}.$$

Just as according to Maxwell the æther is the seat of energy and mass so we obtain here an electric charge (plus current) diffused thinly throughout the world. Variation of the g_{ik}'s gives the gravitational equations

$$\mathbf{R}_i^k - \frac{\mathbf{R} + \lambda\sqrt{g}}{2}\,\delta_i^k = \alpha\mathbf{T}_i^k \tag{85}$$

where

$$\mathbf{T}_i^k = \left\{\mathbf{l} + \tfrac{1}{2}(\phi_r\mathbf{s}^r)\right\}\delta_i^k - f_{ir}\mathbf{f}^{kr} = \phi_i\mathbf{s}^k.$$

The conservation of electricity is expressed in the divergence equation

$$\frac{\partial(\sqrt{g}\phi^i)}{\partial x_i} = 0. \tag{86}$$

This follows, on the one hand, from Maxwell's equations, but must, on the other hand, be derivable from the gravitational equations according to our general results. We actually find, by contracting the latter equations with respect to i, k, that

$$R + 2\lambda = \tfrac{3}{2}(\phi_i\phi^i),$$

and this in conjunction with $-F = 2\lambda$ again gives (86). We get for the pseudo-tensor-density of the energy-momentum, as is to be expected

$$\mathbf{S}_i^k = \alpha\mathbf{T}_i^k + \left\{\mathbf{G} + \tfrac{1}{2}\lambda\sqrt{g}\delta_i^k - \tfrac{1}{2}\frac{\partial g_{\alpha\beta}}{\partial x_i}\mathbf{G}^{\alpha\beta,k}\right\}.$$

From the equation $\delta' \int \mathbf{V}\, dx = 0$ for a variation δ' which is produced by the displacement in the true sense [from formula (76) with $\xi^i = \text{const.}$, $\pi = 0$], we get

$$\frac{\partial({}^*\mathbf{S}_i^k \xi^i)}{\partial x_k} = 0, \tag{87}$$

where

$${}^*\mathbf{S}_i^k = \mathbf{V}\delta_i^k - \tfrac{1}{2}\frac{\partial g_{\alpha\beta}}{\partial x_i}\mathbf{G}^{\alpha\beta,k} + \alpha\frac{\partial \phi}{\partial x_i}\mathbf{f}^{kr}.$$

To obtain the conservation theorems, we must, according to our earlier remarks, write Maxwell's equations in the form

$$\frac{\partial\left(\pi\mathbf{s}^i + \dfrac{\partial \pi}{\partial x_k}\mathbf{f}^{ik}\right)}{\partial x_i} = 0,$$

then set $\pi = -(\phi_i \xi^i)$, and, after multiplying the resulting equation by α, add it to (87). We then get, in fact,

$$\frac{\partial(\mathbf{S}_i^k \xi^i)}{\partial x_k} = 0.$$

The following terms occur in $\mathbf{S}_i^k$: the Maxwell energy-density of the electromagnetic field

$$\mathbf{l}\delta_i^k - f_{ir}\mathbf{f}^{kr},$$

the gravitational energy

$$\mathbf{G}\delta_i^k - \tfrac{1}{2}\frac{\partial g_{\alpha\beta}}{\partial x_i}\mathbf{G}^{\alpha\beta,k},$$

and the supplementary cosmological terms

$$\tfrac{1}{2}(\lambda\sqrt{g} + \phi_r\mathbf{s}^r)\delta_i^k - \phi_i\mathbf{s}^k.$$

The statical world is by its own nature calibrated. The question arises whether $F = \text{const.}$ for this calibration. The answer is in the affirmative. For if we re-calibrate the statical world in accordance with the postulate $F = -1$ and distinguish the resulting quantities by a horizontal bar, we get

$$\bar{\phi}_i = -\frac{F_i}{F}, \quad \text{where we set } F_i = \frac{\partial F}{\partial x_i} \quad (i = 1, 2, 3),$$

$$\bar{g}_{ik} = -Fg_{ik}, \quad \text{that is, } \bar{g}^{ik} = -\frac{g^{ik}}{F}, \qquad \sqrt{\bar{g}} = F^2\sqrt{g},$$

and equation (86) gives

$$\sum_{i=1}^{3}\frac{\partial \mathbf{F}^i}{\partial x_i} = 0 \qquad (\mathbf{F}^i = \sqrt{g}F^i).$$

From this, however, it follows that $F = \text{const.}$

From the fact that a further electrical term becomes added to Einstein's cosmological term, the existence of a material particle becomes possible without a mass horizon becoming necessary. The particle is necessarily charged electrically. If, in order to determine the radially symmetrical solutions for the statical case, we again use the old terms of § 31, and take ϕ to mean the electrostatic potential, then the integral whose variation must vanish, is

$$\int \mathbf{V} r^2\, dr = \int \left\{ w\Delta' - \frac{\alpha r^2 \phi'^2}{2\Delta} + \frac{\lambda r^2}{2}\left(\Delta - \frac{3h^2\phi^2}{2\Delta}\right)\right\} dr$$

(the accent denotes differentiation with respect to r). Variation of w, Δ, and ϕ, respectively, leads to the equations

$$\Delta\Delta' = \frac{3\lambda}{4} h^4 \phi^2 r,$$

$$w' = \frac{\lambda r^2}{2}\left(1 + \tfrac{3}{2}\frac{h^2\phi^2}{\Delta^2}\right) + \frac{\alpha}{2}\frac{r^2\phi'^2}{\Delta^2},$$

$$\left(\frac{r^2\phi'}{\Delta}\right)' = \frac{3}{2\alpha}\frac{h^2 r^2 \phi}{\Delta}.$$

As a result of the normalisations that have been performed, the spatial co-ordinate system is fixed except for a Euclidean rotation, and hence h^2 is uniquely determined. In f and ϕ, as a result of the free choice of the unit of time, a common constant factor remains arbitrary (a circumstance that may be used to reduce the order of the problem by 1). If the equator of the space is reached when $r = r_0$, then the quantities that occur as functions of $z = \sqrt{r_0^2 - r^2}$ must exhibit the following behaviour for $z = 0$: f and ϕ are regular, and $f \neq 0$; h^2 is infinite to the second order, Δ to the first order. The differential equations themselves show that the development of $h^2 z^2$ according to powers of z begins with the term h_0^2, where

$$h_0^2 = \frac{2r_0^2}{\lambda r_0^2 - 2}$$

—this proves, incidentally, that λ must be positive (the curvature F negative) and that $r_0^2 > \frac{2}{\lambda}$—whereas for the initial values f_0, ϕ_0, of f and ϕ we have

$$f_0^2 = \frac{3\lambda}{4} h_0^2 \phi_0^2.$$

If diametral points are to be identified, ϕ must be an even function of z, and the solution is uniquely determined by the initial values for $z = 0$, which satisfy the given conditions (*vide* note 37). It cannot remain regular in the whole region $0 \leq r \leq r_0$, but must, if we let r decrease from r_0, have a singularity at least ultimately when $r = 0$. For otherwise it would follow, by multiplying the differential equation of ϕ by ϕ, and integrating from 0 to r_0, that

$$\int_0^{r_0} \frac{r^2}{\Delta} \left(\phi'^2 + \frac{3}{2\alpha} h^2 \phi^2 \right) dr = 0.$$

Matter is accordingly a true singularity of the field. The fact that the phase-quantities vary appreciably in regions whose linear dimensions are very small in comparison with $\frac{1}{\sqrt{l}}$ may be explained, perhaps, by the circumstance that a value must be taken for r_0^2 which is enormously great in comparison with $\frac{1}{\lambda}$. The fact that all elementary particles of matter have the same charge and the same mass seems to be due to the circumstance that they are all embedded in the same world (of the same radius r_0); this agrees with the idea developed in § 32, according to which the charge and the mass are determined from infinity.

In conclusion, we shall set up the mechanical equations that govern the motion of a material particle. In actual fact we have not yet derived these equations in a form which is admissible from the point of view of the general theory of relativity; we shall now endeavour to make good this omission. We shall also take this opportunity of carrying out the intention stated in § 32, that is, to show that in general the inertial mass is the flux of the gravitational field through a surface which encloses the particle, even when the matter has to be regarded as a singularity which limits the field and lies, so to speak, outside it. In doing this we are, of course, debarred from using a substance which is in motion; the hypotheses corresponding to the latter idea, namely (§ 27):

$$dm\,ds = \mu\,dx, \qquad \mathbf{T}_i^k = \mu u_i u^k$$

are quite impossible here, as they contradict the postulated properties of invariance. For, according to the former equation, μ is a scalar-density of weight $\frac{1}{2}$, and, according to the latter, one of weight 0, since $\mathbf{T}_i^k$ is a tensor-density in the true sense. And we see that these initial conditions are impossible in the new theory for the same reason as in Einstein's Theory, namely, because they lead to a false value for the mass, as was mentioned at the end of § 33. This is obviously intimately connected with the circumstance that the integral $\int dm\,ds$ has now no meaning at all, and hence cannot be introduced as "substance-action of gravitation". We took the first step towards giving a real proof of the mechanical equations in § 33. There we considered the special case in which the body is completely isolated, and no external forces act on it.

From this we see at once that we must start from the laws of conservation

$$\frac{\partial \mathbf{S}_i^k}{\partial x_k} = 0 \tag{89}$$

which hold for the **total energy**. Let a volume Ω, whose dimensions are great compared with the actual essential nucleus of the particle, but small compared with those dimensions of the external field which alter appreciably, be marked off around the material particle. In the course of the motion Ω describes a canal in the world, in the interior of which the current filament of the material particle flows along. Let the co-ordinate system consisting of the "time-co-ordinate" $x_0 = t$ and the "space-co-ordinates" x_1, x_2, x_3, be such that the spaces $x_0 =$ const. intersect the canal (the cross-section is the volume Ω mentioned above). The integrals

$$\int_{\Omega} \mathbf{S}_i^0 \, dx_1 \, dx_2 \, dx_3 = J_i,$$

which are to be taken in a space $x_0 =$ const. over Ω, and which are functions of the time alone, represent the energy ($i = 0$) and the momentum ($i = 1, 2, 3$) of the material particle. If we integrate the equation (89) in the space $x_0 =$ const. over Ω, the first member ($k = 0$) gives the time-derivative $\dfrac{dJ_i}{dt}$; the integral sum over the three last terms, however, becomes transformed by Gauss' Theorem into an integral K_i which is to be taken over the surface of Ω. In this way we arrive at the mechanical equations

$$\frac{dJ_i}{dt} = K_i. \tag{90}$$

On the left side we have the components of the "inertial force," and on the right the components of the external "field-force". Not only the field-force but also the four-dimensional momentum J_i may be represented, in accordance with a remark at the end of § 35, as a flux through the surface of Ω. If the interior of the canal encloses a real singularity of the field the momentum must, indeed, be defined in the above manner, and then the device of the "fictitious field," used at the end of § 35, leads to the mechanical equations proved above. *It is of fundamental importance to notice that in them only such quantities are brought into relationship with one another as are determined by the course of the field outside the particle* (on the surface of Ω), *and have nothing to do with the singular states or phases in its interior.* The antithesis of kinetic and potential which receives expression in the fundamental law of mechanics does not, indeed, depend actually on the separation of energy-momentum into one part belonging to the external field and another belonging to the particle (as we pictured it in § 25), but rather on this juxtaposition, conditioned by the resolution into space and time, of the first and the three last members of the divergence equations which make up the laws of conservation, that is, on the circumstance that the singularity canals of the material particles have an infinite extension in only **one** dimension, but are very limited in **three** other dimensions. This stand was taken most definitely by Mie in the third part of his epoch-making *Foundations of a Theory of Matter*, which deals with "Force and Inertia" (*vide* note 38). Our next object is to work out the full consequences of this view for the principle of action adopted in this chapter.

To do this, it is necessary to ascertain exactly the meaning of the electromagnetic and the gravitational equations. If we discuss Maxwell's equations first, we may disregard gravitation entirely and take the point of view presented by the special theory of relativity. We should be reverting to the notion of substance if we were to interpret the Maxwell-Lorentz equation

$$\frac{\partial f^{ik}}{\partial x_k} = \rho u^i$$

so literally as to apply it to the volume-elements of an electron. Its true meaning is rather this: Outside the Ω-canal, the homogeneous equations

$$\frac{\partial f^{ik}}{\partial x_k} = 0 \tag{91}$$

hold. The only statical radially symmetrical solution $\bar{f}^{ik}$ of (91) is that derived from the potential $\frac{e}{r}$; it gives the flux e (and not 0, as it would be in the case of a solution of (91) which is free from singularities) of the electric field through an envelope Ω enclosing the particle. On account of the linearity of equations (91), these properties are not lost when an arbitrary solution f_{ik} of equations (91), free from singularities, is added to $\bar{f}_{ik}$; such a one is given by $f_{ik} = \text{const}$. **The field which surrounds the moving electron must be of the type:** $f_{ik} + \bar{f}_{ik}$, if we introduce at the moment under consideration a co-ordinate system in which the electron is at rest. This assumption concerning the constitution of the field outside Ω is, of course, justified only when we are dealing with quasi-stationary motion, that is, when the world-line of the particle deviates by a sufficiently small amount from a straight line. The term ρu^i in Lorentz's equation is to express the general effect of the charge-singularities for a region that contains many electrons. But it is clear that this assumption comes into question only for **quasi-stationary motion**. Nothing at all can be asserted about what happens during rapid acceleration. The opinion which is so generally current among physicists nowadays, that, according to classical electrodynamics, a greatly accelerated particle emits radiation, seems to the author quite unfounded. It is justified only if Lorentz's equations are interpreted in the too literal fashion repudiated above, and if, also, it is assumed that the constitution of the electron is not modified by the acceleration. **Bohr's Theory of the Atom** has led to the idea that there are individual stationary orbits for the electrons circulating in the atom, and that they may move permanently in these orbits without emitting radiations; only when an electron jumps from one stationary orbit to another is the energy that is lost by the atom emitted as electromagnetic energy of vibration (*vide* note 39). If matter is to be regarded as a boundary-singularity of the field, our field-equations make assertions only about **the possible states of the field**, and **not about the conditioning of the states of the field by the matter**. This gap is filled by the **Quantum Theory** in a manner of which the underlying principle is not yet fully grasped. The above assumption about the singular component $\bar{f}$ of the field surrounding the particle is, in our opinion, true for a quasi-stationary electron. We may, of course, work out other assumptions. If, for example, the particle is a radiating atom, the $\bar{f}^{ik}$'s will have to be represented as the field of an oscillating Hertzian dipole. (This is a possible state of the field which is caused by matter in a manner which, according to Bohr, is quite different from that imagined by Hertz.)

As far as gravitation is concerned, we shall for the present adopt the point of view of the original Einstein Theory. In it the (homogeneous) gravitational equations have (according to § 31) a statical radially symmetrical solution, which depends **on a single constant m, the mass**. The flux of a gravitational field through a sufficiently great sphere described about the centre is not equal to 0, as it should be if the solution were free from singularities, but equal to m. We assume that this solution is characteristic of the moving particle in the following sense: We consider the values traversed by the g_{ik}'s outside the canal to be extended over the canal, by supposing the narrow deep furrow, which the path of the material particle cuts out in the metrical picture of the world, to be smoothed out, and by treating the stream-filament of the particle as a line in this smoothed-out metrical field. Let ds be the corresponding proper-time differential. For a point of the stream-filament we may introduce a ("normal") co-ordinate system such that, at that point,

$$ds^2 = dx_0^2 - (dx_1^2 + dx_2^2 + dx_3^2),$$

the derivatives $\dfrac{\partial g_{\alpha\beta}}{\partial x_i}$ vanish, and the direction of the stream-filament is given by

$$dx_0 : dx_1 : dx_2 : dx_3 = 1 : 0 : 0 : 0.$$

In terms of these co-ordinates the field is to be expressed by the above-mentioned statical solution (only, of course, in a certain neighbourhood of the world-point under consideration, from which the canal of the particle is to be cut out). If we regard the normal co-ordinates x_i as Cartesian co-ordinates in a four-dimensional Euclidean space, then the picture of the world-line of the particle becomes a definite curve in the Euclidean space. Our assumption is, of course, admissible again only if the motion is quasi-stationary, that is, if this picture-curve is only slightly curved at the point under consideration. (The transformation of the homogeneous gravitational equations into non-homogeneous ones, on the right side of which the tensor $\mu u_i u_k$ appears, takes account of the singularities, due to the presence of masses, by fusing them into a continuum; this assumption is legitimate only in the quasi-stationary case.)

To return to the derivation of the mechanical equations! We shall use, once and for all, the calibration normalised by $F =$ const., and we shall neglect the cosmological terms outside the canal. The influence of the charge of the electron on the gravitational field is, as we know from § 32, to be neglected in comparison with the influence of the mass, provided the distance from the particle is sufficiently great. Consequently, if we base our calculations on the normal co-ordinate system, we may assume the gravitational field to be that mentioned above. The determination of the electromagnetic field is then, as in the gravitational case, a linear problem; it is to have the form $f_{ik} + \bar{f}_{ik}$ mentioned above (with $f_{ik} =$ const. on the surface of Ω). But this assumption is compatible with the field-laws only if $e =$ const. To prove this, we shall deduce from a fictitious field that fills the canal regularly and that links up with the really existing field outside, that

$$\frac{\partial \mathbf{f}^{ik}}{\partial x_k} = \mathbf{s}^i, \qquad \int_{\Omega} \mathbf{s}^0 \, dx_1 \, dx_2 \, dx_3 = e^*$$

in any arbitrary co-ordinate system; e^* is independent of the choice of the fictitious field, inasmuch as it may be represented as a field-flux through the surface of Ω. Since (if we neglect the cosmological terms) the $\mathbf{s}^i$'s on this surface vanish, the equation of definition gives us, if $\dfrac{\partial \mathbf{s}^i}{\partial x_i} = 0$ is integrated, $\dfrac{de^*}{dt} = 0$; moreover, the arguments set out in § 33 show that e^* is independent of the co-ordinate system chosen. If we use the normal co-ordinate system at one point, the representation of e^* as a field-flux shows that $e^* = e$.

Passing on from the charge to the momentum, we must notice at once that, with regard to the representation of the energy-momentum components by means of field-fluxes, we may not refer to the general theory of § 35, because, by applying the process of partial integration to arrive at (84), we sacrificed the co-ordinate invariance of our *Action*. Hence we must proceed as follows. With the help of the fictitious field which bridges the canal regularly, we define $\alpha\mathbf{S}_i^k$ by means of

$$(\mathbf{R}_i^k - \tfrac{1}{2}\delta_i^k\mathbf{R}) + \left(\mathbf{G}\delta_i^k - \tfrac{1}{2}\frac{\partial g_{\alpha\beta}}{\partial x_i}\mathbf{G}^{\alpha\beta,k}\right).$$

The equation

$$\frac{\partial \mathbf{S}_i^k}{\partial x_k} = 0 \tag{92}$$

is an identity for it. By integrating (92) we get (90), whereby

$$J_i = \int_\Omega \mathbf{S}_i^0 \, dx_1 \, dx_2 \, dx_3.$$

K_i expresses itself as the field-flux through the surface Ω. In these expressions the fictitious field may be replaced by the real one, and, moreover, in accordance with the gravitational equations, we may replace

$$\frac{1}{\alpha}(\mathbf{R}_i^k - \tfrac{1}{2}\delta_i^k\mathbf{R}) \quad \text{by} \quad \mathbf{l}\delta_i^k - f_{ir}\mathbf{f}^{kr}.$$

If we use the normal co-ordinate system the part due to the gravitational energy drops out; for its components depend not only linearly but also quadratically on the (vanishing) derivatives $\dfrac{\partial g_{\alpha\beta}}{\partial x_i}$.

We are, therefore, left with only the electromagnetic part, which is to be calculated along the lines of Maxwell. Since the components of Maxwell's energy-density depend quadratically on the field $f + \bar{f}$, each of them is composed of three terms in accordance with the formula

$$(f + \bar{f})^2 = f^2 + 2f\bar{f} + \bar{f}^2.$$

In the case of each, the first term contributes nothing, since the flux of a constant vector through a closed surface is 0. The last term is to be neglected since it contains the weak field $\bar{f}$ as a square; the middle term alone remains. But this gives us

$$K_i = ef_{0i}.$$

Concerning the momentum-quantities we see (in the same way as in § 33, by using identities (92) and treating the cross-section of the stream-filament as infinitely small in comparison with the external field) (1) that, for co-ordinate transformations that are to be regarded as linear in the cross-section of the canal, the J_i's are the co-variant components of a vector which is independent of the co-ordinate system; and (2) that if we alter the fictitious field occupying the canal (in § 33 we were concerned, not with this, but with a charge of the co-ordinate system in the canal) the quantities J_i retain their values. In the normal co-ordinate system, however, for which the gravitational field that surrounds the particle has the form calculated in § 31, we find that, since the fictitious field may be chosen as a statical one, according to page 408: $J_1 = J_2 = J_3 = 0$, and $J_0 =$ the flux of a spatial vector-density through the surface of Ω, and hence $= m$. On account of the property of co-variance possessed by J_i, we find that not only at the point of the canal under consideration, but also just before it and just after it

$$J_i = mu_i \qquad \left(u^i = \frac{dx_i}{ds}\right).$$

Hence the equations of motion of our particle expressed in the normal co-ordinate system are

$$\frac{d(mu_i)}{dt} = ef_{0i}. \tag{93}$$

The 0th of these equations gives us: $\dfrac{dm}{dt} = 0$; thus the field equations require that the mass be constant. But in any arbitrary co-ordinate system we have:

$$\frac{d(mu_i)}{ds} - \tfrac{1}{2}\frac{\partial g_{\alpha\beta}}{\partial x_i} mu^\alpha u^\beta = e \cdot f_{ki}u^k. \tag{94}$$

For the relations (94) are invariant with respect to co-ordinate transformations, and agree with (93) in the case of the normal co-ordinate system. *Hence, according to the field-laws, a necessary condition for a singularity canal, which is to fit into the remaining part of the field, and in the immediate neighbourhood of which the field has the required structure, is that the quantities* e *and* m *that characterise the singularity at each point of the canal remain constant along the canal, but that the world-direction of the canal satisfy the equations*

$$\frac{du_i}{ds} - \tfrac{1}{2}\frac{\partial g_{\alpha\beta}}{\partial x_i} u^\alpha u^\beta = \frac{e}{m} \cdot f_{ki}u^k.$$

In the light of these considerations, it seems to the author that the opinion expressed in § 25 stating that mass and field-energy are identical is a premature inference, and the whole of Mie's view of matter assumes a fantastic, unreal complexion. It was, of course, a natural result of the special theory of relativity that we should come to this conclusion. It is only when we arrive at the general theory that we find it possible to represent the mass as a field-flux, and to ascribe to the world relationships such as obtain in Einstein's *Cylindrical World* (§ 34), when there are cut out of it canals of circular cross-section which stretch to infinity in both directions. This view of m states not only that inertial and gravitational masses are identical in nature, but also that mass as the **point of attack** of the metrical field is identical in nature with mass as the **generator** of the metrical field. That which is physically important in the statement that energy has inertia still persists in spite of this. For example, a radiating particle loses inertial mass of exactly the same amount as the electromagnetic energy that it emits. (In this example Einstein first recognised the intimate relationship between energy and inertia.) This may be proved simply and rigorously from our present point of view. Moreover, the new standpoint in no wise signifies a relapse to the old idea of substance, but it deprives of meaning the problem of the cohesive pressure that holds the charge of the electron together.

With about the same reasonableness as is possessed by Einstein's Theory we may conclude from our results that a **clock** in quasi-stationary motion indicates the proper time $\int ds$ which corresponds to the normalisation $F = \text{const.}$* If during the motion

*The invariant quadratic form $F \cdot ds^2$ is very far from being distinguished from all other forms of the type $E \cdot ds^2$ (E being a scalar of weight -1) as is the ds^2 of Einstein's Theory, which does not contain the derivatives of the

of a clock (e.g. an atom) with infinitely small period, the world-distance traversed by it during a period were to be transferred congruently from period to period in the sense of our world-geometry, then two clocks which set out from the same world-point A with the same period, that is, which traverse congruent world-distances in A during their first period will have, in general, different periods when they meet at a later world-point B. The orbital motion of the electrons in the atom can, therefore, certainly not take place in the way described, independently of their previous histories, since the atoms emit spectral lines of definite frequencies. Neither does a measuring rod at rest in a statical field undergo a congruent transference; for the measure $l = d\sigma^2$ of a measuring rod at rest does not alter, whereas for a congruent transference it would have to satisfy the equation $\frac{dl}{dt} = -l \cdot \phi$. What is the source of this discrepancy between the conception of congruent transference and the behaviour of measuring rods, clocks, and atoms? We may distinguish two modes of determining a quantity in nature, namely, that of **persistence** and that of **adjustment**. This difference is illustrated in the following example. We may prescribe to the axis of a rotating top any arbitrary direction in space; but once this arbitrary initial direction has been fixed the direction of the axis of the top when left to itself is determined from it for all time by a **tendency of persistence** which is active from one moment to another; at each instant the axis experiences an infinitesimal par-

---

potentials at all. For this reason the inference made in our calculation of the **displacement towards the infra-red** (page 367), that similar atoms radiate the same frequency measured in the proper time ds corresponding to the normalisation $F =$ const., is by no means as convincing as in the theory of Einstein: it loses its validity altogether if a principle of action other than that here discussed holds.

allel displacement. Diametrically opposed to this is the case of a magnet needle in the magnetic field. Its direction is determined at every moment, independently of the state of the system at other moments, by the fact that the system, in virtue of its constitution, **adjusts** itself to the field in which it is embedded. There is no *a priori* ground for supposing a pure transference, following the tendency of persistence, to be integrable. But even if this be the case, as, for example, for rotations of the top in Euclidean space, nevertheless two tops which set out from the same point with axes in the same position, and which meet after the lapse of a great length of time, will manifest any arbitrary deviations in the positions of the axes, since they can never be fully removed from all influences. Thus although, for example, Maxwell's equations for the charge e of an electron make necessary the equation of conservation $\frac{de}{dt} = 0$, this does not explain why an electron itself after an arbitrarily long time still has the same charge, and why this charge is the same for all electrons. This circumstance shows that the charge is determined not by persistence but by adjustment: there can be only **one** state of equilibrium of negative electricity, to which the corpuscle adjusts itself afresh at every moment. The same reason enables us to draw the same conclusion for the spectral lines of the atoms, for what is common to atoms emitting equal frequencies is their constitution and not the equality of their frequencies at some moment when they were together far back in time. In the same way, obviously, the length of a measuring rod is determined by adjustment; for it would be impossible to give to **this** rod at **this** point of the field any length, say two or three times as great as the one that it now has, in the way that I can prescribe its direction arbitrarily. The world-curvature makes it theoretically possible to determine a length by adjustment. In consequence of its constitution the rod assumes a length which has such and such a value in relation to the radius of curvature of the world. (Perhaps the time of rotation of a top gives us an example of a time-length that is determined by persistence; if what we assumed above is true for direction then at each moment of the motion of the top the rotation vector would experience a parallel displacement.) We may briefly summarise as follows: The affine and metrical relationship is an *a priori* datum telling us how vectors and lengths alter, **if they happen to follow the tendency of persistence**. But to what extent this is the case in nature, and in what proportion persistence and adjustment modify one another, can be found only by starting from the physical laws that hold, i.e. from the principle of action.

The subject of the above discussion is the principle of action, compatible with the new axiom of calibration invariance, which most nearly approaches the Maxwell-Einstein theory. We have seen that it accounts equally well for all the phenomena which are explained by the latter theory and, indeed, that it has decided advantages so far as the deeper problems, such as the cosmological problems and that of matter are concerned. Nevertheless, I doubt whether the Hamiltonian function (83) corresponds to reality. We may certainly assume that **W** has the form $W\sqrt{g}$, in which W is an invariant of weight -2 formed in a perfectly rational manner from the components of curvature. Only **four** of these invariants may be set up, from which every other may be built up linearly by means of numerical co-efficients (*vide* note 40). One of these is Maxwell's:

$$l = \tfrac{1}{4} f_{ik} f^{ik}; \tag{95}$$

another is the F^2 used just above. But curvature is by its nature a linear matrix-tensor of the second order: $F_{ik}\, dx_i\, \delta x_k$. According to the same law by which (95), the square of the numerical value, is produced from the distance-curvature f_{ik} we may form

$$\tfrac{1}{4} F_{ik} F^{ik} \tag{96}$$

from the total curvature. The multiplication is in this case to be interpreted as a composition of matrices; (96) is therefore itself again a matrix. But its trace L is a scalar of weight -2. The two quantities L and l seem to be invariant and of the kind sought, and they can be formed most naturally from the curvature; invariants of this natural and simple type, indeed, exist only in a four-dimensional world at all. It seems more probable that W is a linear combination of L and l. Maxwell's equations become then as above: (when the calibration has been normalised by $F =$ const.) $\mathbf{s}^i =$ a constant multiple of $\sqrt{g}\phi^i$, and $\mathbf{h}^{ik} = \mathbf{f}^{ik}$. The gravitational laws in the statical case here, too, agree to a first approximation with Newton's laws. Calculations by Pauli (*vide* note 41) have indeed disclosed that the field determined in § 31 is not only a rigorous solution of Einstein's equations, but also of those favoured here, so that the amount by which the perihelion of Mercury's orbit advances and the amount of the deflection of light rays owing to the proximity of the sun at least do not conflict with these equations. But in the question of the mechanical equations and of the relationship holding between the results obtained by measuring-rods and clocks on the one hand and the quadratic form on the other, the connecting link with the old theory seems to be lost; here we may expect to meet with new results.

One serious objection may be raised against the theory in its present state: it does not account for the **inequality of positive and negative electricity** (*vide* note 42). There seem to be two ways out of this difficulty. Either we must introduce into the law of action a square root or some other irrationality; in the discussion on Mie's theory, it was mentioned how the desired inequality could be caused in this way, but it was also pointed out what obstacles lie in the way of such an irrational *Action*. Or, secondly, there is the following view which seems to the author to give a truer statement of reality. We have here occupied ourselves only with the **field** which satisfies certain generally invariant functional laws. It is quite a different matter to inquire into the **excitation** or **cause** of the field-phases that appear to be possible according to these laws; it directs our attention to the reality lying beyond the field. Thus in the æther there may exist convergent as well as divergent electromagnetic waves; but only the latter event can be brought about by an atom, situated at the centre, which emits energy owing to the jump of an electron from one orbit to another in accordance with Bohr's hypothesis. This example shows (what is immediately obvious from other considerations) that the idea of causation (in contra-distinction to functional relation) is intimately connected with the **unique direction of progress characteristic of Time**, namely **Past → Future**. This oneness of sense in Time exists beyond doubt—it is, indeed, the most fundamental fact of our perception of Time—but *a priori* reasons exclude it from playing a part in physics of the field, But we saw above (§ 33) that the sign, too, of an isolated system is fully determined, as soon as a definite sense of flow, Past → Future, has been prescribed to the world-canal swept out by the system. This connects the inequality of positive and negative electricity with the inequality of Past and Future; but the roots of this problem are not in the field, but lie outside it. Examples of such regularities of structure that concern, not the field, but the causes of the field-phases are instanced: by the existence of cylindrically shaped boundaries of the field: by our assumptions above concerning the constitution of the field in their immediate neighbourhood: lastly, and above all, by the facts of the quantum theory. But the way in which these regularities have hitherto been formulated are, of course, merely provisional in character. Nevertheless, it seems that the **theory of statistics** plays a part in it which is fundamentally necessary. We must here state in unmistakable language that physics at its present stage can in no wise be regarded as lending support to the belief that there is a causality of physical nature which is founded on rigorously exact laws. The extended field, "æther," is merely the *transmitter* of effects and is, of itself, powerless; it plays a part that is in no wise different from that which space with its rigid Euclidean metrical structure plays, according to the old view; but now the rigid motionless character has become transformed into one which gently yields and adapts itself. But freedom of action in the world is no more restricted by the rigorous laws of field physics than it is by the validity of the laws of Euclidean geometry according to the usual view.

If Mie's view were correct, we could recognise the field as objective reality, and physics would no longer be far from the goal of giving so complete a grasp of the nature of the physical world, of matter, and of natural forces, that logical necessity would extract from this insight the unique laws that underlie the occurrence of physical events. For the present, however, we must reject these bold hopes. The laws of the metrical field deal less with real ity itself than with the shadow-like extended medium that serves as a link between material things, and with the formal constitution of this medium that gives it the power of transmitting effects. **Statistical physics**, through the quantum theory, has already reached a deeper stratum of reality than is accessible to field physics; but the problem of matter is still wrapt in deepest gloom. But even if we recognise the limited range of field physics, we must gratefully acknowledge the insight to which it has helped us. Whoever looks back over the ground that has been traversed, leading from the Euclidean metrical structure to the mobile metrical field which depends on matter, and which includes the field phenomena of gravitation and electromagnetism; whoever endeavours to get a complete survey of what could be represented only successively and fitted into an articulate manifold, must be overwhelmed by a feeling of freedom won—the mind has cast off the fetters which have held it captive. He must feel transfused with the conviction that reason is not only a human, a too human, makeshift in the struggle for existence, but that, in spite of all disappointments and errors, it is yet able to follow the intelligence which has planned the world, and that the consciousness of each one of us is the centre at which the One Light and Life of Truth comprehends itself in Phenomena. Our ears have caught a few of the fundamental chords from that harmony of the spheres of which Pythagoras and Kepler once dreamed.

APPENDIX I

(Pp. 266 and 341)

To distinguish "normal" co-ordinate systems among all others in the special theory of relativity, and to determine the metrical groundform in the general theory, we may dispense with not only rigid bodies but also with clocks.

In the *special* theory of relativity the postulate that, for the transformation corresponding to the co-ordinates x_i of a piece of the world to an Euclidean "picture" space, the world-lines of points moving freely under no forces are to become **straight** lines (Galilei's and Newton's Principle of Inertia), fixes this picture space **except for an affine transformation**. For the theorem, that affine transformations of a portion of space are the only con-

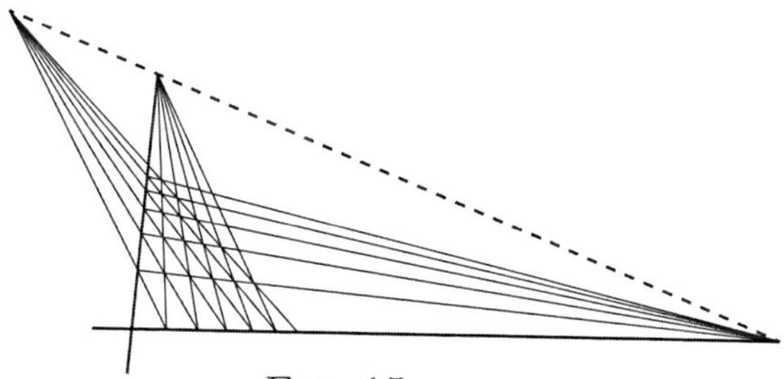

FIG. 15.

tinuous ones which transform straight lines into straight lines, holds. This is immediately evident if, in Möbius' mesh construction (Fig. 12), we replace infinity by a straight line intersecting our portion of space (Fig. 15). The phenomenon of light propagation then fixes **infinity** and the **metrical structure** in our four-dimensional projective space; for its (three dimensional) "plane at infinity" E is characterised by the property that the light-cones are projections, taken from different world-points, of one and the same two-dimensional conic section situated in E.

In the *general* theory of relativity these deductions are best expressed in the following form. The four-dimensional Riemann space, which Einstein imagines the world to be, is a particular case of general metrical space (§ 16). If we adopt this view we may say that the phenomenon of light propagation determines the **quadratic** groundform ds^2, whereas the **linear** one remains unrestricted. Two different choices of the linear groundform which differ by $d\phi = \phi_i \, dx_i$ correspond to two different values of the affine relationship. Their difference is, according to formula (49), § 16, given by

$$[\Gamma^i_{\alpha\beta}] = \tfrac{1}{2}(\delta^i_\alpha \phi_\beta + \delta^i_\beta \phi_\alpha - g_{\alpha\beta}\phi^i).$$

The difference between the two vectors that are derived from a world-vector u^i at the world-point O by means of an infinitesimal parallel displacement of u^i in its own direction (by the same amount $dx_i = \epsilon \cdot u^i$), is therefore ϵ times

$$u^i(\phi_\alpha u^\alpha) - \tfrac{1}{2}\phi^i, \qquad (*)$$

whereby we assume $g_{\alpha\beta}u^\alpha u^\beta = 1$. If the geodetic lines passing through O in the direction of the vector u^i coincide for the two fields, then the above two vectors derived from u^i by parallel displacement must be coincident in direction; the vector $(*)$, and hence ϕ^i, must have the same direction as the vector u^i. If this agreement holds for **two** geodetic lines passing through O in different directions, we get $\phi^i = 0$. Hence if we know the world-lines of two point-masses passing through O and moving only under the influence of the guiding field, then the linear groundform, as well as the quadratic groundform, is uniquely determined at O.

APPENDIX II

(Page 345)

Proof of the Theorem that, in Riemann's space, R is the sole invariant that contains the derivatives of the g_{ik}'s only to the second order, and those of the second order only linearly.

According to hypothesis, the invariant J is built up of the derivatives of the second order:

$$g_{ik,rs} = \frac{\partial^2 g_{ik}}{\partial x_r \, \partial x_s};$$

thus

$$J = \sum \lambda_{ik,rs} g_{ik,rs} + \lambda.$$

The λ's denote expressions in the g_{ik}'s and their first derivatives; they satisfy the conditions of symmetry:

$$\lambda_{ki,rs} = \lambda_{ik,rs}, \qquad \lambda_{ik,sr} = \lambda_{ik,rs}.$$

At the point O at which we are considering the invariant, we introduce an orthogonal geodetic co-ordinate system, so that, at that point, we have

$$g_{ik} = \delta_i^k, \qquad \frac{\partial g_{ik}}{\partial x_r} = 0.$$

The λ's become **absolute constants**, if these values are inserted. The unique character of the co-ordinate system is not affected by:

(1) linear orthogonal transformations;

(2) a transformation of the type

$$x_i = x'_i + \frac{1}{6} \alpha^i_{krs} x'_k x'_r x'_s$$

which contains no quadratic terms; the co-efficients α are symmetrical in k, r, and s, but are otherwise arbitrary.

Let us therefore consider in a Euclidean-Cartesian space (in which arbitrary orthogonal linear transformations are allowable) the biquadratic form dependent on two vectors $x = (x_i)$, $y = (y_i)$, namely

$$G = g_{ik,rs} x_i x_k y_r y_s$$

with arbitrary co-efficients $g_{ik,rs}$ that are symmetrical in i and k, as also in r and s; then

$$\lambda_{ik,rs} g_{ik,rs} \tag{1}$$

must be an invariant of this form. Moreover, since as a result of the transformation (2) above, the derivatives $g_{ik,rs}$ transform themselves, as may easily be calculated, according to the equation

$$g'_{ik,rs} = g_{ik,rs} + \tfrac{1}{2}(\alpha^i_{krs} + \alpha^k_{irs}),$$

we must have

$$\lambda_{ik,rs} \alpha^i_{krs} = 0 \tag{2}$$

for every system of numbers α symmetrical in the three indices k, r, s.

Let us operate further in the Euclidean-Cartesian space; (x, y) is to signify the scalar product $x_1 y_1 + x_2 y_2 + \ldots + x_n y_n$. It will suffice to use for G a form of the type

$$G = (a, x)^2 (b, y)^2$$

in which a and b denote arbitrary vectors. If we now again write x and y for a and b, then (1) expresses the postulate that

$$\Lambda = \Lambda_x = \sum \lambda_{ik,rs} x_i x_k y_r y_s \tag{1*}$$

is an orthogonal invariant of the two vectors x, y. In (2) it is sufficient to choose

$$\alpha^i_{krs} = x_i \cdot y_k y_r y_s$$

and then this postulate signifies that the form which is derived from Λ_x by converting an x into a y, namely,

$$\Lambda_y = \sum \lambda_{ik,rs} x_i y_k y_r y_s \tag{2*}$$

vanishes identically. (It is got from Λ_x by forming first the symmetrical bilinear form $\Lambda_{x,x'}$ in x, x' (it is related quadratically to y), which, if the series of variables x' be identified with x, resolves into Λ_x, and by then replacing x' by y.) I now assert that it follows from (1*) that Λ is of the form

$$\Lambda = \alpha(x, x)(y, y) - \beta(x, y)^2 \tag{I}$$

and from (2*) that

$$\alpha = \beta. \tag{II}$$

This will be the complete result, for then we shall have

$$J = \alpha(g_{ii,kk} - g_{ik,ik}) + \lambda$$

or since, in an orthogonal geodetic co-ordinate system, the Riemann scalar of curvature is

$$R = g_{ik,ik} - g_{ii,kk}$$

we shall get

$$J = -\alpha R + \lambda. \tag{*}$$

Proof of I: We may introduce a Cartesian co-ordinate system such that x coincides with the first co-ordinate axis, and y with the $(1,2)$th co-ordinate plane, thus;

$$x = (x_1, 0, 0, \ldots, 0), \qquad y = (y_1, y_2, 0, \ldots, 0),$$
$$\Lambda = x_1^2(ay_1^2 + 2by_1y_2 + cy_2^2),$$

whereby the sense of the second co-ordinate axis may yet be chosen arbitrarily. Since Λ may not depend on this choice, we must have $b = 0$, therefore

$$\Lambda = cx_1^2(y_1^2 + y_2^2) + (a - c)(x_1y_1)^2 = c(x,x)(y,y) + (a-c)(x,y)^2.$$

Proof of II: From the $\Lambda = \Lambda_x$ which are given under I, we derive the forms

$$\Lambda_{x,x'} = \alpha(x,x')(y,y) - \beta(x,y)(x',y),$$
$$\Lambda_y = (\alpha - \beta)(x,y)(y,y).$$

If Λ_y is to vanish then α must equal β.

We have tacitly assumed that the metrical groundform of Riemann's space is definitely positive; in case of a different index of inertia a slight modification is necessary in the "Proof of I". In order that the second derivatives be excluded from the volume integral J by means of partial integration, it is necessary that the $\lambda_{ik,rs}$'s depend only on the g_{ik}'s and not on their derivatives; we did not, however, require this fact at all in our proof. Concerning the physical meaning entailed by the possibility, expressed in $(*)$, of adding to a multiple of R also a universal constant λ, we refer to § 34. Concerning the theorem here proved, cf. Vermeil, *Nachr. d. Ges. d. Wissensch. zu Göttingen*, 1917, pp. 334–344.

In the same way it may be proved that g_{ik}, Rg_{ik}, R_{ik} are the only tensors of the second order that contain derivatives of the g_{ik}'s only to the second order, and these, indeed, only linearly.

BIBLIOGRAPHY

(The number of each note is followed by the number of the page on which reference is made to it)

INTRODUCTION AND CHAPTER I

Note 1. (**7**) The detailed development of these ideas follows very closely the lines of Husserl in his "Ideen zu einer reinen Phänomenologie und phänomenologischen Philosophie" (Jahrbuch f. Philos. u. phänomenol. Forschung, Bd. 1, Halle, 1913).

Note 2. (**21**) Helmholtz in his dissertation, "Über die Tatsachen, welche der Geometrie zugrunde liegen" (Nachr. d. K. Gesellschaft d. Wissenschaften zu Göttingen, math.-physik. Kl., 1868), was the first to attempt to found geometry on the properties of the group of motions. This "Helmholtz space-problem" was defined more sharply and solved by S. Lie (Berichte d. K. Sachs. Ges. d. Wissenschaften zu Leipzig, math.-phys. Kl., 1890) by means of the theory of transformation groups, which was created by Lie (cf. Lie-Engel, Theorie der Transformationsgruppen, Bd. 3, Abt. 5). Hilbert then introduced great restrictions among the assumptions made by applying the ideas of the theory of aggregates (Hilbert, Grundlagen der Geometrie, 3 Aufl., Leipzig, 1909, Anhang IV).

Note 3. (**27**) The systematic treatment of affine geometry not limited to the dimensional number 3 as well as of the whole subject of the geometrical calculus is contained in the epoch-making work of Grassmann, Lineale Ausdehnungslehre (Leipzig, 1844). In forming the conception of a manifold of more than three dimensions, Grassmann as well as Riemann was influenced by the philosophic ideas of Herbart.

Note 4. (**78**) The systematic form which we have here given to the tensor calculus is derived essentially from Ricci and Levi-Civita: Méthodes de calcul différentiel absolu et leurs applications, Math. Ann.,

Bd. 54 (1901).

CHAPTER II

Note 1. (**113**) For more detailed information reference may be made to Die Nicht-Euklidische Geometrie, Bonola and Liebmann, published by Teubner.

Note 2. (**118**) F. Klein, Über die sogenannte Nicht-Euklidische Geometrie, Math. Ann., Bd. 4 (1871), p. 573. Cf. also later papers in the Math. Ann., Bd. 6 (1873), p. 112, and Bd. 37 (1890), p. 544.

Note 3. (**121**) Sixth Memoir upon Quantics, Philosophical Transactions, t. 149 (1859).

Note 4. (**133**) Mathematische Werke (2 Aufl., Leipzig, 1892), Nr. XIII, p. 272. Als besondere Schrift herausgegeben und kommentiert vom Verf. (2 Aufl., Springer, 1920).

Note 5. (**137**) Saggio di interpretazione della geometria non euclidea, Giorn. di Matem., t. 6 (1868), p. 204; Opere Matem. (Höpli, 1902), t. 1, p. 374.

Note 6. (**137**) Grundlagen der Geometrie (3 Aufl., Leipzig, 1909), Anhang V.

Note 7. (**142**) Cf. the references in Chap. I.[2] Christoffel, Über die Transformation der homogenen Differentialausdrücke zweiten Grades, Journ. f. d. reine und angew. Mathemathik, Bd. 70 (1869): Lipschitz, in the same journal, Bd. 70 (1869), p. 71, and Bd. 72 (1870), p. 1.

Note 8. (**150**) Christoffel (l.c.[7]). Ricci and Levi-Civita, Méthodes de calcul différentiel absolu et leurs applications, Math. Ann., Bd. 54 (1901).

Note 9. (**151**) The development of this geometry was strongly influenced by the following works which were created in the light of Einstein's Theory of Gravitation: Levi-Civita, Nozione di parallelismo in una varietà qualunque ..., Rend. del Circ. Mat. di Palermo, t. 42 (1917), and Hessenberg, Vektorielle Begründung der Differentialgeometrie, Math. Ann., Bd. 78 (1917). It assumed a perfectly definite form in the dissertation by Weyl, Reine Infinitesimalgeometrie, Math. Zeitschrift, Bd. 2 (1918).

Note 10. (**166**) The conception of parallel displacement of a vector was set up for Riemann's geometry in the dissertation quoted in Note 9; to derive it, however, Levi-Civita assumed that Riemann's space is embedded in a Euclidean space of higher dimensions. A direct explanation of the conception was given by Weyl in the first edition of this book with the help of the geodetic co-ordinate system; it was elevated to the rank of a fundamental axiomatic conception, which is characteristic of the degree of the affine geometry, in the paper "Reine Infinitesimalgeometrie," mentioned in Note 9.

Note 11. (**198**) Hessenberg (l.c.[9]), p. 190.

Note 12. (**213**) Cf. the large work of Lie-Engel, Theorie der Transformationsgruppen, Leipzig, 1888–93; concerning this so-called "second fundamental theorem" and its converse, *vide* Bd. 1, p. 156, Bd. 3, pp. 583, 659, and also Fr. Schur, Math. Ann., Bd. 33 (1888), p. 54.

Note 13. (**218**) A second view of the problem of space in the light of the theory of groups forms the basis of the investigations of Helmholtz and Lie quoted in Chapter I.[2]

CHAPTER III

Note 1. (**221**) All further references to the special theory of relativity will be found in Laue, Die Relativitätstheorie I (3 Aufl., Braunschweig, 1919).

Note 2. (**238**) Helmholtz, Monatsber. d. Berliner Akademie, Marz, 1876, or Ges. Abhandlungen, Bd. 1 (1882), p. 791. Eichenwald, Annalen der Physik, Bd. 11 (1903), p. 1.

Note 3. (**250**) This is true, only subject to certain limitations; *vide* A. Korn, Mechanische Theorie des elektromagnetischen Feldes, Phys. Zeitschr., Bd. 18, 19 and 20 (1917–19).

Note 4. (**253**) A. A. Michelson, Sill. Journ., Bd. 22 (1881), p. 120. A. A. Michelson and E. W. Morley, *idem*, Bd. 34 (1887), p. 333. E. W. Morley and D. C. Miller, Philosophical Magazine, vol. viii (1904), p. 753, and Bd. 9 (1905), p. 680. H. A. Lorentz, Arch. Néerl., Bd. 21 (1887), p. 103, or Ges. Abhandl., Bd. 1, p. 341. Since the enunciation of the theory of relativity by Einstein, the experiment has been discussed repeatedly.

Note 5. (**255**) Cf. Trouton and Noble, Proc. Roy. Soc., vol. lxxii (1903), p. 132. Lord Rayleigh, Phil. Mag., vol. iv (1902), p. 678. D. B. Brace, *idem* (1904), p. 317, vol. x (1905), pp. 71, 591. B. Strasser, Annal. d. Physik, Bd. 24 (1907), p. 137. Des Coudres, Wiedemanns Annalen, Bd. 38 (1889), p. 71. Trouton and Rankine, Proc. Roy. Soc., vol. viii. (1908), p. 420.

Note 6. (**256**) Zur Elektrodynamik bewegter Körper, Annal. d. Physik, Bd. 17 (1905), p. 891.

Note 7. (**257**) Minkowski, Die Grundgleichungen für die elektromagnetischen Vorgänge in bewegten Körpern, Nachr. d. K. Ges. d. Wissensch. zu Göttingen, 1908, p. 53, or Ges. Abhandl., Bd. 2, p. 352.

Note 8. (**265**) Möbius, Der barycentrische Calcul (Leipzig, 1827; or Werke, Bd. 1), Kap. 6 u. 7.

Note 9. (**276**) In taking account of the dispersion it is to be noticed that q' is the velocity of propagation for the frequency ν' in water at rest, and not for the frequency ν (which exists inside and outside the water). Careful experimental confirmations of the result have been given by Michelson and Morley, Amer. Jour. of Science, **31** (1886), p. 377, Zeeman, Versl. d. K. Akad. v. Wetensch., Amsterdam, **23** (1914), p. 245; **24** (1915), p. 18. There is a new interference experiment by Zeeman similar to that performed by Fizeau: Zeeman,

Versl. Akad. v. Wetensch., Amsterdam, **28** (1919), p. 1451; Zeeman and Snethlage, *idem*, p. 1462. Concerning interference experiments with rotating bodies, *vide* Laue, Annal. d. Physik, **62** (1920), p. 448.

Note 10. (**285**) Wilson, Phil. Trans. (A), vol. 204 (1904), p. 121.

Note 11. (**292**) Röntgen, Sitzungsber. d. Berliner Akademie, 1885, p. 195; Wied. Annalen, Bd. 35 (1888), p. 264, and Bd. 40 (1890), p. 93. Eichenwald, Annalen d. Physik, Bd. 11 (1903), p. 421.

Note 12. (**292**) Minkowski (l.c.[7]).

Note 13. (**296**) W. Kaufmann, Nachr. d. K. Gesellsch. d. Wissensch. zu Göttingen, 1902, p. 291; Ann. d. Physik, Bd. 19 (1906), p. 487, and Bd. 20 (1906), p. 639. A. H. Bucherer, Ann. d. Physik, Bd. 28 (1909), p. 513, and Bd. 29 (1919), p. 1063. S. Ratnowsky, Determination experimentale de la variation d'inertie des corpuscules cathodiques en fonction de la vitesse, Dissertation, Geneva, 1911. E. Hupka, Ann. d. Physik, Bd. 31 (1910), p. 169. G. Neumann, Ann. d. Physik, Bd. 45 (1914), p. 529, mit Nachtrag von C. Schaefer, *ibid.*, Bd. 49, p. 934. Concerning the atomic theory, *vide* K. Glitscher, Spektroskopischer Vergleich zwischen den Theorien des starren und des deformierbaren Elektrons, Ann. d. Physik, Bd. 52 (1917), p. 608.

Note 14. (**305**) Die Relativitätstheorie I (3 Aufl., 1919), p. 229.

Note 15. (**305**) Einstein (l.c.[6]). Planck, Bemerkungen zum Prinzip der Aktion und Reaktion in der allgemeinen Dynamik, Physik. Zeitschr., Bd. 9 (1908), p. 828; Zur Dynamik bewegter Systeme, Ann. d. Physik, Bd. 26 (1908), p. 1.

Note 16. (**307**) Herglotz, Ann. d. Physik, Bd. 36 (1911), p. 453.

Note 17. (**307**) Ann. d. Physik, Bd. 37, 39, 40 (1912–13).

CHAPTER IV

Note 1. (**325**) Concerning this paragraph, and indeed the whole chapter up to § 34, *vide* A. Einstein, Die Grundlagen der allgemeinen Relativitätstheorie (Leipzig, Joh. Ambr. Barth, 1916); Über die spezielle und die aligemeine Relativitätstheorie (gemeinverständlich; Sammlung Vieweg, 10 Aufl., 1910). E. Freundlich, Die Grundlagen der Einsteinschen Gravitationstheorie (4 Aufl., Springer, 1920). M. Schlick, Raum und Zeit in der gegenwärtigen Physik (3 Aufl., Springer, 1920). A. S. Eddington, Space, Time, and Gravitation (Cambridge, 1920), an excellent, popular, and comprehensive exposition of the general theory of relativity, including the development described in §§ 35, 36. Eddington, Report on the Relativity Theory of Gravitation (London, Fleetway Press, 1919). M. Born, Die Relativitätstheorie Einsteins (Springer, 1920). E. Cassirer, Zur Einsteinschen Relativitätstheorie (Berlin, Cassirer, 1921). E. Kretschmann, Über den physikalischen Sinn der Relativitätspostulate, Ann. Phys., Bd. 53 (1917), p. 575. G. Mie, Die Einsteinsche Gravitationstheorie und das Problem der Materie, Phys. Zeitschr., Bd. 18 (1917), pp. 551–56, 574–80 and 596–602. F. Kottler, Über die physikalischen Grundlagen der allgemeinen Relativitätstheorie, Ann. d. Physik, Bd. 56 (1918), p. 401. Einstein, Prinzipielles zur allgemeinen Relativitätstheorie, Ann. d. Physik, Bd. 55 (1918), p. 241.

Note 2. (**326**) Even Newton felt this difficulty; it was stated most clearly and emphatically by E. Mach. Cf. the detailed references in A. Voss, Die Prinzipien der rationellen Mechanik, in der Mathematischen Enzyklopädie, Bd. 4, Art. 1, Absatz 13–17 (phoronomische Grundbegriffe).

Note 3. (**335**) Mathematische und naturwissenschaftliche Berichte aus Ungarn VIII (1890).

Note 4. (**339**) Concerning other attempts (by Abraham, Mie, Nordström) to adapt the theory of gravitation to the results arising from the special theory of relativity, full references are given in M. Abraham, Neuere Gravitationstheorien, Jahrbuch der Radioaktivität und Elektronik, Bd. 11 (1915), p. 470.

Note 5. (**347**) F. Klein, Über die Differentialgesetze für die Erhaltung von Impuls und Energie in der Einsteinschen Gravitationstheorie, Nachr. d. Ges. d. Wissensch. zu Göttingen, 1918. Cf., in the same periodical, the general formulations given by E. Noether, Invariante Variationsprobleme.

Note 6. (**355**) Following A. Palatini, Deduzione invariantiva delle equazioni gravitazionali dal principio di Hamilton, Rend. del Circ. Matem. di Palermo, t. 43 (1919), pp. 203–12.

Note 7. (**357**) Einstein, Zur allgemeinen Relativitätstheorie, Sitzungsber. d. Preuss. Akad. d. Wissenschaften, 1915, **44**, p. 778, and an appendix on p. 799. Also Einstein, Die Feldgleichungen der Gravitation, *idem*, 1915, p. 844.

Note 8. (**357**) H. A. Lorentz, Het beginsel van Hamilton in Einstein's theorie der zwaartekracht, Versl. d. Akad. v. Wetensch. te Amsterdam, XXIII, p. 1073: Over Einstein's theorie der zwaartekracht I, II, III, *ibid.*, XXIV, pp. 1389, 1759, XXV, p. 468. Trestling, *ibid.*, Nov., 1916; Fokker, *ibid.*, Jan., 1917, p. 1067. Hilbert, Die Grundlagen der Physik, 1 Mitteilung, Nachr. d. Gesellsch. d. Wissensch. zu Göttingen, 1915, 2 Mitteilung, 1917. Einstein, Hamiltonsches Prinzip und allgemeine Relativitätstheorie, Sitzungsber. d. Preuss. Akad. d. Wissensch., 1916, **42**, p. 1111. Klein, Zu Hilberts erster Note über die Grundlagen der Physik, Nachr. d. Ges. d. Wissensch. zu Göttingen, 1918, and the paper quoted in Note 5, also Weyl, Zur Gravitationstheorie, Ann. d. Physik, Bd. 54 (1917), p. 117.

Note 9. (**359**) Following Levi-Civita, Statica Einsteiniana, Rend. della R. Accad. dei Linceï, 1917, vol. xxvi., ser. 5a, 1° sem., p. 458.

Note 10. (**364**) Cf. also Levi-Civita, La teoria di Einstein e il principio di Fermat, Nuovo Cimento, ser. 6, vol. xvi. (1918), pp. 105–14.

Note 11. (**367**) F. W. Dyson, A. S. Eddington, C. Davidson, A Determination of the Deflection of Light by the Sun's Gravitational Field, from Observations made at the Total Eclipse of May 29th, 1919; Phil. Trans. of the Royal Society of London, Ser. A, vol. 220 (1920), pp. 291–333. Cf. E. Freundlich, Die Naturwissenschaften, 1920, pp. 667–73.

Note 12. (**368**) Schwarzschild, Sitzungsber. d. Preuss. Akad. d. Wissenschaften, 1914, p. 1201. Ch. E. St. John, Astrophys. Journal, **46** (1917), p. 249 (vgl. auch die dort zitierten Arbeiten von Halm und Adams). Evershed and Royds, Kodaik. Obs. Bull., **39**. L. Grebe and A. Bachem, Verhandl. d. Deutsch. Physik. Ges., **21** (1919), p. 454; Zeitschrift für Physik, **1** (1920), p. 51. E. Freundlich, Physik. Zeitschr., **20** (1919), p. 561.

Note 13. (**369**) Einstein, Sitzungsber. d. Preuss. Akad. d. Wissensch., 1915, **47**, p. 831. Schwarzschild, Sitzungsber. d. Preuss. Akad. d. Wissensch., 1916, **7**, p. 189.

Note 14. (**369**) The following hypothesis claimed most favour. H. Seeliger, Das Zodiakallicht und die empirischen Glieder in der Bewegung der inneren Planeten, Münch. Akad., Ber. 36 (1906). Cf. E. Freundlich, Astr. Nachr., Bd. 201 (June, 1915), p. 48.

Note 15. (**371**) Einstein, Sitzungsber. d. Preuss. Akad. d. Wissensch., 1916, p. 688; and the appendix: Über Gravitationswellen, *idem*, 1918, p. 154. Also Hilbert (l.c.[8]), 2 Mitteilung.

Note 16. (**376**) Phys. Zeitschr., Bd. 19 (1918), pp. 33 and 156. Cf. also de Sitter, Planetary motion and the motion of the moon according to Einstein's theory, Amsterdam Proc., Bd. 19, 1916.

Note 17. (**377**) Cf. Schwarzschild (l.c.[12]); Hilbert (l.c.[8]), 2 Mitt.; J. Droste, Versl. K. Akad. v. Wetensch., Bd. 25 (1916), p. 163.

Note 18. (**387**) Concerning the problem of n bodies, *vide* J. Droste, Versl. K. Akad. v. Wetensch., Bd. 25 (1916), p. 460.

Note 19. (**388**) Cf. A. S. Eddington, Report, §§ 29, 30.

Note 20. (**389**) L. Flamm, Beiträge zur Einsteinschen Gravitationstheorie, Physik. Zeitschr., Bd. 17 (1916), p. 449.

Note 21. (**389**) H. Reistner, Ann. Physik, Bd. 50 (1916), pp. 106–20. Weyl (l.c.[8]). G. Nordström, On the Energy of the Gravitation Field in Einstein's Theory, Versl. d. K. Akad. v. Wetensch., Amsterdam, vol. xx., Nr. 9, 10 (Jan. 26th, 1918). C. Longo, Legge elettrostatica elementare nella teoria di Einstein, Nuovo Cimento, ser. 6, vol. xv. (1918). p. 191.

Note 22. (**399**) Sitzungsber. d. Preuss. Akad. d. Wissensch., 1916, **18**, p. 424. Also H. Bauer, Kugelsymmetrische Lösungssysteme der Einsteinschen Feldgleichungen der Gravitation für eine ruhende, gravitierende Flüssigkeit mit linearer Zustandsgleichung, Sitzungsber. d. Akad. d. Wissensch. in Wien, math.-naturw. Kl., Abt. IIa, Bd. 127 (1918).

Note 23. (**399**) Weyl (l.c.[8]), §§ 5, 6. And a remark in Ann. d. Physik, Bd. 59 (1919).

Note 24. (**402**) Levi-Civita: ds^2 einsteiniani in campi newtoniani, Rend. Accad. dei Linceï, 1917–19.

Note 25. (**402**) A. De-Zuani, Equilibrio relativo ed equazioni gravitazionali di Einstein nel caso stazionario, Nuovo Cimento, ser. v, vol. xviii. (1819), p. 5. A. Palatini, Moti Einsteiniani stazionari, Atti del R. Instit. Veneto di scienze, lett. ed arti, t. 78 (2) (1919), p. 589.

Note 26. (**405**) Einstein, Grundlagen [(l.c.[1])] S. 49. The proof here is according to Klein (l.c.[5]).

Note 27. (**405**) For a discussion of the physical meaning of these equations, *vide* Schrödinger, Phys. Zeitschr., Bd. 19 (1918), p. 4; H. Bauer, *idem*, p. 163; Einstein, *idem*, p. 115, and finally, Einstein, Der Energiesatz in der allgemeinen Relativitätstheorie, in den Sitzungsber. d. Preuss. Akad. d. Wissensch., 1918, p. 448, which cleared away the difficulties, and which we have followed in the text. Cf. also F. Klein, Über die Integralform der Erhaltungssätze und die Theorie der räumlich geschlossenen Welt, Nachr. d. Ges. d. Wissensch. zu Göttingen, 1918.

Note 28. (**408**) Cf. G. Nordström, On the mass of a material system according to the Theory of Einstein, Akad. v. Wetensch., Amsterdam, vol. xx., No. 7 (Dec. 29th, 1917).

Note 29. (**411**) Hilbert (l.c.[8]), 2 Mitt.

Note 30. (**414**) Einstein, Sitzungsber. d. Preuss. Akad. d. Wissensch., 1917 **6**, p. 142.

Note 31. (**420**) Weyl, Physik. Zeitschr., Bd. 20 (1919), p. 31.

Note 32. (**422**) Cf. de Sitter's Mitteilungen im Versl. d. Akad. v. Wetensch. te Amsterdam, 1917, as also his series of concise articles: On Einstein's theory of gravitation and its astronomical consequences (Monthly Notices of the R. Astronom. Society); also F. Klein (l.c.[27]).

Note 33. (**422**) The theory contained in the two following articles were developed by Weyl in the Note "Gravitation und Elektrizität," Sitzungsber. d. Preuss. Akad. d. Wissensch., 1918, p. 465. Cf. also Weyl, Eine neue Erweiterung der Relativitätstheorie, Ann. d. Physik, Bd. 59 (1919). A similar tendency is displayed (although obscure to the present author in essential points) in E. Reichenbächer (Grundzüge zu einer Theorie der Elektrizität und Gravitation, Ann. d. Physik, Bd. 52 [1917], p. 135; also Ann. d. Physik, Bd. 63 [1920], pp. 93–144). Concerning other attempts to derive Electricity and Gravitation from a common root cf. the articles of Abraham quoted in Note 4; also G. Nordström, Physik. Zeitschr., **15** (1914), p. 504; E. Wiechert, Die Gravitation als elektrodynamische Erscheinung, Ann. d. Physik, Bd. 63 (1920), p. 301.

Note 34. (**428**) This theorem was proved by Liouville: Note IV in the appendix to G. Monge, Application de l'analyse à la géométrie (1850), p. 609.

Note 35. (**429**) This fact, which here appears as a self-evident result, had been previously noted: E. Cunningham, Proc. of the London Mathem. Society (2), vol. viii. (1910), pp. 77–98; H. Bateman, *idem*, pp. 223–64.

Note 36. (**442**) Cf. also W. Pauli, Zur Theorie der Gravitation und der Elektrizität von H. Weyl, Physik. Zeitschr., Bd. 20 (1919), pp. 457–67. Einstein arrived at partly similar results by means of a further modification of his gravitational equations in his essay: Spielen Gravitationsfelder im Aufbau der materiellen Elementarteilchen eine wesentliche Rolle? Sitzungsber. d. Preuss. Akad. d. Wissensch., 1919, pp. 349–56.

Note 37. (**449**) Concerning such existence theorems at a point of singularity, *vide* Picard, Traité d'Analyse, t. 3, p. 21.

Note 38. (**452**) Ann. d. Physik, Bd. 39 (1913).

Note 39. (**454**) As described in the book by Sommerfeld, Atombau and Spektrallinien, Vieweg, 1919 and 1921.

Note 40. (**463**) This was proved by R. Weitzenböck in a letter to the present author; his investigation will appear soon in the Sitzungsber. d. Akad. d. Wissensch. in Wien.

Note 41. (**464**) W. Pauli, Merkur-Perihelbewegung und Strahlenablenkung in Weyl's Gravitationstheorie, Verhandl. d. Deutschen physik. Ges., Bd. 21 (1919), p. 742.

Note 42. (**464**) Pauli (l.c.[36]).

INDEX

Printed at Repro India Ltd.